D0086511

Physiology of Crop Plants

FRANKLIN P. GARDNER
PROFESSOR OF AGRONOMY, UNIVERSITY OF FLORIDA

R. BRENT PEARCE
PROFESSOR OF AGRONOMY, IOWA STATE UNIVERSITY

ROGER L. MITCHELL
DEAN OF THE COLLEGE OF AGRICULTURE, UNIVERSITY OF MISSOURI-COLUMBIA

IOWA STATE UNIVERSITY PRESS: AMES

Physiology of Crop Plants

Composed and printed by The Iowa State University Press, Ames, Iowa 50010

First edition, 1985

Gardner, Franklin P. (Franklin Pierce), 1924–
Physiology of crop plants.

Includes index.
1. Field crops—Physiology. 2. Crops—Physiology. I. Pearce, R. Brent (Robert Brent), 1936– . II. Mitchell, Roger L., 1932– . III. Title.
SB185.5.G37 1985 582.13′041 84–15667
ISBN 0–8138–1376–X

CONTENTS

Foreword vii

Acknowledgments viii

1. Photosynthesis 3

2. Carbon Fixation by Crop Canopies 31

3. Transport and Partitioning 58

4. Water Relations 76

5. Mineral Nutrition 98

6. Biological Nitrogen Fixation 132

7. Plant Growth Regulation 156

8. Growth and Development 187

9. Seeds and Germination 209

10. Root Growth 246

11. Vegetative Growth 271

12. Flowering and Fruiting 296

Index 321

FOREWORD

THIS SECOND EDITION of a textbook focused on crop physiology reflects the many changes and expanded efforts occurring since the first edition was published in 1970, at which time it was acceptable for agronomists to use only a broad title like *Crop Growth and Culture.* In the ensuing years, the discipline of crop physiology has become widely recognized; thus this second edition can be titled *Physiology of Crop Plants.*

As noted in the first edition, the unique contribution of agronomy as a discipline, represented by the subdivisions crops, soils, and climatology, is the integration of biological, chemical, and physical phenomena into useful crop management systems. With the basic biological scientist's continuing emphasis on molecular biology (the reductionist approach), it remains imperative for the agronomist and the crop physiologist to integrate information, synthesize new levels of knowledge, and develop systems for problem solving, all the while interfacing with biologists, chemists, physicists, and researchers in other basic areas of science.

Our treatment continues to break from the tradition of organization on a crop basis; the emphasis is on physiological concepts and the factors influencing metabolism, growth, and reproduction. While crop plant examples are a key part of the discussion, the first order of business is to identify the basic principles that apply across species. As in the first edition, specialized terminology has been kept at a moderate level and illustrations have been used liberally in order to enhance readability and understanding for undergraduates in advanced crop science courses and to provide a text or reference for introductory graduate courses in crop physiology. Crop physiology peers have indicated the desirability of such a dual-level approach.

This discussion thus has two major purposes: to develop an understanding of the important principles underlying the practices used in the culture of crop plants and to develop the ability to apply these principles in production strategies.

The second edition expands on crop physiology and omits specific chapters on seeding; winter and drought survival; weed, insect and disease problems; and harvest and storage. We concluded these topics were more appropriately treated in other courses.

This approach to crop physiology was developed in outline form by Frank Gardner and Roger Mitchell in 1963 and prepared as a first edition by Roger Mitchell. In this second edition, Frank Gardner took a primary role in a major rewriting and Brent Pearce contributed extensively to the expanded focus on crop physiology.

Roger L. Mitchell

ACKNOWLEDGMENTS

TO S. L. Albrecht, D. D. Baltensperger, D. W. Beatty, J. M. Bennett, W. G. Blue, W. G. Duncan, M. H. Gaskins, D. J. Hume, M. J. Kasperbauer, and R. M. Shibles for their most helpful suggestions in the writing of this book.

TO Randy Compton, Lynn Breitenbeck, Beverly Harvey, An Nguyen, and Peg Pearce for their patience and skillful typing.

TO Opal, Peg, and Joyce.

Physiology of Crop Plants

1 Photosynthesis

AGRICULTURE is basically a system of exploiting solar energy through *photosynthesis*. The primary source of energy for humans, photosynthesis has supplied the energy for food, feed, and the fossil fuels that power electrical generating plants and many machines. A study of crop physiology soon leads to the discovery that the yield of crop plants ultimately depends on the size and efficiency of this photosynthetic system. Crop management practices proceed from this assumption. Because photosynthesis is the cornerstone of crop production, it is important to be aware of the energy available to drive photosynthesis and to consider how the anatomical features and biochemical processes in the plant interact to capture and store radiant energy.

Light Used in Photosynthesis

PROPERTIES

Visible light, the source of energy used by the plant for photosynthesis, is part of the radiant energy spectrum (Fig. 1.1). Radiant energy has unique characteristics that can be explained by using two related theories, the electromagnetic wave theory and the quantum theory. The *electromagnetic wave theory* states that light travels through space as a wave. The number of waves passing a given point in a certain interval of time is a *frequency*.

$$\upsilon = c/\lambda$$

where υ = frequency (wavelengths/sec), c = speed of light (3×10^{10} cm/sec), and λ = wavelength. If we divide the speed of light by the frequency, we obtain the *wavelength*.

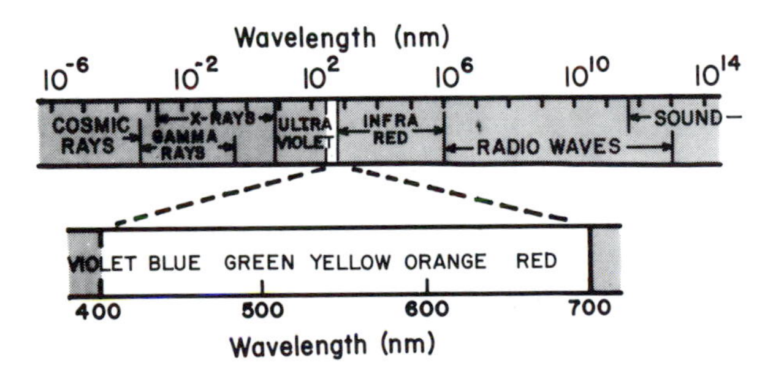

Fig. 1.1. Radiant energy spectrum. Photons in the 400- to 700-nm range are used in photosynthesis.

The *quantum theory* states that light travels in a stream of particles called *photons.* The energy present in one photon is called a *quantum.* Because the energy present in one photon is proportional to the frequency, the quantum can be expressed in terms of wavelength and the energy per photon is inversely proportional to the wavelength (Fig. 1.2).

$$E = hv = c/\lambda$$

where E = photon energy (quantum), h = Planck's constant (662×10^{-7} erg/s), c = speed of light (3×10^{10} cm/sec), and λ = wavelength. The light reaction of photosynthesis is a direct result of photon absorption by pigment molecules such as chlorophyll. Not all photons have the proper energy level to excite leaf pigments. Above 760 nm the photons do not have enough energy; below 390 nm, the photons (if absorbed by leaf pigments) have too much energy, causing ionization and pigment degradation. Only the photons with wavelengths between 390 and 760 nm (corresponding to visible light) have the proper energy level for photosynthesis.

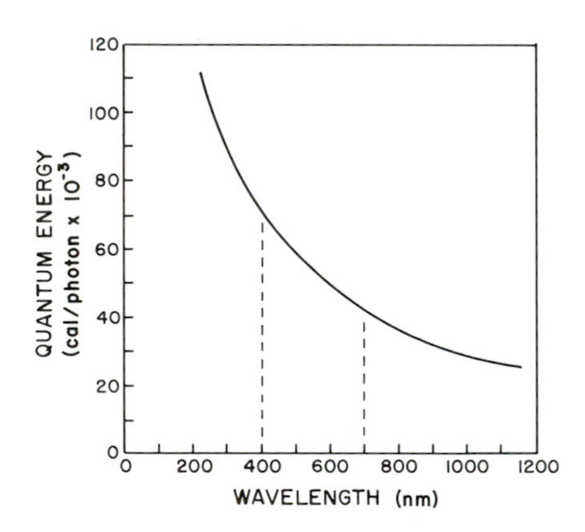

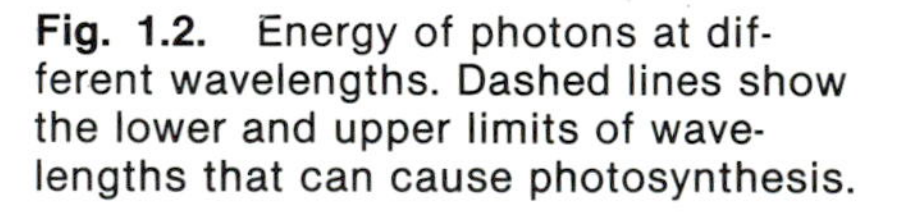
Fig. 1.2. Energy of photons at different wavelengths. Dashed lines show the lower and upper limits of wavelengths that can cause photosynthesis.

Because pigment excitation is a direct result of interaction between a photon and the pigment, a measure of light used in photosynthesis is often based on photon flux density rather than on energy. *Photon flux density* is the number of photons striking a given surface area per unit of time. Since wavelengths between 400 and 700 nm are most efficiently used in photosynthesis, light measurement for photosynthesis is usually based on photon flux density within those wavelengths. These measurements are called photosynthetically active radiation (PAR) or photosynthetic photon flux density (PPFD). The term *Einstein* (E) is defined as one mole of photons, so PAR is often listed in terms of $\mu E \cdot m^{-2} \cdot sec^{-1}$, or under the international system of units as simply $\mu mol \cdot m^{-2} \cdot s^{-1}$.

SOLAR RADIATION

The radiant energy available for photosynthesis on earth comes from the sun. Every energy source used by humans, directly or indirectly, results from solar radiation, with the exception of atomic energy and possibly geothermal energy. For crop growth and development, the sun is the only source of energy.

The sun is a blackbody radiator, and according to Wein's law, the maximum wavelength is inversely proportional to the heat of the body and

$$\max \lambda = 2.88 \times 10^6/K$$

where 2.88×10^6 is Wein's displacement constant and K is the temperature. For example, the temperature of the sun is believed to be 5750 K, so

$$\max \text{ of sun} = (2.88 \times 10^6)/5750 = 500 \text{ nm (green)}$$

Thus, the solar radiation spectrum has a peak at λ of 500 nm (Fig. 1.3). Plants have apparently adapted to solar radiation because the visible light of λ between 400 and 700 nm corresponds to 44 to 50% of the total solar radiation entering the earth's atmosphere.

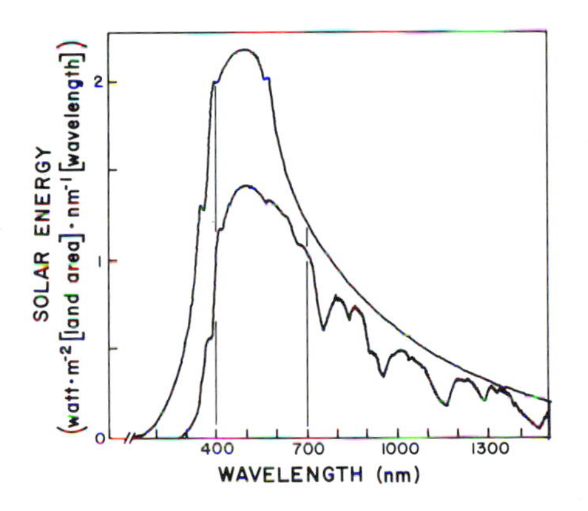

Fig. 1.3. Energy at different wavelengths of solar radiation at solar noon. The top line is the energy just outside the earth's atmosphere, the lower line is the solar energy hitting the earth's surface.

The *solar constant* is 2.00 cal • cm^{-2} • min^{-1} (1395 W • m^{-2}). It is the amount of energy received on a flat surface that is perpendicular to the sun's rays and immediately outside the earth's atmosphere. The solar radiation level decreases as it passes through the earth's atmosphere due to absorption and scattering. The solar radiation at the earth's surface, when that surface is perpendicular to the sun's rays, is reduced from 2.0 to between 1.4 and 1.7 cal • cm^{-2} • min^{-1} on a clear day.

Figure 1.4 illustrates that the axis the earth spins around is tilted in rela-

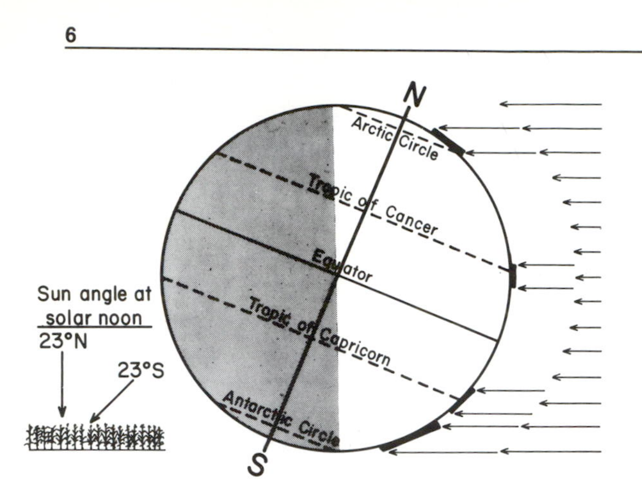

Fig. 1.4. Earth's relationship to the sun on June 22. Earth is angled 23° toward the sun, so the northern hemisphere has days longer than and the southern hemisphere days shorter than 12 hr. The Arctic has constant sunlight (mainly from the horizon), and Antarctic has no direct sun rays. The sun's angle at solar noon is 90° from horizontal at the Tropic of Cancer but only 46° at the Tropic of Capricorn. The situation is reversed on Dec. 22, when the South Pole is oriented 23° toward the sun.

tion to the sun. Therefore annual cycles (Fig. 1.5) and diurnal (daily) cycles (Fig. 1.6) of solar radiation are governed primarily by latitude. Because of this latitude effect, the following factors influence the amount of solar radiation received in one day:

1. Angle of the sun's rays directed on that spot. When solar radiation comes in at smaller and smaller angles from perpendicular to the earth's surface, the light spreads out over a larger ground area, reducing the light level per unit of ground area.
2. Day length.
3. The amount of atmosphere the radiation passed through as a function of the angle of the sun's rays. If the sun is 90° overhead, the number of atmospheres light must pass through equals one; at 60° it is equal to two atmospheres, and at 30° it is equal to five atmospheres.
4. The number of particles (e.g., dust or condensed water particles such as fog or clouds) in the atmosphere. In many tropical regions much less light hits the earth's surface in the cloudy monsoon season than in the cloudless dry season.
5. Other minor factors, such as fluctuations of solar output, distance of the earth from the sun, and the earth's reflecting ability.

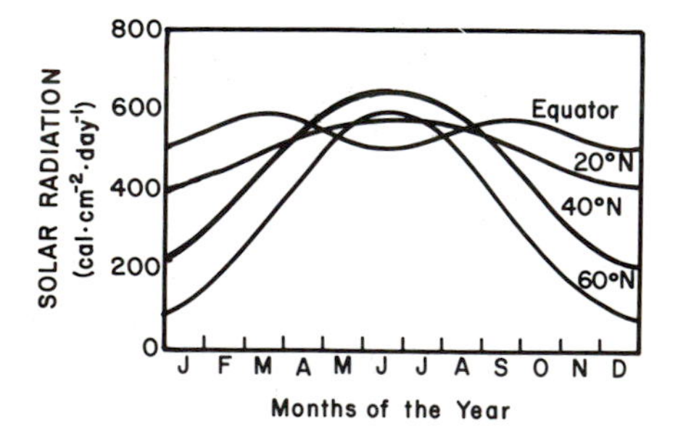

Fig. 1.5. Annual variation in solar radiation energy at different latitudes on cloudless days.

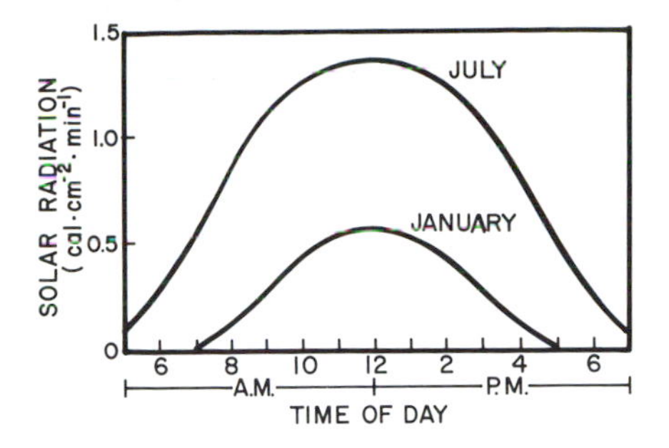

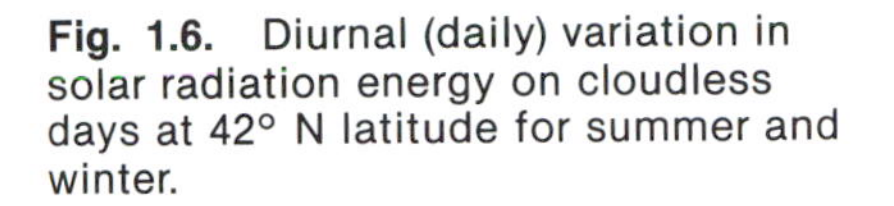
Fig. 1.6. Diurnal (daily) variation in solar radiation energy on cloudless days at 42° N latitude for summer and winter.

Of the solar radiation absorbed during the daytime by a crop surface, 75 to 85% is used to evaporate water; 5 to 10% goes into sensible heat storage in the soil; 5 to 10% goes into sensible heat exchange with the atmosphere by convection processes; and 1 to 5% goes into photosynthesis.

Since the maximum solar radiation level occurs in June and July for the northern hemisphere, the untutored observer might expect agriculturalists to always have their crops ready to make their peak growth at that time (e.g., to have sorghum at the grain-filling stage). However, the opportunity to utilize this radiation peak is limited by seasonal temperature boundaries and the fact that most crops must develop from small seeds or other small organs before the *economic yield* (the harvested portion of the dry matter) can be produced. The challenge to crop physiologists and plant breeders is to develop crops and crop management practices that will place the crop in the appropriate growth cycle to take maximum advantage of this radiation peak.

The Photosynthetic Apparatus

LIGHT REACTION

Electron microscopy has made it possible to look more closely at the chloroplast, which is the photosynthetic apparatus of the plant. The chloroplast, a lens-shaped organelle 1 to 10 μm across, displays two key areas: (1) the *lamellae* (membranes), consisting of stroma lamellae (a double lamella) and grana lamellae (stacked lamella), both of which are concentrated areas of photosynthetic pigments, and (2) the *stroma*, a less dense, fluid area where the reduction of carbon dioxide (dark reaction) occurs (Fig. 1.7). The transformation of light energy to chemical energy (photophosphorylation) occurs in lamellae and consists of the oxidation of water and production of chemical potential, or reduced nicotinamide adenine dinucleotide phosphate (NADPH) and the phosphorylation of adenosine diphosphate (ADP) to adenosine triphosphate (ATP) (Fig. 1.8). The NADPH is one of the most powerful *reductants* (acceptors of electrons and suppliers of hydrogen ions) known in biological systems. ATP is synonymous with available energy in the biological

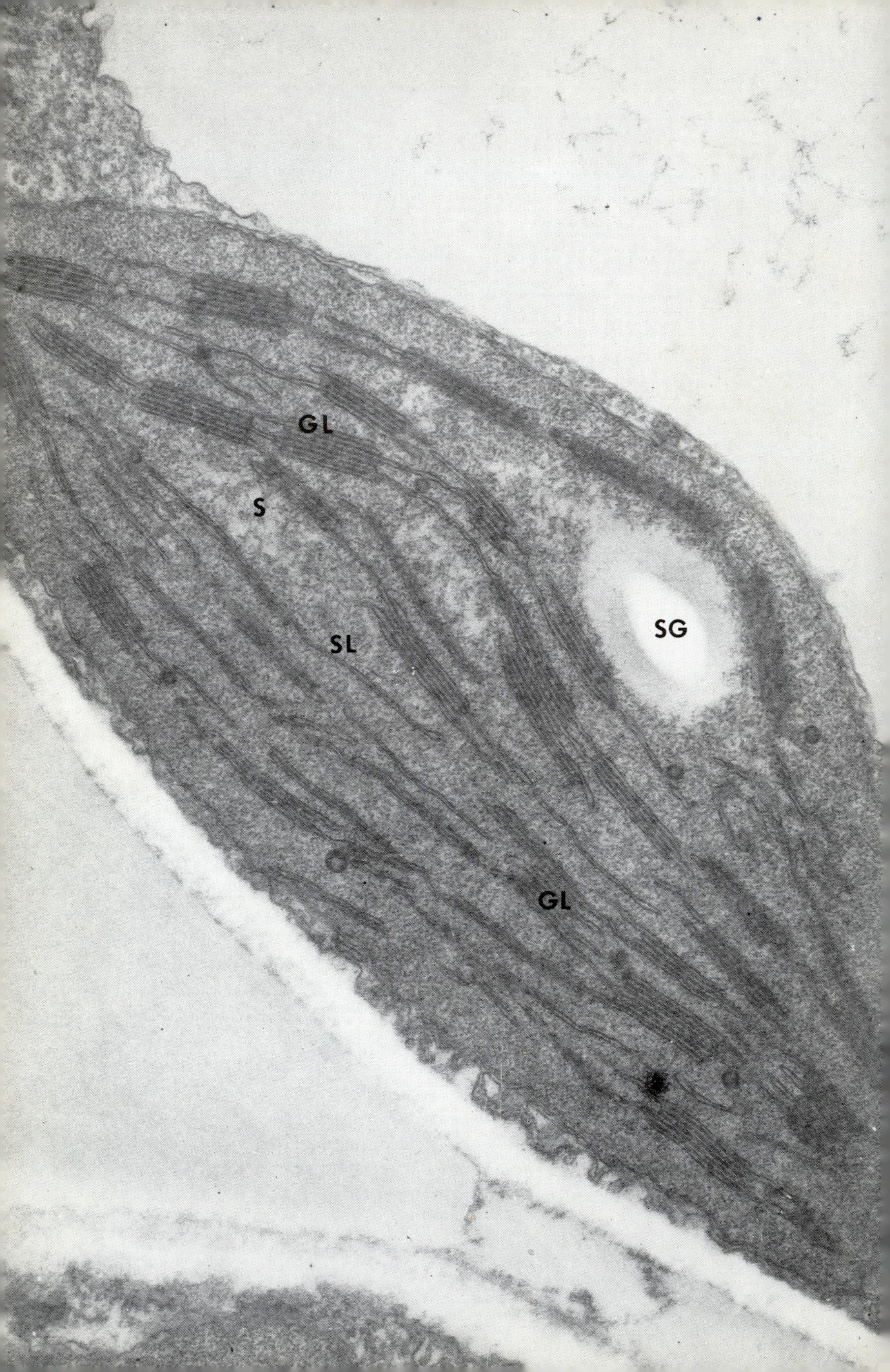
GL
S
SG
SL
GL

Fig. 1.7. Photomicrograph of an alfalfa chloroplast enlarged 64,500 times (Stifel et al. 1968): stroma (*S*), grana lamella (*GL*), stroma lamella (*SL*), starch granule (*SG*).

system; when a phosphate group is released from ATP, energy is also released. The released phosphate, attaching to some molecule (*phosphorylation*) by an energy input, raises the energy of the molecule and allows it to undergo even more chemical reactions. Both NADPH and ATP are needed to convert carbon dioxide (CO_2) to organic molecules.

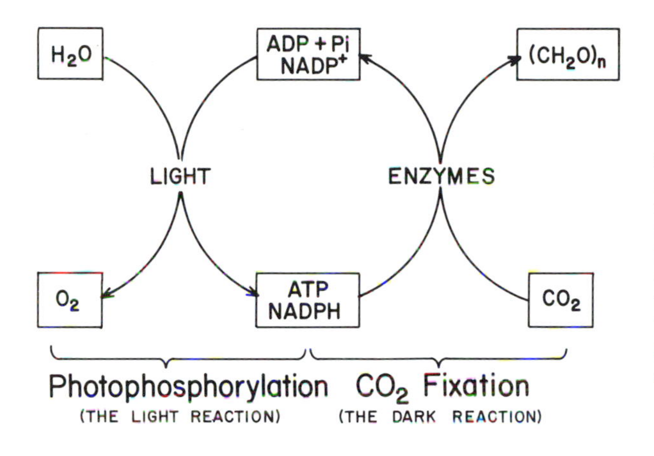

Fig. 1.8. The light and dark reactions that make up photosynthesis. The energy flows from light (irradiance) to high-energy intermediate compounds (ATP and NADPH) and then to long-term energy in bonds connecting carbon atoms of organic molecules.

The electron transport system is fairly well understood (Fig. 1.9). There are two reaction centers where energy from absorbed photons are used to drive the system. These reaction centers have many pigment molecules. When a pigment like chlorophyll or a carotenoid absorbs a photon, the energy lifts an electron (e^-) from a lower (ground) energy state to a higher (excited) state. While in this excited state the pigment molecule can donate and accept electrons from other molecules. Photosystem II catalyzes the removal of electrons from water molecules, and these electrons are accepted by a substance labeled Q. Photosystem I, using more energy from absorbed photons, catalyzes the removal of electrons from Q. This sets up the energy needed for *photophosphorylation* (ATP formation) and the reduction of $NADP^+$ (Fig. 1.9). The chloroplast lamellae are specialized membranes containing the pigments, proteins, and lipid materials that facilitate electron transport (Fig. 1.10).

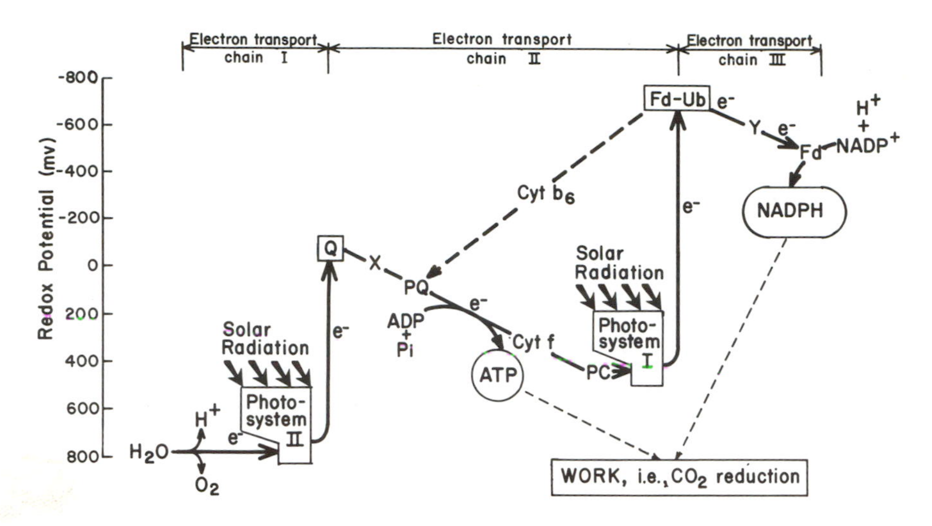

Fig. 1.9. Energy diagram of the photosynthetic electron transport system plotted in terms of standard oxidation-reduction potentials. Electron transport chains I and III are connected by a central chain (II) between the primary electron acceptors of photosystem I and photosystem II, which causes the phosphorylation of ADP to ATP. Noncyclic electron flow employs both systems, starting with water and ending with NADPH. Cyclic electron flow requires only photosystem I (see heavy dashed line) mediated by cytochrome b_6 and generates the production of ATP in electron transport chain II.

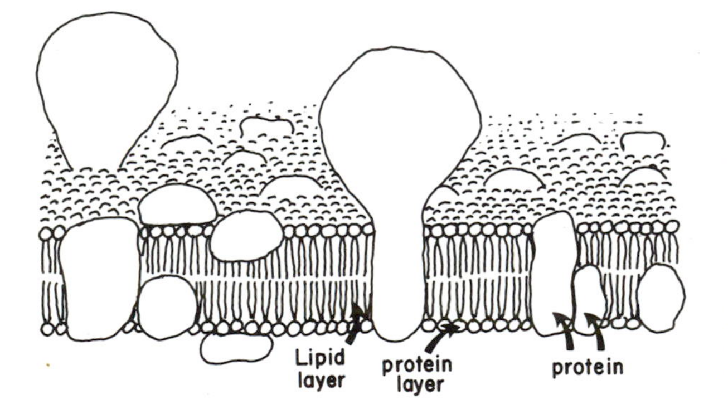

Fig. 1.10. Chloroplast lamella. The internal lipid layer has an external protein layer on both sides. Embedded in the membrane are enzymes and other components of photosynthesis. The pigments are located in the lipid layer with any *polar groups* (i.e., porphyrin ring of chlorophyll) being associated with the protein layer. The lamellar membrane is where the light absorbed by lamella pigments causes electron transport and proton gradients, resulting in the phosphorylation of ADP to ATP and the reduction of $NADP^+$ to NADPH.

The pigments in the chloroplast lamellae consist largely of two kinds of *chlorophyll* (a and b) and two kinds of yellow to orange pigments classified as *carotenoids* (*carotenes* and *xanthopylls*). The structures of some of these pigments are shown in Figure 1.11. Experiments indicate that the *porphyrin ring* of chlorophyll is associated with the protein component of membranes, and the phytol tail as well as the hydrophobic carotenoids are probably associated with the lipid interior of the lamella. Carotenoids serve as auxiliary pigments in light absorption. Some are inactive; others absorb light and transfer excited electrons to chlorophyll as well as from one photosystem to another, a phenomenon known as Emerson enhancement. In addition, they appear to have the capacity to slow the rate of chlorophyll photodestruction (Anderson 1975).

Through adaptation the spectrum of light absorbed by chlorophyll and other leaf pigments corresponds to the visible light range of the human eye. The light absorption by the leaf is quite different from light absorption by chlorophyll in ether. Figure 1.12 shows that *quantum efficiency* (moles of CO_2 reduced per mole of photons) in field beans ranges from 8 to 12% in monochromatic (single wavelength) light from 400 to 700 nm. The red range is most efficient, blue next, and green the least. There is not as much variation in either the action spectrum or the adsorption spectrum of leaves from 400 to 700 nm as one might expect when looking at the variation in light absorption by individual leaf pigments.

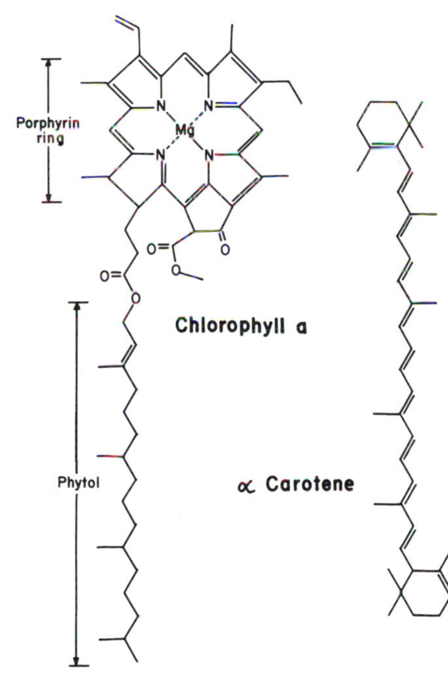

Fig. 1.11. Structures of chlorophyll a and α carotene.

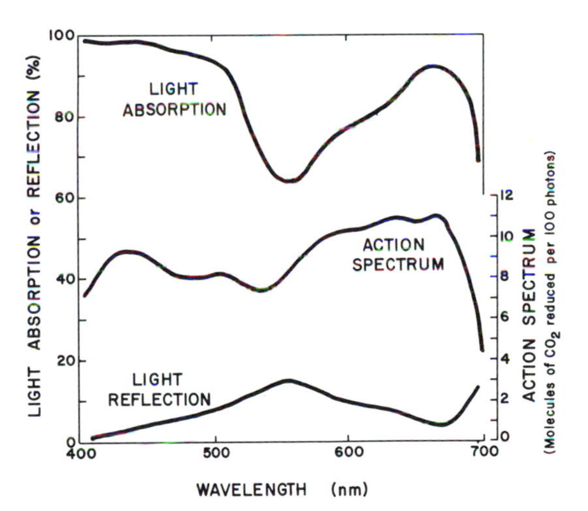

Fig. 1.12. Light absorption and high reflection of bean leaves, compared with their action spectrum (Balegh and Biddulph 1970, by permission).

Acceptors and donors of e^- other than pigments are associated with the lamella proteins. One of these compound types are *cytochromes*, which are proteins that have porphyrin rings similar to chlorophyll. However, the central metallic element is iron (Fe), not magnesium, that gives up or accepts the e^-. In other compounds copper (Cu) donates or accepts e^-.

$$e^- + Cu^{++} = Cu^+$$
$$e^- + Fe^{+++} = Fe^{++}$$

CARBON DIOXIDE FIXATION

Agriculture is based on the yield or weight of crop products. Since the harvested weight is usually calibrated at a specific moisture content, yield is equated to *dry matter production* by the plant, which is the balance between CO_2 uptake (photosynthesis) and CO_2 evolution (respiration). During growth daily respiration for most crop species under cropping environments is 25 to 30% of total photosynthesis, so the plant increases in dry weight. When respiration is more than photosynthesis, the plant loses dry weight, as can be shown by putting a plant in the dark, preventing photosynthesis.

Light reactions transform light energy to the short-term chemical energy of NADPH and ATP. These compounds are then used to reduce CO_2 to stable organic compounds from which dry weight results.

Carbon Dioxide Fixation—C_3 Species. The path of carbon in photosynthesis, which forms the basis of our present knowledge, has been worked out by Calvin and co-workers (Bassham and Calvin 1957). This Calvin cycle is outlined in Figure 1.13. The CO_2 fixation portion of the Calvin cycle is catalyzed by the enzyme ribulose bis-phosphate (RuBP) carboxylase. Notice that the ATP produced in photophosphorylation is used to convert ribulose-5-phosphate to RuBP (Fig. 1.13). After CO_2 fixation ATP, along with reduced nucleotides from the light process, change 3-phosphoglyceric acid (3-PGA) to 3-phosphoglyceraldehyde (3-PGald). Species with this pathway are called C_3 pathway species because the first product that can be measured after adding radioactive CO_2 ($^{14}CO_2$) is a three-carbon molecule, 3-PGA.

Carbon Dioxide Fixation—C_4 Species. From 1954 to 1966 the Calvin cycle was considered the only pathway for CO_2 fixation in higher plants. Then Hatch and Slack (1966), working in Australia, presented detailed evidence that another pathway for CO_2 fixation exists in some species. This pathway incorporates CO_2 using phosphoenol pyruvate (PEP) carboxylase enzyme. Notice in Figure 1.14 that the ATP produced in photophosphorylation is used to convert pyruvate to PEP. The PEP, a three-carbon molecule, is carboxylated to three four-carbon acids (oxaloacetate, malate, and aspartate). These acids

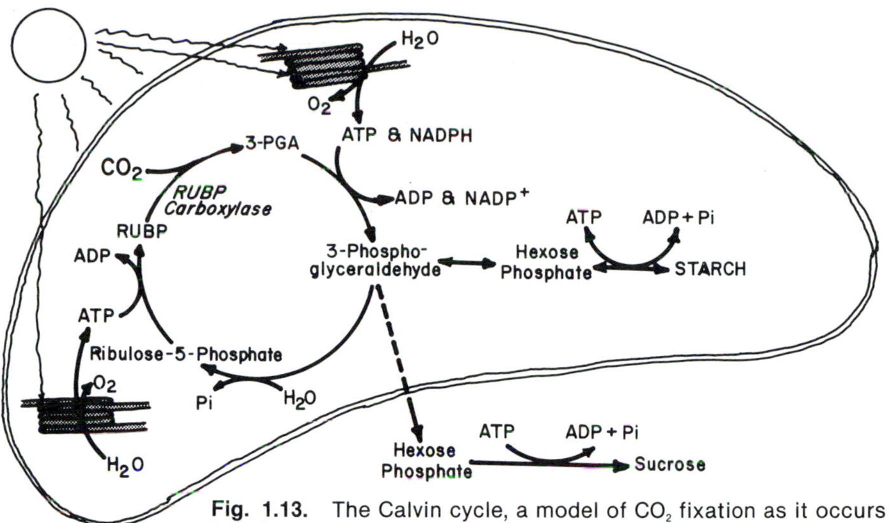

Fig. 1.13. The Calvin cycle, a model of CO_2 fixation as it occurs in a C_3 photosynthesizing chloroplast.

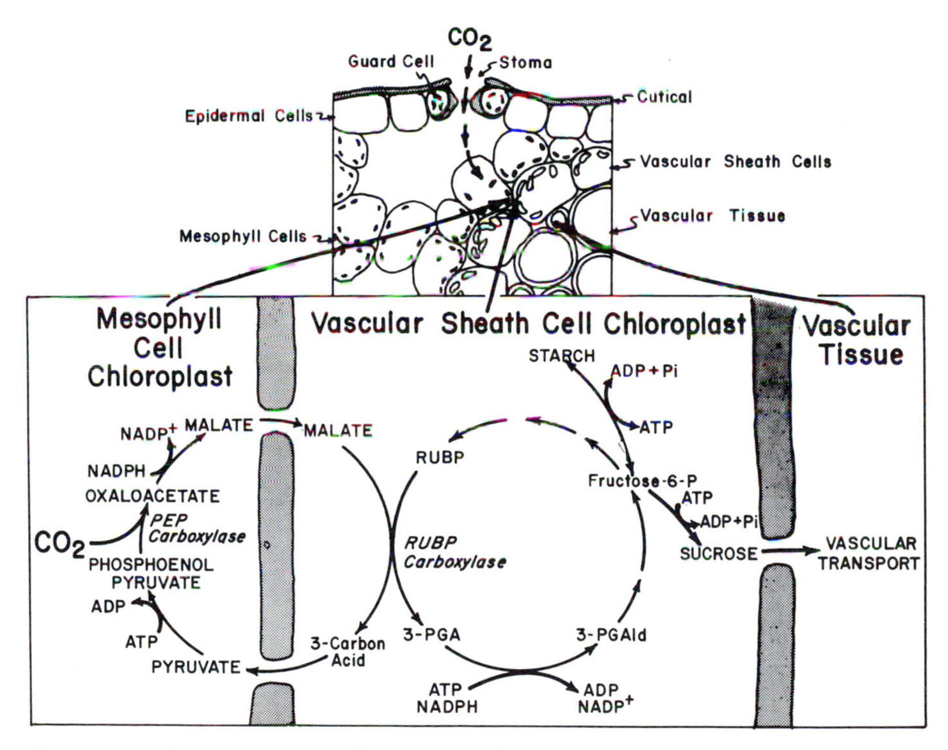

Fig. 1.14. CO_2 movement and fixation in a C_4 photosynthesizing species.

are translocated to vascular sheath cells where they are converted to pyruvate. In the change to pyruvate, a carbon is released that is converted, either by addition to RuBP or by addition to a two-carbon molecule, to 3-PGA by RuBP carboxylase. After 3-PGA is produced, the Calvin cycle is operative. Species with the Hatch and Slack pathway are called C_4 species because the first product of photosynthesis in the mesophyll is a four-carbon molecule.

Comparing C_3 and C_4 Species. Comparisons between the species using these two pathways show many differences.

1. Anatomical differences (The leaf anatomy of C_4 species is called the *Kranz anatomy.*)
   a. The C_4 species have chloroplasts in the vascular sheath cells; C_3 species do not.
   b. Chloroplasts in the mesophyll of C_3 and C_4 species look similar (they usually have double external membranes and well-developed grana) but are very different biochemically. In C_3 species, CO_2 is fixed by RuBP carboxylase, the Calvin cycle is operative, and starch is accumulated (Fig. 1.13). In C_4 species, CO_2 is fixed by PEP carboxylase, which forms four-carbon acids that translocate to vascular sheath cells. Starch is not formed in these mesophyll cells, only four-carbon acids.
   c. Chloroplasts in vascular sheath cells of C_4 species are anatomically different. They are larger and have less-developed grana than in mesophyll cell chloroplasts; and since the Calvin cycle is operative, they store starch.
2. PEP carboxylase enzyme has a greater affinity for CO_2 than does the RuBP carboxylase enzyme, so it can operate more efficiently at low CO_2 concentrations.
3. Species with the C_4 pathway generally have higher photosynthetic rates than do C_3 species, especially at high light intensities (see Fig. 1.20).
4. C_4 species may use more energy than C_3 species to fix one CO_2 molecule. This point has not been proven but seems probable, as ATP is required to form PEP.
5. Ribulose bis-phosphate carboxylase is present at much lower levels in C_4 species than in C_3 species (i.e., around 10%). C_3 species apparently do not have PEP carboxylase present.
6. Differences in adaptation occur for species with different CO_2 fixation mechanisms: C_3 species are adapted to cool and moist to hot and moist conditions while C_4 species are adapted to hot, dry, or moist conditions.
7. The main factor causing the increased photosynthetic efficiency of C_4 species is the lack of measurable *photorespiration* (respiration in the light). It results in CO_2 loss in photosynthetic tissue and is a major source of CO_2 evolution by C_3 species in the light. It occurs as a by-product of the Calvin

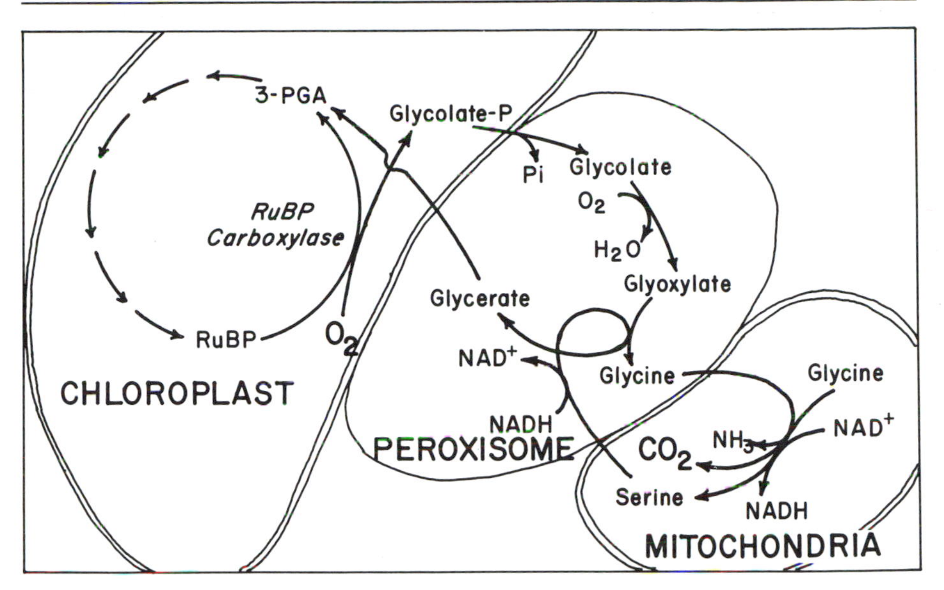

Fig. 1.15. Pathway of photorespiration in a leaf mesophyll cell of a C_3 photosynthesizing plant.

cycle (Fig. 1.15); since RuBP carboxylase is also RuBP oxygenase, the O_2 and CO_2 compete for the same enzyme and for the same ribulose bisphosphate substrate. Photorespiration is not significant in C_4 species; this is believed to be the major factor giving C_4 species higher photosynthetic efficiency than C_3 species. The C_4 species are believed to have little or no photorespiration because movement of four-carbon acids into the vascular sheath cells concentrates CO_2 in these cells, which would favor the RuBP carboxylase reaction over RuBP oxygenase. However, any CO_2 evolved from the vascular sheath cells would probably not leave the leaf because of the great affinity of PEP carboxylase for CO_2 in mesophyll cells; thus any photorespiration that did occur could not be measured (Goldsworthy 1970). Photorespiration evolves CO_2 without seemingly coupling the energy acquired to a useful purpose. Photorespiration does, however, provide amination for amino acid synthesis and keeps cycling inorganic phosphate, which may be beneficial under low light and cool temperatures. Special techniques must be used to measure photorespiration:

a. Air without CO_2 is passed over an illuminated leaf. If CO_2 is evolved, it is a measure of photorespiration (Fig. 1.16).
b. Plants or leaves in a gas-tight container under illumination will pull down the CO_2 concentration in the air until it comes to an equilibrium (compensation concentration), which is a measure of photorespiration (Fig. 1.16).

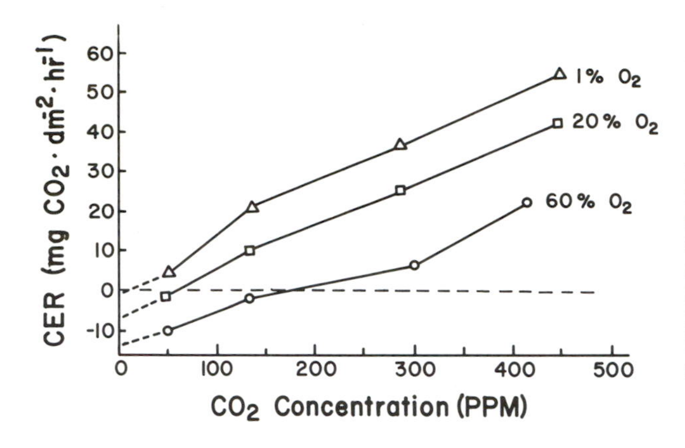

Fig. 1.16. CO_2 exchange rates (CERs) of soybean (C_3 species) leaves at different CO_2 concentrations and different O_2 concentrations. Note that photorespiration is near zero at 1% O_2 and increases with the O_2 concentration. (From Hitz 1978)

c. If a leaf is suddenly put into darkness, photosynthesis will stop but photorespiration will continue a short while to use up the glycolic acid. This causes a postillumination burst of CO_2 evolution much greater than the equilibrium of CO_2 evolution of dark respiration.

d. Oxygen (O_2) is required for the conversion of glycolic acid to glyoxylic acid. If O_2 is reduced in the air from 21 to 1% or less, the photorespiration is stopped. Therefore, the difference in photosynthesis at O_2 concentrations of 21 and 1% is a measure of photorespiration (Fig. 1.16).

Both types of CO_2 fixation occur in crop species.

C_3 species	C_4 species
Cool-season grasses (e.g., wheat, oats, barley, rice, rye, bluegrass, fescue, bromegrass)	Warm-season grasses (e.g., corn, sorghum, sudangrass, sugarcane, millets, Bermudagrass, warm-season prairie grasses)
Dicot species (e.g., legumes, cotton, sugar beets, flax, tobacco, potatoes)	Dicot species (no major crop species but found in several weed species, e.g., *amaranthus,* pigweed)

Crassulation Acid Metabolism Plants. A third type of CO_2 fixation, called crassulation acid metabolism (CAM), occurs primarily in *succulent* plants, which have fleshy leaves or stems. Such plants are adapted to arid conditions where low *transpiration* (evaporation from plant surfaces) is a survival necessity. Under low moisture conditions they open their stomata at night to absorb CO_2 and close them in the day, thus reducing the transpiration load of the plant. There are only a few CAM plants classified as domestic crops; these include pineapple, *Agave* (sisal, henequen, and others), and prickly pear.

The CAM species fix CO_2 into four-carbon acids with PEP carboxylase as do the C_4 species, only it occurs at night when stomata are open and the energy required comes from glycolysis. Solar radiation causes stomatal closure and irradiates the leaf; this light energy is used to drive the Calvin cycle, taking CO_2 from the four-carbon acids as in the reaction in the bundle sheath cells of C_4 species. The chloroplasts of CAM plants are more similar to those of the C_3 species. Under favorable moisture conditions, many CAM species, change stomatal function and carboxylation is similar to that of C_3 species.

Thus, CAM plants have developed an ingenious physiological method of reducing moisture loss and escaping drought. They often are important crop plants where moisture availability for crops remains low.

The Leaf as a Photosynthetic Organ

The leaf serves as the major photosynthetic organ of higher plants. Leaf evolution has provided a structure that will withstand environmental rigors and yet provide both effective light absorption and rapid CO_2 uptake for photosynthesis. Most crop leaves have (1) a large, flat external surface; (2) upper and lower protective surfaces; (3) many stomata per unit area; (4) extensive internal surface and interconnecting air spaces; (5) an abundance of chloroplasts in each cell; and (6) a close relationship between the vascular and photosynthetic cells. A leaf ideal for gaseous exchange and light interception would be only one cell thick, but the rigors of the natural environment demand several layers of cells and a protective surface for survival.

The large, flat external surface of the leaf allows for maximum light interception per unit of volume and minimizes the distance CO_2 must travel from leaf surface to chloroplast, a distance of around 0.1 mm for the leaves of most crops (Fig. 1.17). The epidermis serves as a barrier to gaseous exchange pri-

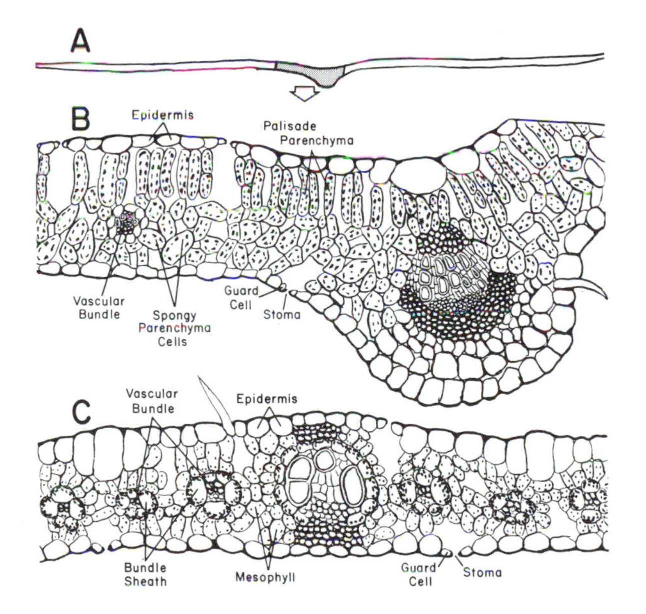

Fig. 1.17. Leaf cross-sections: *A.* Outline of an alfalfa leaflet; *B.* cross-section of an alfalfa (C_3 dicot) leaf; *C.* cross-section of a maize (C_4 grass) leaf. The dark areas in cells represent chloroplasts.

TABLE 1.1. Stomatal number and size of several crop species

		Stomata (number/mm^2)		Open
Common Name	Scientific Name	Adaxial (upper) epidermis	Abaxial (lower) epidermis	Stomatal Size (μm)
Alfalfa	*Medicago sativa* L.	169	138	. . .
Apple	*Pyrus malus* L.	0	294	. . .
Bean	*Phaseolus vulgaris* L.	40	281	7 × 3
Cabbage	*Brassica oleracea* L.	141	226	. . .
Castor bean	*Ricinus communis* L.	64	176	10 × 4
Maize	*Zea mays* L.	52	68	19 × 5
Oat	*Avena sativa* L.	25	23	38 × 8
Potato	*Solanum tuberosum* L.	51	161	. . .
Sunflower	*Helianthus annus* L.	85	156	22 × 8
Tomato	*Lycopersicon esculentum* Mill.	12	130	13 × 6
Wheat	*Triticum sativum* L.	33	14	38 × 7

Source: From Meyer et al., 1960.

marily because epidermal cells are covered by a waxy layer called the cuticle. Both the cuticle and the epidermis are nearly transparent and readily allow visible light to enter the leaf. The cuticle prevents gaseous exchange between leaf and atmosphere, which is important in preventing excess water loss. Most of the gaseous exchange in leaves occurs through stomata. There are many stomata on the leaf surface (12 to 281 stomata • mm^{-2}), which allows for maximum CO_2 diffusion into the leaf when stomata are open (Table 1.1). Guard cells surrounding the stomatal opening control opening and closing (Fig. 1.18). Stomatal closing is important in preventing water loss when water is limiting, but at the same time it limits CO_2 uptake for photosynthesis. Most crop species are grown under full solar radiation and have stomata on both sides of the leaf. Most shade species have stomata only on the *abaxial* (lower) epidermis.

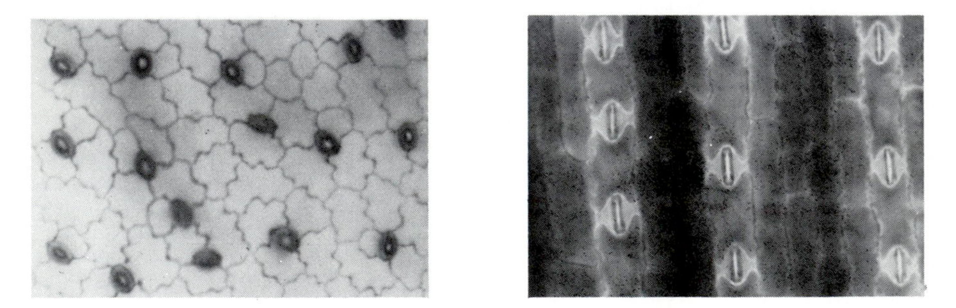

Fig. 1.18. Epidermis of (*left*) a clover (dicot) leaf and (*right*) a maize (grass) leaf. Differences in guard cells and stomata arrangement in the two species can be noted.

Inside the leaf are many mesophyll cells and intercellular spaces. Dicot and grass types have different leaf anatomies, but there is little indication that any one structure is more efficient in intercepting light or in CO_2 diffusion. However, anatomical differences among C_3, C_4, and CAM species do affect photosynthetic efficiency.

The many mesophyll cells in leaves increase the total internal surface area (6 to 10 times the exterior area) to allow CO_2 to come into more contact with cell walls. The intercellular spaces allow for rapid CO_2 diffusion from stomata to cell surfaces. The pathway of CO_2 into the leaf is from stomata to the cell walls, where it dissolves in water and then diffuses to the chloroplast due to a gradient established by CO_2 fixation.

Most mesophyll cells contain a large number of chloroplasts (20 to 100 per cell) where the light reaction for photosynthesis takes place. When light illuminates the leaf, the chloroplasts often congregate along the side of the cell wall, orienting themselves to intercept the most light under dim conditions or sometimes to intercept the least light under high-light conditions. Being close to cell walls also facilitates rapid CO_2 diffusion from cell walls into chloroplasts.

Leaf cells are not far from vascular tissue, which allows for rapid movement of water and minerals to the photosynthetic cells and of photosynthetic products from the cells and from the leaf (Fig. 1.17). Reduction in the movement of raw materials to the chloroplasts or of products from the chloroplasts can reduce the photosynthetic rate.

Factors Essential for Photosynthesis

Light, CO_2, and the proper temperatures, the factors directly affecting photosynthesis, will be discussed in this chapter. Water and mineral elements also influence photosynthesis and will be discussed in greater detail in subsequent chapters.

LIGHT

The use of light by leaf surfaces has been discussed. Light response curves for leaves are illustrated in Figures 1.19 and 1.20. If there is no light, there is dark respiration, which for a leaf is usually 5 to 10% of CO_2 uptake in bright light. As the light level gradually increases, photosynthesis increases to the *light compensation level*, which is the light level at which CO_2 uptake is equal to CO_2 evolution (the carbon exchange rate, or CER = 0). If the light level continues to increase, there is less increase in CER for each unit increase in light level until the *light saturation level* is reached. Any increase in light level after this level will not significantly increase CER; therefore leaves are more efficient at utilizing light energy at low irradiance levels.

Species differ in their responses to light levels. Most C_4 species (Fig. 1.20, curve *A*) are able to increase photosynthesis even at light levels equal to full

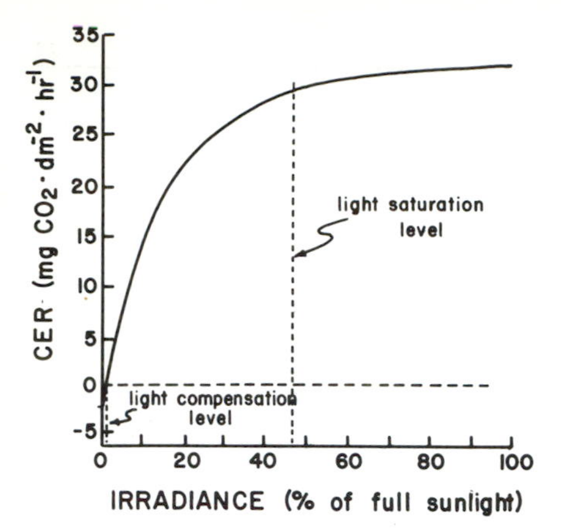

Fig. 1.19. Light-response curve for a CO_2 exchange rate (CER) measurement on a red clover (C_3 species) leaf. The light compensation level is the irradiance level at which CO_2 uptake due to photosynthesis is equal to CO_2 evolution due to respiration. The light saturation level is the irradiance level at which an irradiance increase would not result in a significant CER increase.

sunlight, whereas most C_3 species reach light saturation before full sunlight. Figure 1.20 illustrates that usually the lower the maximum CER, the lower the light level at which light saturation occurs. It should be noted that even though C_4 species often do not become light saturated and do use high light levels better than C_3 species, they use dimmer light more efficiently (CO_2 uptake per unit of light) than bright light. For example, at 50 and 10% of full sun the CER is approximately 72 and 17%, respectively, of that at full sun; the most efficient use of light by CER is always at the lowest light levels. Efficiency is the slope of the light response curve.

CARBON DIOXIDE

Concentration in the Atmosphere. Carbon dioxide is a gaseous component of air. Dry air contains 78% nitrogen (N_2), 21% oxygen (O_2), 0.93% argon (Ar),

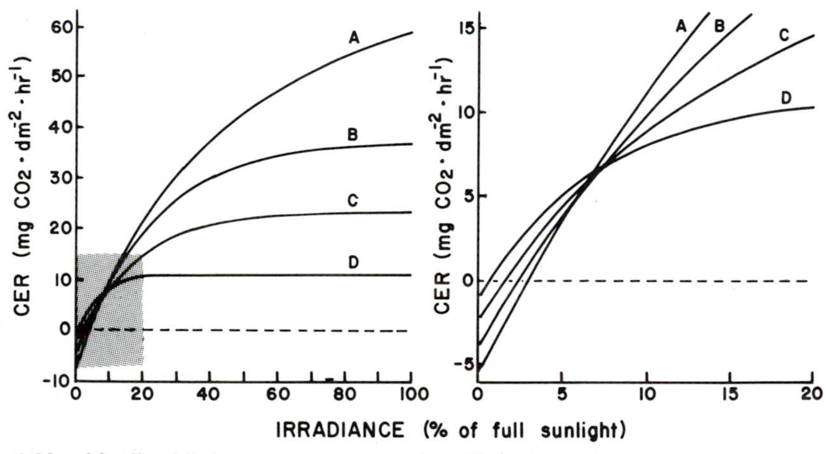

Fig. 1.20. Idealized light-response curves for different species. The graph on the right is an enlargement of the shaded area of the graph on the left. The letters represent the following species: (*A*) C_4 species (e.g., corn, sorghum, sugarcane, bermuda grass); (*B*) efficient C_3 sun species (e.g., soybean, cotton, alfalfa); (*C*) less efficient C_3 sun species (e.g., tobacco, red clover, orchard grass); and (*D*) C_3 shade species (e.g., hardwood trees, house plants).

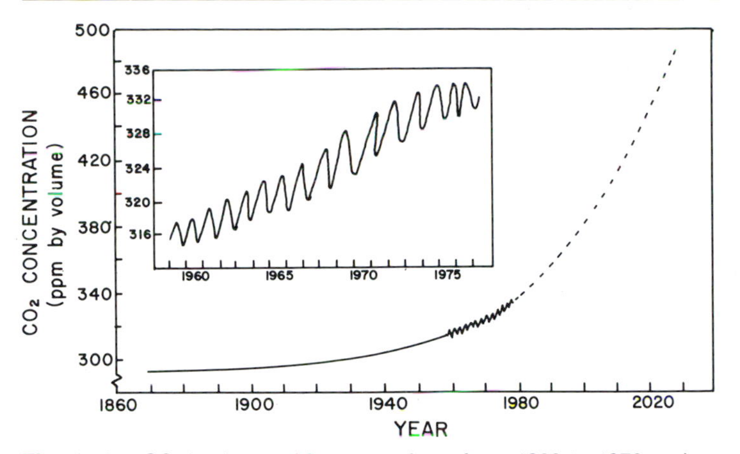

Fig. 1.21. CO_2 in the earth's atmosphere from 1860 to 1976 and projected to 2030 on predicted fossil fuel utilization. (From Kellogg 1977) The inset, which represents the sawtoothed area of the curve, shows CO_2 measurements taken at the Scripps Institute of Oceanography, in Hawaii. These seasonal fluctuations reflect CO_2 removal by photosynthesis during the growing season in the northern hemisphere and CO_2 evolution during fall and winter months. (From Woodwell 1978)

0.034% (340 ppm) CO_2, and traces of other gases. Although CO_2 is at a low concentration, 85 to 92% of a plant's dry weight is derived from CO_2 uptake in photosynthesis.

Because of the burning of fossil fuels (which represent photosynthetic production millions of years ago) and the burning of forests, there has been an increasing CO_2 concentration in the atmosphere (Fig. 1.21). The projections for fossil fuel (primarily coal) utilization indicate an even larger increase in this concentration in the future. Since CO_2 causes a greenhouse effect by absorbing infrared bands of light, the higher concentration will cause the earth to retain more heat, which could increase mean global temperatures. Such an increase could influence the global weather pattern enough to change rainfall patterns and crop productive capabilities in many regions of the earth (Williams 1979).

Under high-light conditions, most crop species show a linear response for leaf photosynthesis to CO_2 levels above the current atmospheric concentration of 340 ppm (Fig. 1.22). Crop yields could be increased considerably in an atmosphere enriched with CO_2 up to 1500 ppm. Although there is no current practical means of doing this under field conditions, CO_2 enrichment has shown great benefits in greenhouses, not only increasing dry matter yield but also hastening plant development. It would be interesting to know how much the increase in CO_2 concentration in the atmosphere in the past 100 years (from approximately 290 to 340 ppm) has increased crop yields and influenced crop maturation.

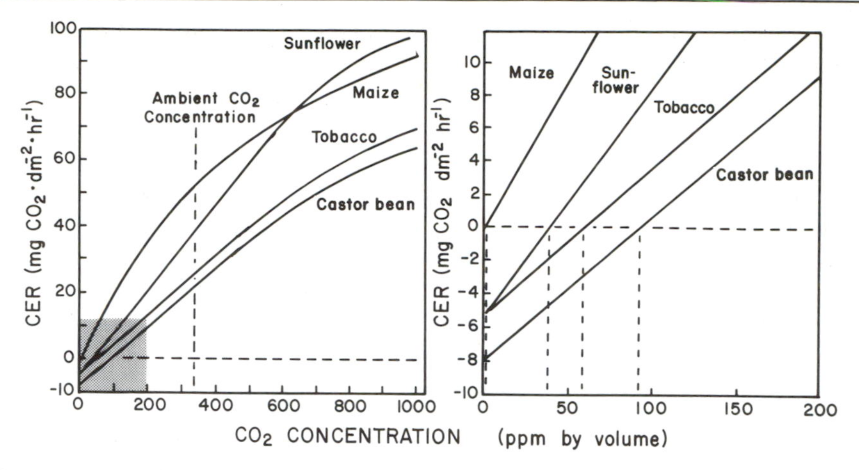

Fig. 1.22. The CO_2 exchange rates (CERs) of four species in response to CO_2 concentration at a photosynthetically active radiation level equal to full sunlight. (*Right*) A close-up of the CO_2 compensation levels from the graph on the left. (From Hesketh 1963)

Leaf Resistances to CO_2 Assimilation. Carbon dioxide gets to the chloroplast by diffusion from the air through the stomata to the cell and then to the chloroplast.

Impediments to CO_2 movement into and through the leaf do occur, and scientists have termed them *resistances* and quantified them.

$$r_{CO_2} = r_a + r_s + r_m \tag{1.1}$$

where: r_{CO_2} = CO_2 exchange rate, r_a = laminar resistance; r_s = stomatal resistance; and r_m = mesophyll resistance. *Laminar resistance* (r_a) is the CO_2 concentration at the leaf surface (also called the *boundary layer effect*); the lower the concentration, the higher the resistance. Since ambient CO_2 concentration is between 300 and 360 ppm, the factors that cause reduced concentration would increase r_a. In the field, turbulence is the major factor influencing r_a. If no air movement occurs, CO_2 uptake by the leaf causes a CO_2 diffusion gradient that reduces CO_2 concentration at the leaf surface. As wind turbulence increases, it will eventually reduce r_a to the minimum level at most leaf surfaces within a plant canopy.

Stomatal resistance (r_s) is the resistance of CO_2 diffusion from outside the leaf through the stomata. Crop leaves generally have the stomatal frequency necessary for efficient CO_2 diffusion. The major factor affecting r_s is the degree to which stomata are open. To calculate r_s, crop physiologists measure the water loss from the leaf, which is a measure of stomatal impedance and diffusion. The r_s is easily measured assuming that the relative humidity inside the leaf remains near saturation and any water loss is due to stomatal opening and r_a.

Mesophyll resistance (r_m) is calculated as residual resistance to CO_2 uptake by the leaf:

$$r_m = r_{CO_2} - r_a - r_s$$

Mesophyll resistance is a measure of everything about the leaf that affects CO_2 uptake except for r_a and r_s. This is because anything that influences CO_2 fixation will affect CO_2 concentration at the chloroplast, which in turn influences the total diffusion rate of CO_2 from air to chloroplast.

The resistance formula (1.1) is used by crop physiologists as a method to determine if CO_2 uptake by a crop plant is affected by resistance to CO_2 diffusion into the leaf (r_a and r_s) or by CO_2 fixation in the leaf (r_m).

Temperature. Photosynthesis must be separated into its component parts to establish its response to temperature. The light reaction, or photophosphorylation, is independent of temperature in the temperature range in which plants grow. Carbon dioxide fixation is an enzymatically controlled reaction and increases at an increasing rate with increases in temperature until temperature reaches a level that favors enzymatic denaturization. Respiration rates will continue to increase as temperature increases. Measurement of net CERs show a minimal response of CER to temperature (Fig. 1.23). Photorespiration also increases with temperature, since it is also enzymatically controlled, resulting in lower CER rates for C_3 species than for C_4 species at the higher temperatures of plant growth.

Water. Water is a substrate for photosynthesis, but only about 0.1% of the total water is used by the plant for photosynthesis. Transpiration accounts for 99% of the water used by plants; approximately 1% is used to hydrate the plant, maintain turgor pressure, and make growth possible. The primary influence of water stress on CER is an increase in r_s due to stomatal closure. If

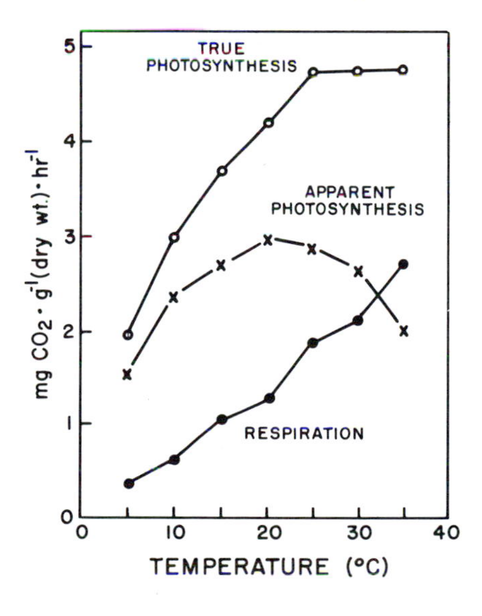

Fig. 1.23. Photosynthesis as affected by temperature in bryophyllum. Note that respiration (CO_2 evolution) increases 8-fold, CO_2 uptake increases 2.5-fold, and CO_2 exchange rate increases 2-fold to a peak and then decreases, which shows that apparent photosynthesis remains fairly stable over the temperature range of growth. (From Stalfelt 1937)

water stress becomes severe, r_m will also increase because of permanent damage to the photosynthetic apparatus. The effects of water on photosynthesis will be discussed in further detail in Chapter 4.

Leaf Age and Mineral Status. Leaf age has an effect on photosynthesis: senescence causes a reduction in the process. The major factor that influences the rate of senescence is the mineral nutrient status of the leaf. Adequate mineral nutrient supplies allow both old and young leaves to meet their nutrient needs. However, limited nutrients are preferentially distributed to young leaves and reduce the photosynthetic rate of older leaves.

In maize, Peaslee and Moss (1966) measured lower photosynthetic rates for lower leaves. The lower rates were correlated with lower levels of potassium, phosphorus, magnesium, and nitrogen (Table 1.2). Apparently if these nutrients are in short supply they are translocated from older to younger leaves, causing what appears to be more rapid ageing in lower leaves. Other nutrients that are less mobile in the plant (e.g., calcium and iron) can reduce photosynthesis in younger leaves, while photosynthesis increases in older leaves due to the steady increase in calcium and iron content over time.

Reduced nutrient levels influence photosynthesis primarily by influencing the photosynthetic apparatus. For example, chlorophyll contains both nitrogen and magnesium; if they are limited in supply, chlorophyll may not form. Precursor molecules for chlorophyll synthesis include iron, and if it is not present, chlorophyll cannot be synthesized. Nutrient effects will be covered in more detail in Chapter 5.

Differences in Photosynthetic Rates among and within Species

Figure 1.20 shows CO_2 exchange rates (photosynthesis) in response to light. Response *A* is typical for crop species with the C_4 pathway. Responses *B*

TABLE 1.2. Potassium content and photosynthesis of maize leaves

Leaf Number (from top)	K Content (μg/g fresh wt)	Photosynthesis (mg $CO_2 \cdot dm^{-2} \cdot h^{-1}$)
Well fertilized		
2	6,100	40
4	5,500	38
7	5,000	36
11	4,350	36
K-stressed		
2	2,150	33
6	800	15
7	600	14
11	250	1

Source: From Peaslee and Moss 1966.
Note: Environment is high light, 30°C.

and *C* represent crop species with the C_3 pathway. Response *D* represents C_3 plants that show adaptation to shade conditions; these include certain hardwood trees and house plants. Plants with response *D* have stomata only on lower sides of leaves (this does not account completely, however, for their low photosynthetic rates) and are inefficient at dry matter production. Few if any crop species have response *D* photosynthetic rates.

Many studies have shown that varieties within a crop species have different CO_2 exchange rates (Table 1.3), with 2-fold to 3-fold ranges in CO_2 uptake between the lowest and highest examples. This has encouraged speculation that yields might be increased by selecting for, and developing populations with, higher CO_2 uptake rates.

TABLE 1.3. Photosynthetic variation within selected species

Species	Location	Photosynthesis ($mg\ CO_2 \cdot dm^{-2} \cdot h^{-1}$)
Maize, *Zea mays* L.	New York[a]	21–59
	Philippines[a]	28–85
	Iowa[b]	22–52
Soybean, *Glycine max* L.	Iowa[c]	29–43
	Illinois[d]	12–24
Alfalfa, *Medicago sativa* L.	Maryland[e]	28–60

Note: Within each of these species there is at least a 3-fold difference in photosynthesis.
[a]Heichel and Musgrave 1969.
[b]Crosbie et al. 1977.
[c]Dornhoff and Shibles 1970.
[d]Curtis et al. 1969.
[e]Pearce et al. 1969.

Since C_4 species have a high CO_2 uptake rate and are among the most productive crop species (e.g., maize, sorghum, and sugarcane), it would appear desirable to introduce the C_4 mechanism into C_3 crop species. Several attempts have been conducted with soybeans, barley, and a few other crops to determine if any C_4 plants occur in C_3 crop species, but without success. In the future other methods may be tried to change a C_3 to a C_4 species.

Photosynthate Utilization by the Plant

FOR STORAGE AND STRUCTURE

Although it is convenient to regard photosynthesis as ending with the formation of hexose sugar, many further changes may occur. The hexose may immediately interconvert from glucose to fructose, or combine to form sucrose for translocation to other cells, or polymerize to starch for temporary storage in the chloroplast. The sucrose may go to enlarging cell walls, where it may be transformed to structural components such as cellulose. The sucrose might also be transported to other areas of the plant where there is active growth (meristems) or where it is converted to polysaccharides as storage or structural compounds.

RESPIRATION AND GROWTH

Hexose can also enter into the respiratory system of the cell where it is broken down to release energy or be converted to organic components used in important structural, metabolic, and storage compounds (Fig. 1.24). The first step is the anaerobic respiratory process called glycolysis, in which reduced nucleotides and ATP are formed to do work in the cells by splitting a hexose phosphate sugar to pyruvic acid. The pyruvic acid then loses a carbon through oxidation to CO_2, reduces NAD^+ (since the reduced form of nicotinamide adenine dinucleotide [NADH] is used to reduce O_2 to H_2O, the production and utilization of NADH is called *aerobic respiration*), and forms acetyl-coA. Acetyl-coA enters the Krebs cycle by combining with a four-carbon molecule from the Krebs cycle to make a six-carbon molecule. In the Krebs cycle, more carbon is oxidized to CO_2, which is coupled to the reduction of NAD^+. At the same time, the compounds and energy produced in the Krebs cycle are used to form and transport amino acids and nucleic acids for polymer synthesis (i.e., proteins, RNA, and DNA). This energy comes from NADH oxidation, which is coupled to the reduction of O_2 to H_2O and the phosphorylation of ADP plus inorganic phosphorus (Pi) to ATP (Fig. 1.24). Since O_2 is used in this process, it is called *oxidative phosphorylation*. The Krebs cycle occurs in the mitochondria between membranes, and oxidative phosphorylation occurs in the inner membrane of mitochondria. This membrane is very similar to chloroplast membranes (Fig. 1.10) except that it does not contain the photosynthesizing pigments. Oxidative phosphorylation utilizes an electron transport system similar to photophosphorylation; the major proteins involved are cytochromes.

Photosynthesis and respiration, although very similar, are in many ways opposing reactions (Table 1.4). Both use energy for synthesis, but respiration must catabolize organic molecules to obtain energy for its processes (i.e., for the synthesis of storage, structure, and metabolic compounds and for processes such as translocation and nutrient transport across membranes). Respiration uses energy from photosynthate to do its work. Photosynthesis results in increased plant dry weight due to CO_2 uptake, while the catabolic processes of respiration cause CO_2 release, reducing dry weight. Both processes are essential: photosynthesis to fix CO_2 for hexose production and respiration to transform hexose to the structural, storage, and metabolic substances required in plant growth and development. The concern of the crop physiologist is to make both processes as efficient as possible, that is, for photosynthesis to use light energy as efficiently as possible and for respiration to use that captured energy to form the productive crop plant as efficiently as possible.

How efficient is respiration? To answer this question, the production and utilization of ATP must be examined. Respiration is like a storage battery with

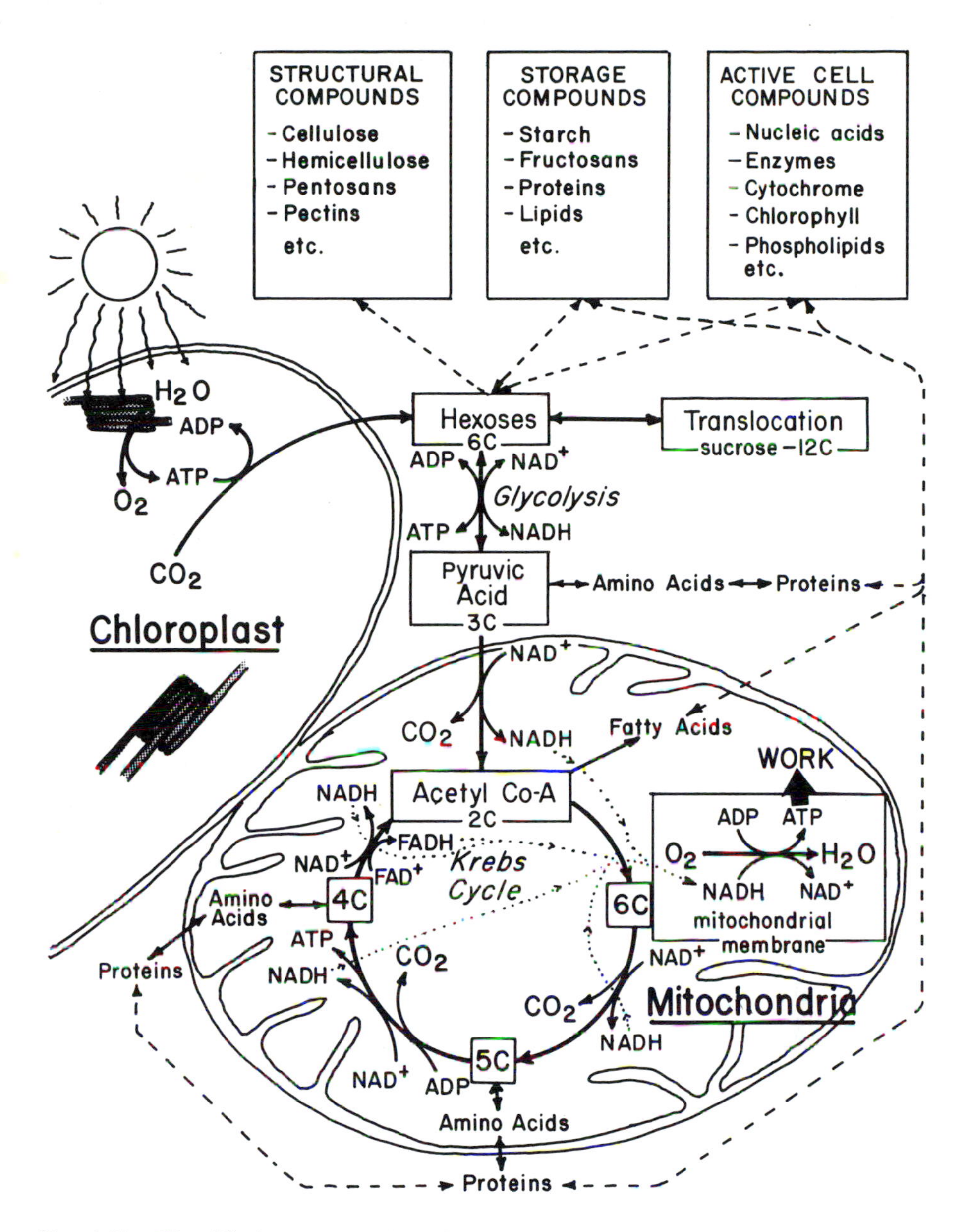

Fig. 1.24. Simplified representation of plant metabolism emphasizing three aspects of respiration: glycolysis, Krebs cycle, and oxidative phosphorylation. Besides producing ATP and NADH, respiration provides compounds for building larger, more complex compounds for structure, metabolism, and storage.

TABLE 1.4. Comparison of photosynthesis and respiration

Subject	Photosynthesis	Respiration
Phosphorylation	Photophosphorylation using light energy	Oxidative phosphorylation using chemical energy
Nucleotide reduction	NADPH formed with light energy, used for CO_2 reduction	NADH formed with oxidation of carbon, used for O_2 reduction
Carbon dioxide	Substrate	Product
Water	Substrate	Product
Oxygen	Product	Substrate
Organic compounds	Product	Substrate and product

energy that may be released to other molecules, which in turn become energized and more active.

The rapid combustion of stored energy in a mechanical engine gives energy release as heat and converts stored energy to useful work at approximately 35% efficiency. In contrast, respiration is slow energy release. Substrates are energized by phosphorylation, and the entire process is done in many small steps that are catalyzed by enzymes under isothermal conditions. For example, glucose is completely converted to CO_2 in approximately 30 steps.

$$C_6H_{12}O_6 + 6O_2 \xrightarrow{(30\ \text{steps})} 6CO_2 + 6H_2O + 637\ \text{kcal (potential energy)}$$

These steps include glycolysis and the Krebs cycle. Together with oxidative phosphorylation, these cycles are known to phosphorylate 38 ADP molecules. Since 1 ATP = 12 kcal, the 38 ATP molecules generate 456 kcal, or 456 (actual)/673 (potential) = 68% efficiency.

The goal of the agriculturalist is to couple the energy of ATP production with the most efficient plant growth and development system. The end result is the most agricultural product possible per unit land area.

Estimating Maximum Crop Growth Rates

Patterns of solar radiation levels in an area, which remain fairly constant from year to year, are considered the ultimate limitation to crop yield. Loomis and Williams (1963) present a thoughtful analysis of the maximum possible dry matter production using solar energy levels as the limiting factor (Table 1.5) (total solar radiation based on a 100-day period from June 1 to Sept. 8, 1960, in a location in the United States). They conservatively chose 500 cal • cm^{-2} • day^{-1} and converted it to micro-Einsteins. They used several other assumptions: (1) 82% of the visible light was intercepted by chloroplasts, (2) the maximum quantum efficiency was 10% (10 photons to reduce one CO_2 molecule), and (3) respiration would evolve 33% of the CO_2 reduced in photosynthesis. With these assumptions they were able to set an estimated maximum for daily dry weight increases at 77 g • m^{-2} • day^{-1}, which is called *crop growth rate* (CGR). Their estimate set the efficiency of energy conversion at 5.3% of the total, and 12% of the visible, solar radiation.

TABLE 1.5. Calculation of potential daily productivity by a crop surface

1. Total solar radiation	500 cal • cm^{-2}
2. Visible solar radiation (400–700 nm) = 44%	222 cal • cm^{-2}
3. Total quanta in the visible range (approx. 19.5 μE/cal)	4320 μE • cm^{-2}
a. Albedo (reflection loss) 6–12% in visible spectrum	–360 μE • cm^{-2}
b. Inactive absorption loss = 10% (e.g., cell walls)	–432 μE • cm^{-2}
4. Total quanta usefully absorbed in the visible spectrum and available for photosynthesis	3528 μE • cm^{-2}
5. Amount of CO_2 reduced (10 quanta [photons] per CO_2 molecule reduced)	353 μmol CO_2 • cm^{-2}
6. Respiratory loss[a] of CO_2 (33%)	–116 μmol CO_2 • cm^{-2}
7. Net production of CH_2O (1 CH_2O produced/CO_2 reduced)	237 μmol CH_2O • cm^{-2}
8. Conversion of μmoles • cm^{-2} to g • m^{-2}	
a. 237 μmol • cm^{-2} = 0.000237 mol • cm^{-2}	2.37 mol • m^{-2}
b. CH_2O = 30 g • mol^{-1} × 2.37 mol • m^{-2}	71 g • m^{-2} • day^{-1}
9. If CH_2O is 92% of dry weight and inorganic constituents are 8%, total dry matter = 71 g • m^{-2} • day^{-1}/0.92	77 g • m^{-2} • day^{-1}
77 g • m^{-2} • day^{-1} = 687 lb • $acre^{-1}$ • day^{-1} = 34.35 T (2000 lb • T^{-1}) in 100-day season	

Source: From Loomis and Williams 1963.
[a]Respiration loss is an estimate. Measured values range widely, often 25–50%.

When the maximum CGR of 77 g • m^{-2} • day^{-1} is compared with actual measurements of short-term CGRs, we find that under ideal conditions some crops are able to come within 60% of the estimated maximum (Table 1.6).

TABLE 1.6. Maximum short-term crop growth rates recorded for several crop species

Species		Carboxylation Type	Maximum CGR[a]
Alfalfa	*Medicago sativa*	C_3	23
Bermudagrass	*Cynodon dactylon*	C_4	20
Cattail	*Typha latifolia*	C_3	34
Maize	*Zea mays*	C_4	52
Millet	*Pennisetum typhoides*	C_4	54
Pineapple	*Ananas comosus*	CAM	28
Potato	*Solanum tuberosum*	C_3	37
Rice	*Oryza sativa*	C_3	36
Soybean	*Glycine max*	C_3	17
Sudangrass	*Sorghum vulgare*	C_4	51
Sugar beet	*Beta vulgaris*	C_3	31
Sugarcane	*Saccharum officinarum*	C_4	38

Note: These CGRs should be compared with the 77 g • m^{-2} • day^{-1} established by Loomis and Williams as the maximum CGR possible at 500 cal • cm^{-2} • day^{-1} solar radiation. However, some of these crops were grown when the average solar radiation was close to 700 cal • cm^{-2} • day^{-1} for the period of measurement, which increases the possible CGR to 100 g • m^{-2} • day^{-1}.
Source: Selected from Loomis and Williams 1963, Evans 1975, and Monteith 1978.
[a]Crop growth rate.

Summary

Solar radiation is the source of energy for crop plants. Plants take the radiant energy and transform it to chemical energy. The first chemicals formed are reduced nucleotides and ATP. These compounds are short-term intermediates that carboxylate CO_2 into stable organic compounds. There are two carboxylation systems found in plants. The C_3 species seem to be less efficient at photosynthesis than the C_4 species, primarily because they exhibit photorespiration.

Environmental factors such as light, CO_2, temperature, water status, and mineral status directly affect leaf photosynthetic rates by influencing either the light reaction or carboxylation systems of the chloroplast.

Rates of photosynthesis vary widely among species and are often related to the environment to which they are adapted. Crop species usually are among the most efficient of plant species. Leaf photosynthetic rates within species also vary, which suggests the possibility of increasing crop yields and quality by selecting for high photosynthetic rates.

Products of photosynthesis are used for storage, structure, respiration, and growth. How efficiently the plant partitions its photosynthetic supply into these different fractions has an important effect on yield.

With light the ultimate limiting factor, the maximum crop growth rate is estimated at around 77 $g \cdot m^{-2} \cdot day^{-1}$. This represents a 12% efficiency for visible radiant energy.

References

Anderson, J.M. 1975. Biochim. Biophys. Acta 416:191–235.
Balegh, S. E., and O. Biddulph. 1970. Plant Physiol. 46:1–5.
Bassham, J. A., and M. Calvin. 1957. The Path of Carbon in Photosynthesis. Englewood Cliffs, N.J.: Prentice-Hall.
Crosbie, T. M., J. J. Mock, and R. B. Pearce. 1977. Crop Sci. 17:511–14.
Curtis, P. E., W. L. Ogren, and R. H. Hageman. 1969. Crop Sci. 9:323–27.
Dornhoff, G. M., and R. M. Shibles. 1970. Crop Sci. 10:42–45.
Evans, L. T. 1975. In Crop Physiology, ed. L. T. Evans. London: Cambridge University Press.
Goldsworthy, A. 1970. Bot. Rev. 36:321–40.
Hatch, M. D., and C. R. Slack. 1966. Biochem. J. 101:103–11.
Heichel, G. H., and R. B. Musgrave. 1969. Crop Sci. 9:483–86.
Hesketh, J. D. 1963. Crop Sci. 3:493–96.
Hitz, W. D. 1978. Ph.D. diss., Iowa State University, Ames.
Kellogg, W. W. 1977. World Meteorol. Org. Tech. Note 156. Geneva.
Loomis, R. S., and W. A. Williams. 1963. Crop Sci. 3:67–72.
Meyer, B. S., D. B. Anderson, and R. H. Bohning. 1960. Introduction to Plant Physiology. New York: Van Nostrand.
Monteith, J. L. 1978. Exp. Agric. 14:1–5.
Pearce, R. B., G. E. Carlson, D. K. Barnes, R. H. Hart, and C. H. Hanson. 1969. Crop Sci. 9:423–26.
Peaslee, D. E., and D. N. Moss. 1966. Soil Sci. Soc. Am. Proc. 30:220–23.
Stalfelt, M. G. 1937. Planta 27:30–60.
Stifel, F. B., R. L. Vetter, R. S. Allen, and H. T. Horner, Jr. 1968. Phytochemistry 7: 355–64.
Williams, J. 1979. Carbon Dioxide, Climate and Society. New York: Pergamon.
Woodwell, G. M. 1978. Sci. Am. 238:23–43.

Carbon Fixation by Crop Canopies

TOTAL DRY MATTER YIELD of field crops results from accumulation of net CO_2 assimilation throughout the growing season. Because CO_2 assimilation results from solar energy (*irradiance*) absorption and because solar radiation, on a seasonal basis, is distributed uniformly over a land surface, the primary factors affecting total dry matter yield are the solar radiation absorbed and the efficiency of utilizing that energy for CO_2 fixation.

Carbon dioxide assimilation at the subcellular, cellular, and tissue levels of organization was discussed in Chapter 1. Controlled experiments under laboratory conditions have provided detailed information about CO_2 assimilation at these levels, but there is less information on CO_2 assimilation at the crop community level of organization. Crop communities compound the problem because (1) the environmental factors (external micro- and macroenvironments) in crop communities are constantly changing (e.g., seasonal changes in irradiance, day length, temperature, water availability, CO_2 concentration, nutrient availability, oxygen concentration, air turbulence) and (2) plants respond to the complexity of cropping environments in many different ways. The concept of light interception by crop canopies and its relationship to crop productivity is emphasized here.

Leaf Area, Interception of Solar Radiation, and Crop Growth

LEAF AREA AND SOLAR RADIATION INTERCEPTION

For a crop to use solar radiation efficiently, most of the radiation must be absorbed by green, photosynthetic tissue. Leaves, the primary organs for light interception and photosynthesis in crop plants, develop either from embryos in seeds or from meristematic tissue in stems. Some perennial crops maintain a nearly complete *ground cover* (ground area shaded by leaves) in tropical or subtropical climates, but in temperate regions the low winter temperatures terminate it. In the spring when temperatures favor growth, a new leaf canopy is regenerated from quiescent buds supported by stored foods. In perennials the regenerative organs are overwintering buds. In annuals the initial leaf area develops from seedlings and is small for much of the early growth; this results in absorption of most of the solar radiation by the soil surface, producing *sensible energy* (heat). Efficient crop species tend to invest most of their early growth in expansion of leaf area, which results in efficient use of solar radiation. Many agronomic practices such as starter fertilizer, high plant densities, and more uniform plant-spacing arrangements (e.g., narrow rows) are used to hasten ground cover and increase light interception.

Leaf area development in an annual, determinate crop (vegetative growth stops at flowering) is illustrated in Figure 2.1A. As leaf area develops, radiation interception by leaves increases. Early leaf area develops at an exponential rate, but since the initial leaf area is small, significant radiation interception does not occur for several weeks. Since flowering terminates leaf area development, cultural objectives are to maximize photosynthesis by the crop intercepting all or nearly all of the solar radiation. This is an efficient pattern for grain crops, in which the majority of seed weight comes from photosynthesis after flowering.

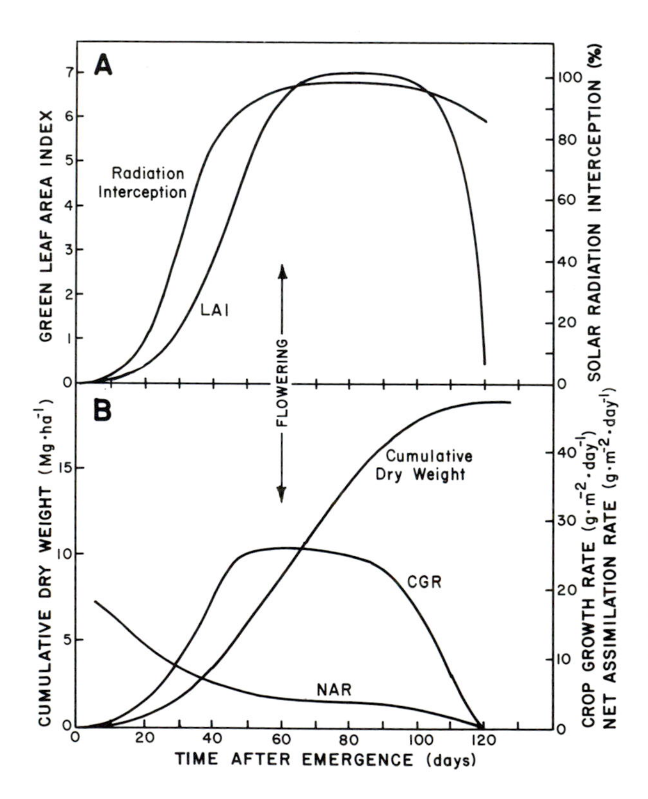

Fig. 2.1. Seasonal development of an annual, determinate grain crop. Note the close relationships among leaf area index (*LAI*), light interception, and crop growth rate (*CGR*). These smooth, idealistic curves approximate those found in the field. Under field conditions, however, environmental factors (especially light, temperature, and moisture) are constantly changing, which results in much more erratic responses. (Some weight conversions: 1.0 Mg • ha^{-1} = 1,000 kg • ha^{-1} = 100 g • m^{-2} = 893 lb • acre^{-1}.)

Early growth of individual plants, with little plant-to-plant competition, is exponential and is described by the *relative growth rate* (see Chap. 8), which is based on the rate of dry matter increase in relation to the dry matter weight of the whole plant or crop. As leaf area develops and there is shading of lower leaves, descriptions of crop growth are based on leaf area or land area rather than on individual plants. Watson (1947) coined the term *leaf area index* (LAI), which is the ratio of the leaf area (one side only) of the crop to the ground area. Because solar radiation is evenly spread over the land surface, LAI is a rough measure of leaf area per unit of available solar radiation.

LEAF AREA INDEX AND RATE OF DRY MATTER PRODUCTION

Crop Growth Rate. The concepts of growth analysis, discussed in Chapter 8, must be introduced here to facilitate discussion of yield in crop canopies. The most meaningful growth analysis term for crop canopies is dry matter accumulation per unit of land area per unit of time, or the *crop growth rate* (CGR). CGR is measured by harvesting (sampling) a crop community at frequent intervals and calculating the increase in dry weight from one sampling to the next. It is usually expressed in units such as $g \cdot m^{-2}$ (land area) $\cdot$ day^{-1}. Ideally, all living tissue of the crops growing in the sampled area should be measured, but the difficulty of sampling roots often excludes their use in some CGR studies. CGRs of a species are usually closely related to interception of solar radiation (Fig. 2.1).

Net Assimilation Rate. Since leaf surfaces are the primary photosynthetic organs of the plant, it is sometimes desirable to express growth on a leaf area basis. The dry matter accumulation rate per unit of leaf area per unit of time is called the *net assimilation rate* (NAR) and is usually expressed in $g \cdot m^{-2}$ (leaf area) $\cdot$ day^{-1}. The NAR is a measure of the average photosynthetic efficiency of leaves in a crop community. It is highest when the plants are small and most of the leaves are exposed to direct sunlight. As the crop grows and LAI increases, more and more leaves become shaded, causing a decrease in NAR as the growing season progresses (Fig. 2.1B). In canopies with a high LAI, the young leaves at the top absorb the most radiation, have a high CO_2 assimilation rate, and translocate large amounts of assimilate to other plant parts. In contrast, older leaves at the bottom of the canopy under shaded conditions have low CO_2 assimilation rates and contribute less assimilate to other plant parts. The net assimilation rate does not take into account *nonlaminar photosynthesis*, which is the photosynthesis of plant parts other than leaves (i.e., petioles, stems, sheaths, and inflorescences), which can significantly contribute to crop yield (see Chap. 3).

The NAR is a measure of the average net CO_2 exchange rate per unit of leaf area in the plant canopy. Thus, when it is multiplied by the LAI, the product is the CGR.

$$CGR = (NAR)(LAI)$$

CRITICAL AND OPTIMUM LEAF AREA INDEXES

Critical Leaf Area Index. Two kinds of relationships have been found between the CGR and LAI. Brougham (1956) in New Zealand hypothesized that if enough leaf area was kept in a pasture to intercept most of the solar radiation, the maximum growth rate should be maintained. To test this hypothesis, Brougham clipped plots of ryegrass-clover mixture to heights of 12.7, 7.6, and 2.5 cm. He measured dry matter, LAI, and light interception of the canopies every 4 days for a total of 32 days after clipping. He found that immediately after clipping, the plots clipped to 12.7 cm intercepted 95% of solar radiation whereas plots clipped to 2.5 cm intercepted less than 20% (Fig. 2.2). Brougham showed that the CGR increased as the LAI increased up to 5, where the canopy intercepted 95% of solar radiation (Fig. 2.2). Leaf area indices above 5 did not change the CGR significantly so Brougham termed the LAI at which the canopy first reached maximum CGR (which occurred at 95% light interception) as the *critical LAI.*

The LAI with 95% solar radiation interception has been adopted as the critical LAI by most crop physiologists for two reasons. First, radiant energy interception approaches a maximum asymptotically, which means the LAI at 100% interception would be impossible to measure. Second, 95% interception under maximum solar radiation of 2300 μmol photons • m^{-2} • sec^{-1} means that the radiation level at the bottom of the canopy is 115 μmol photons • m^{-2} •

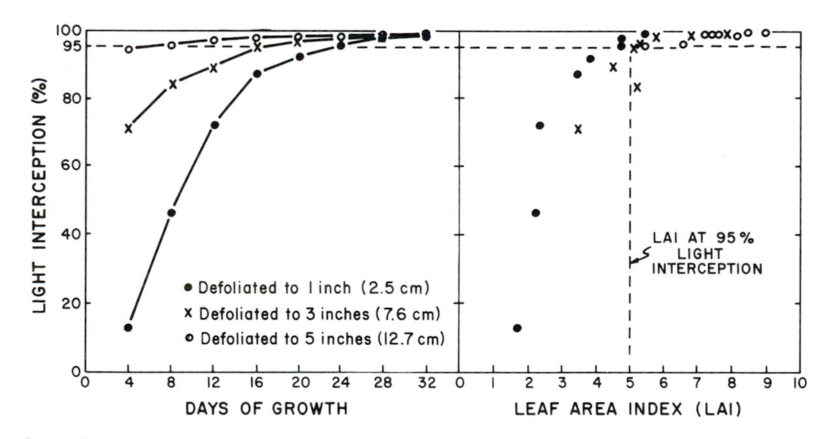

Fig. 2.2. Measurements from a ryegrass-clover pasture. (*Left*) Light interception with days after defoliation at three defoliation levels. (*Right*) CGRs in relation to the LAI resulting from dry matter and leaf area measurements made every 4 days (Brougham 1956, by permission).

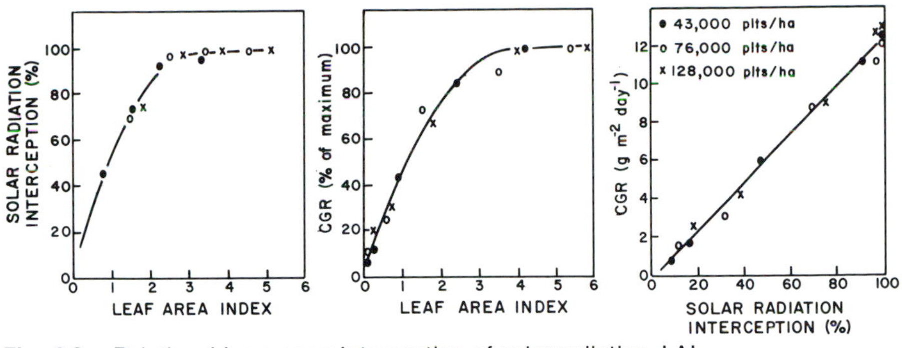

Fig. 2.3. Relationships among interception of solar radiation, LAI, CGR in soybean (Shibles and Weber 1965, by permission).

sec^{-1}; this is the light compensation point for many species. Increasing the LAI above 95% radiation interception would result in insignificant CGR increases. Growth analysis of soybeans by Shibles and Weber (1965) showed the critical LAI response and classical relationships among the interception of solar radiation, LAI, and CGR (Fig. 2.3).

Optimum Leaf Area Index. In England, Watson (1958) ran an experiment similar to that of Brougham's. Watson grew kale and sugar beets in rows and varied the number of plants per row to change the LAI. Measuring the LAI and dry weight at 10-day intervals, Watson showed results similar to Brougham's except that with kale the CGR peaked at a LAI of about 3.5 and then declined at greater LAI values (Fig. 2.4). This finding was similar to theoretical calculations by Kasanaga and Monsi (1954) in Japan, who labeled

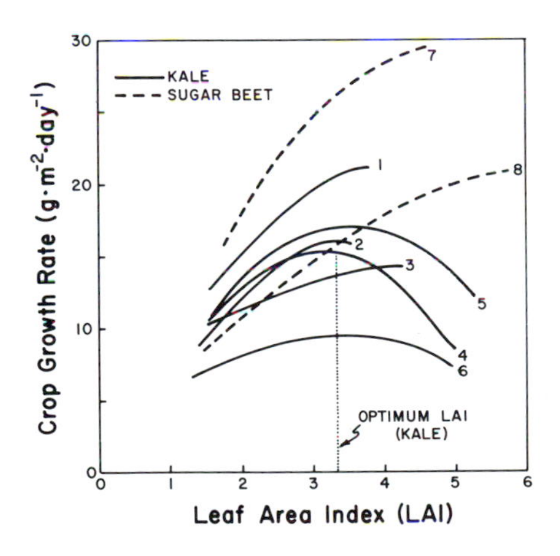

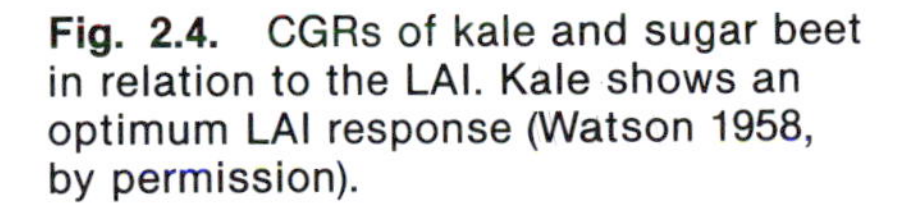

Fig. 2.4. CGRs of kale and sugar beet in relation to the LAI. Kale shows an optimum LAI response (Watson 1958, by permission).

the LAI at maximum CGR as *optimum LAI* because the CGR decreased as the LAI increased beyond the optimum. In Watson's experiments sugar beets were more efficient than kale and did not seem to reach a maximum CGR even at a LAI of 5.

Critical and Optimum LAI Concepts. Canopies exhibiting either optimum or critical LAI concepts show an increase in CGR with an increased LAI up to the LAI at which most of the solar radiation is intercepted. After the maximum CGR is attained, however, the two concepts differ because of respiration (Fig. 2.5). Photosynthesis increases until all the radiation is intercepted by photosynthetic surfaces. Any further increase in leaf area would only shade lower leaves, which would then be unable to produce enough photosynthesis to meet respiration requirements and might use photosynthetic products from other leaves (become parasitic), thus resulting in a reduced CGR. In most species new leaves are produced at the top of the plant and older, bottom leaves are shaded. Fully expanded leaves have not been shown to import photosynthetic products from other leaves (Wolf and Blaser 1971). In addition, as leaves become shaded their respiration is reduced along with reduced photosynthesis (Duncan et al. 1967). In such species we would expect a critical LAI response. Research by King and Evans (1967) in Australia confirmed that increases in respiration are much reduced once the critical LAI is reached in wheat and alfalfa (Fig. 2.6). Therefore, these species showed a critical LAI response.

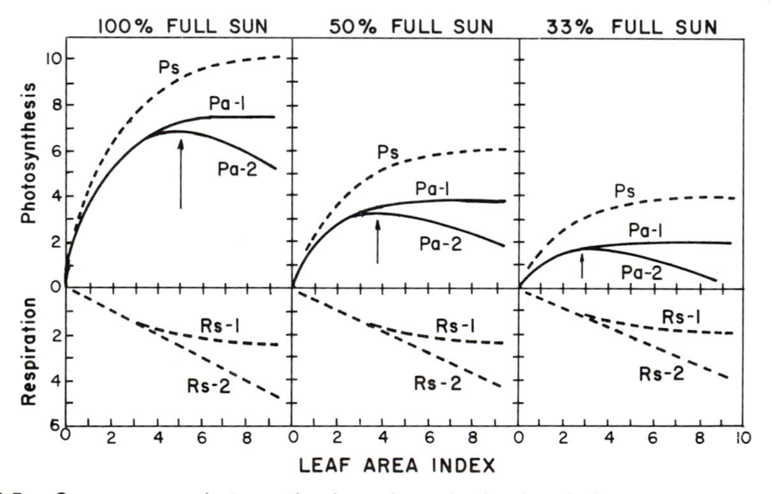

Fig. 2.5. Crop canopy photosynthesis and respiration in relation to LAI. Apparent photosynthesis (*Pa*) is the difference between photosynthesis (*Ps*) and respiration (*Rs*). The two lines for Pa illustrate the difference between the critical LAI response (*Pa-1*) and the optimum LAI response (*Pa-2*), caused by different respiration responses. The Pa is equivalent to the CGR. Note that the maximum Pa (*arrow*) occurs at a lower LAI as the level of solar radiation decreases.

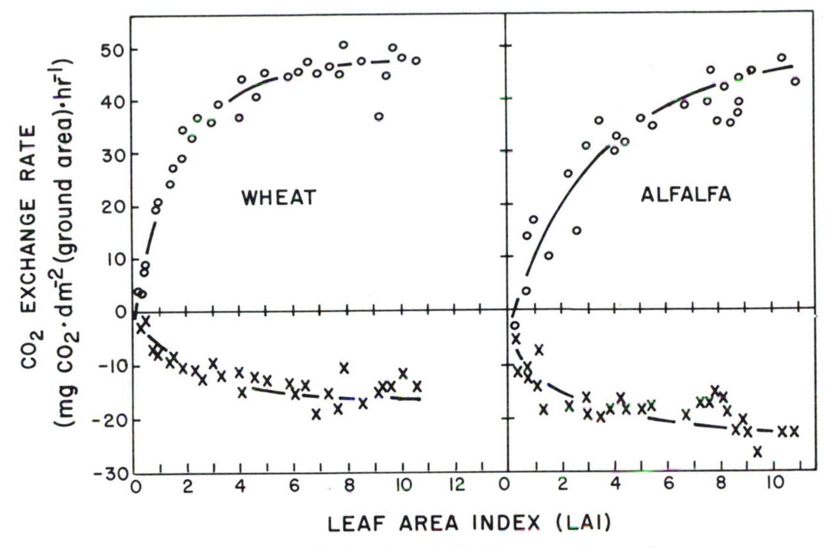

Fig. 2.6. CO_2 exchange rates of wheat and alfalfa canopies under photosynthetically active radiation approximately 1/3 of full sun (*o*) and in the dark (*x*) in relation to the LAI (King and Evans 1967).

The optimum LAI response may occur when young leaf tissue is shaded. Pearce et al. (1965) showed an optimum LAI response in orchardgrass with growth that occurred after flowering. Since orchardgrass leaves elongate from intercalary meristems near the nonelongated stem, at a high LAI older leaf parts at the top of the canopy shade younger leaf parts at the bottom. The young leaf tissues utilize photosynthate from other leaf tissues as a result of their growth function. This increases their respiration, which they cannot match in photosynthesis because of shading, and the optimum LAI response occurs.

Although critical and optimum LAIs differ in definition, the same quantity is being characterized, that is, the minimum LAI to attain the maximum CGR (Loomis and Williams 1963). Both assume a LAI that intercepts most of the light.

RADIATION ATTENUATION THROUGH CROP CANOPIES

Crop communities intercept both direct sunlight and indirect or diffuse skylight. The upper leaves receive both direct and diffuse radiation, while leaves lower in the canopy receive a smaller portion of direct radiation (e.g., sun flecks). Indirect radiation becomes more prominent due to radiation transmitted through leaves and reflected from plant and soil surfaces. Both quantity and quality of radiation change with depth in the canopy because light transmitted by leaves is predominantly infrared. Since plants preferentially absorb energy in the 400- to 700-nm wavelength range, the longer wavelengths become more prominent at lower levels. For that reason, in photosynthetic studies most instruments used to measure radiation in crop communities only measure quantum or energy levels between wavelengths of 400 and 700 nm. The radiation thus measured is called photosynthetic photon flux density (PPFD) as a quantum measurement or photosynthetically active radiation (PAR) as either a quantum or energy measurement. The term *radiation* hereon in the chapter will mean PAR.

The attenuation of radiation down through the plant canopy has been shown to approximate that of colored solutions or algae cell suspensions (Monsi and Saeki 1953). This pattern of extinction conforms to the Lambert-Beer law of absorption, which states that each layer of equal thickness absorbs an equal fraction of radiation that traverses it. For plant canopies the layer of equal thickness is based on units of LAI. The mathematical expression is

$$I_i/I_0 = e^{-kL}$$

where I_0 = photosynthetically active radiation above the canopy, I_i = photosynthetically active radiation below the ith layer of leaves, L = leaf area index of the ith layer of leaves, k = extinction coefficient characteristic of the canopy, and e = base of natural logarithms (2.71828). Thus the amount of sunlight penetrating through the canopy is affected by the LAI and the pattern of leaf display. The extinction coefficient (k) gives a numerical indicator of light attenuation in the canopy. The k is a characteristic of canopy leaf display, which primarily consists of leaf inclination and the way leaves are grouped within the canopy.

Leaf Inclination. Types of leaf inclination have been defined and illustrated (Fig. 2.7) by de Wit (1965). These idealized patterns range from *planophile*, with most leaves nearly horizontal (<35° from horizontal), to *erectophile*, with most leaves nearly vertical (>60° from horizontal). Trenbath and Angus (1975), who tabulated studies showing how different crop species fit different patterns of inclination, cited studies in which plant species have shown all the leaf inclination patterns except *extremophile*.

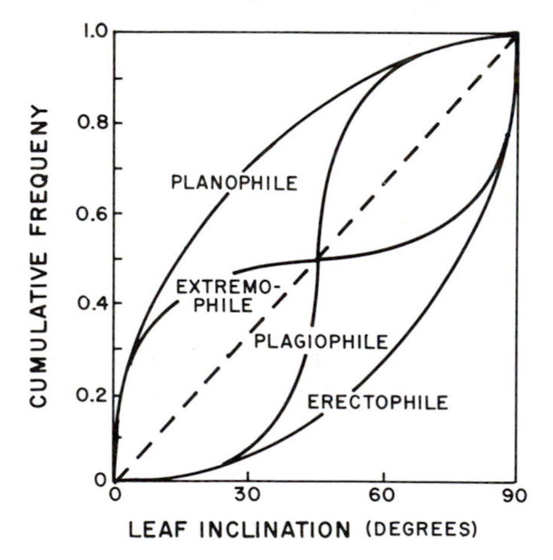

Fig. 2.7. Idealized cumulative frequency distributions of leaf inclination in four types of canopies (de Wit 1965, by permission). The broken line represents a fifth type, labeled *uniform distribution* by Trenbath and Angus (1975).

The inclination of leaves affects radiation interception and distribution in the canopy. The planophile clover canopy needs less leaf area to intercept most of the radiation than the erectophile grass canopy (Fig. 2.8). Approximate *k* values for clover and grass stands are 0.6 for clover and 0.25 for grass (Loomis and Williams 1969). Warren Wilson (1959), using the frequency of foliage contact with vertical and horizontal needles passed through various strata to calculate mean foliage angle (Fig. 2.8), illustrated the planophile and erectophile inclination patterns of clover and ryegrass canopies, respectively. According to Brougham's (1956) theory of critical LAI, the clover canopy in Figure 2.8 intercepts 95% of the radiation at a LAI of 5, so the critical LAI for clover is 5 whereas the CGR of grass continues to increase up to a critical LAI of 9.

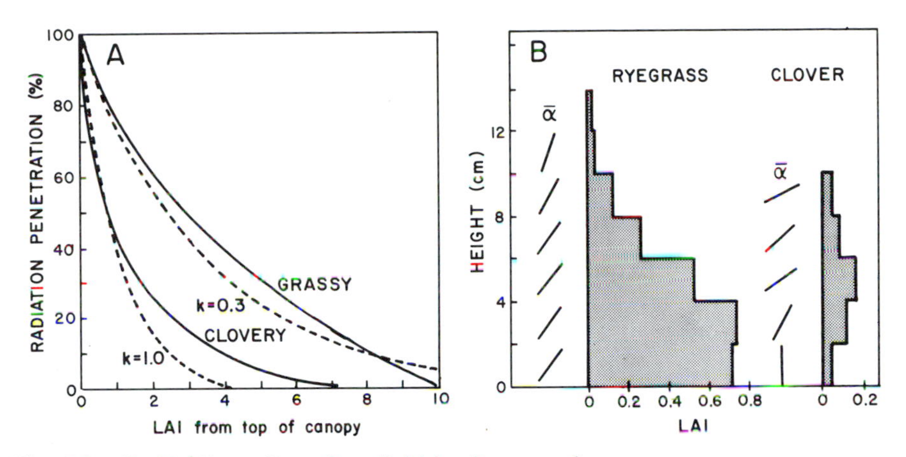

Fig. 2.8. (*Left*) Attenuation of sunlight in clovery and grassy swards as a function of the LAI, compared with extinction coefficients (*k*) of 1.0 and 0.3. (From Stern and Donald 1962) (*Right*) Vertical distribution of LAI and mean leaf angle (α) determined using point quadrates for ryegrass and clover canopies. (From Warren Wilson 1959, as presented by Loomis and Williams 1969, by permission)

A high proportion of the crops studied for leaf inclination are the planophile type (Trenbath and Angus 1975). This could be due to previous selection for competition to weed development in crop stands; most weeds are severely hampered in growth by shading, so crop plants reduce competition for water, nutrients, and radiation by maximum shading of weeds during vegetative development.

Leaf Inclination and Photosynthetic Efficiency. Leaf photosynthesis is most efficient (i.e., CO_2 fixed per unit of light) at low radiation levels. Most individual leaves are radiation saturated in direct sunlight (Fig. 2.9). In a planophile canopy, upper leaves are radiation saturated and lower leaves have reduced photosynthesis due to shading. Theoretically, a planophile canopy would be more efficient if radiation were distributed more evenly over leaf surfaces. Such an equitable distribution could be accomplished by having leaves, at least the upper leaves, at a vertical leaf inclination when the sun is at high elevations.

Radiation Attenuation and Crop Growth Rate. How vertical should leaves be? Figure 2.9 illustrates the theoretical concept of maximizing canopy photosynthesis through more vertical leaf orientation. For example, when leaves are angled 75° from the horizontal with a vertical radiation source they intercept 26% as much radiation as horizontal leaves and the effective radiation level on the leaf is 26% of that on a horizontal leaf. Because the photosynthetic response to radiation is curvilinear and radiation efficiency is greatest at low radiation levels, the vertical leaf is more efficient per unit of radiation intercepted. In the red clover example (Fig. 2.9), when the leaf is angled 75° from horizontal and is intercepting only 26% of the light, leaf photosynthesis is reduced only 21% from that of horizontal leaves. A small reduction in upper leaf photosynthesis because of vertical leaf inclination allows much more radiation to penetrate to lower leaves. Canopy photosynthesis and CGR can theoretically be increased dramatically by vertically orienting leaves at a high LAI (Table 2.1).

The C_4 species usually do not reach radiation saturation in direct sunlight (see Fig. 1.6), which means they use high radiation levels more efficiently than do C_3 species. But they still use radiation more efficiently at low levels than under full sunlight. For example, Figure 1.6 shows that when radiation levels at the leaf surface are 25% and 50% of full sunlight, the CO_2 exchange rate is 45% and 72%, respectively, compared with that at full sunlight.

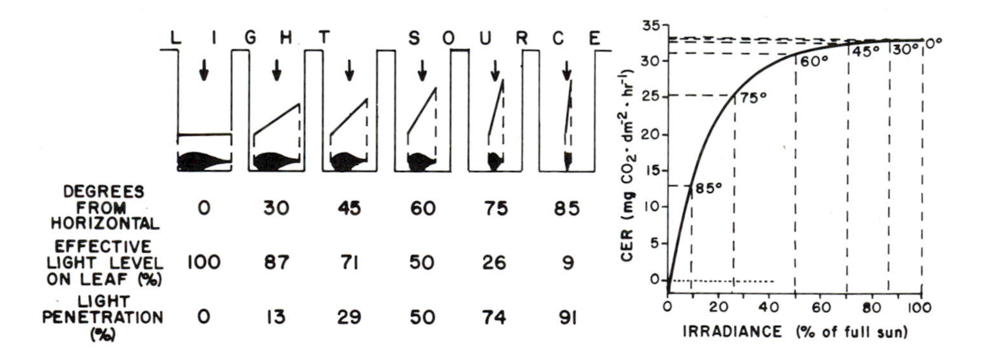

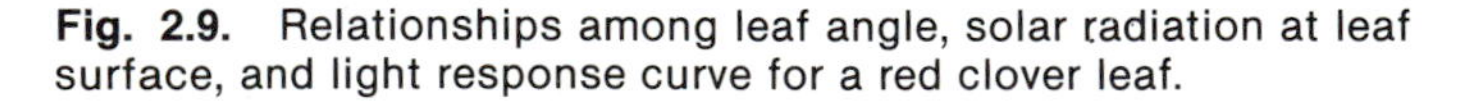
Fig. 2.9. Relationships among leaf angle, solar radiation at leaf surface, and light response curve for a red clover leaf.

TABLE 2.1. Calculated relationships among leaf orientation, leaf photosynthesis, LAI, and total plot photosynthesis

Leaf Angle from Horizontal	Leaf Photosynthetic Rate (mg CO_2 • dm^{-2} • hr^{-1})	LAI to Intercept[a] Most Light	Total Plot Photosynthesis (mg CO_2 • dm^{-2} of ground • hr^{-1})
0	33	1	33
60	31	2	62
75	26	4	104
85	12	10	120

Note: Calculations made from Figure 2.9.
[a]Assuming perfect leaf placement—this would also correspond to critical LAI.

Loomis and Williams (1969), using a computer modeling program, estimated the influence of leaf inclination and leaf amount on the CGR of maize and clover (Fig. 2.10). As discussed previously, the critical LAI (95% radiation interception) is lowest for canopies with horizontal leaves and highest for canopies with vertical leaves. Canopies with horizontal leaves provide the highest CGR at LAIs below 3. A canopy of vertical leaves needed an LAI of 4 or greater to have a distinctly higher CGR than canopies with horizontal leaves. At low LAIs there is little shading among leaves, so canopies with horizontal leaves have a slight advantage over canopies with vertical leaves because of higher irradiance at the leaf surface. At high LAIs canopies with vertical leaves have the advantage because light is more evenly distributed over the canopy leaf area; less radiation interception by upper leaves allows more light to be intercepted by lower leaves.

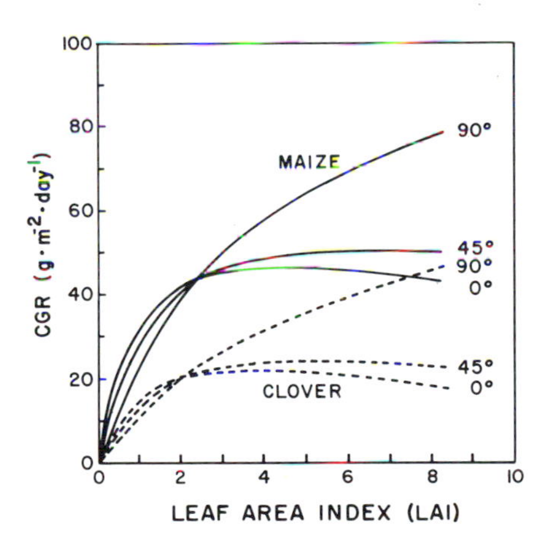

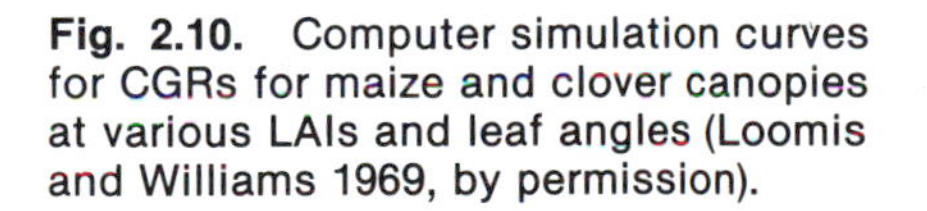
Fig. 2.10. Computer simulation curves for CGRs for maize and clover canopies at various LAIs and leaf angles (Loomis and Williams 1969, by permission).

Solar Angle, Radiation Attenuation, and Crop Growth Rate. The sun is not always vertically overhead; the angle of the sun's rays shining into crop canopies changes seasonally and diurnally. Duncan et al. (1967) calculated the effect of leaf angle and leaf amount on the CGR during the diurnal period (Fig.

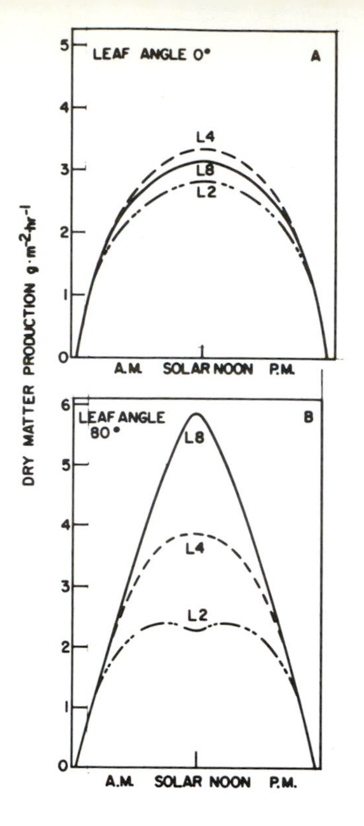

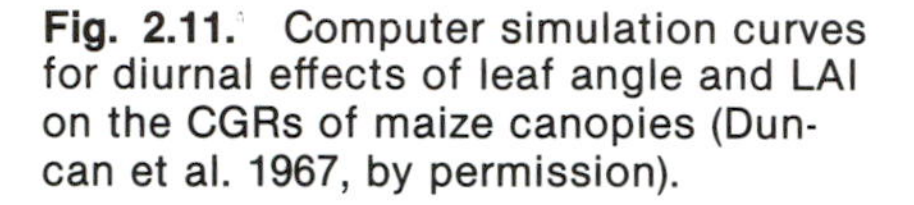

Fig. 2.11. Computer simulation curves for diurnal effects of leaf angle and LAI on the CGRs of maize canopies (Duncan et al. 1967, by permission).

2.11). During the early morning and late evening, when the sun's rays come in at nearly horizontal angles, the leaf angle or LAI had little effect on the CGR. At solar noon horizontal leaves had the advantage at an LAI of 2 and leaves 80° from the horizontal at an LAI of 8. Duncan (1971) estimated that for maize in the U.S. Corn Belt a leaf angle of 80° would be the most productive.

Leaf Inclination Variation within Canopies. Leaf inclination may vary in different strata within the canopy (Fig. 2.8). Canopies with leaves vertically inclined at the top and gradually becoming more horizontal closer to the ground have been termed the ideal foliage display by Trenbath and Angus (1975). Pendleton et al. (1968) showed that maize canopies with the leaves above the ear tied in a vertical inclination yielded more than those with the leaves in their normal planophile position or those with all plant leaves tied into vertical inclinations (Table 2.2). The pattern of vertical upper leaves and more horizontal lower leaves allows vertical leaves in the radiation-rich environment to intercept less radiation, putting them at a more photosynthetically efficient radiation level and allowing more radiation to pass to lower leaves. With radiation more evenly distributed over the total leaf area, the canopy may not require the extremely high LAIs that are necessary for a high CGR in canopies with all leaves vertically inclined (Duncan 1971).

Advantages and Disadvantages of Erectophile Canopies. Trenbath and Angus (1975) cite four studies on sugar beet, barley, rice, and tea in which the rela-

TABLE 2.2. Means of grain yields and barren plants from leaf angle study

Comparisons	Yield (kg/ha)	Plants Barren (%)
Genetic isolines of hybrid C103 × Hy		
Normal leaf	6,202 a	28 a*
Upright leaf	8,769 b	14 b
Mechanical manipulation of leaf angle of 'Pioneer 3306'		
Normal (untreated)	10,683 c	4 c
All leaves positioned upright	11,386 cd	6 bc
Leaves above ear positioned upright	12,202 d	3 c

Source: Pendleton et al. 1968.

*Means followed by the same letter are not significantly different at the 5% level of significance.

tionship between CGR and leaf inclination was measured. Erectophile type CGRs were 19 to 108% greater than planophile type CGRs. They also cited fourteen studies on wheat, barley, rice, and maize in which grain yield was measured in relation to leaf inclination. Three studies showed a yield advantage for planophile types (5–18%), and eleven showed a yield advantage for erectophile types (4–68%). In all cases, erectophile canopies that performed to advantage were planted at plant densities achieving or exceeding the critical LAI.

Most of the studies concerning leaf inclination have been conducted on grasses. Broadleaf (dicotyledonous) plants often change their inclination in response to the sun (*heliotropic* movement). Many crops, including legumes, cotton, and sunflower, have heliotropic responses (Trenbath and Angus 1975). Some orient their leaves perpendicular to direct solar radiation; others may actually angle the leaf surface in relation to direct solar radiation (Ross 1970). Under cloudy conditions soybeans have been found to orient their leaves perpendicular to the brightest part of the sky while maintaining an oblique angle to direct solar radiation (Kawashima 1969). Few studies have been made to determine if this characteristic can be stabilized and used for better radiation interception.

Vertical Separation of Leaves. Vertical leaf density influences the skylight pattern within the canopy. A leaf just beneath a top leaf can receive both direct shade and direct sunlight according to its position; the farther away the leaf is from the top leaf the less prominent the sunflecks and shade because of diffusion of shadow edges (Fig. 2.12). Large but widely separated leaves like those of sunflower may actually create a diffuse light pattern in the plant canopy similar to that of shorter plants such as alfalfa. If leaves are widely spaced vertically or are very narrow (e.g., asparagus and conifers), shadow edges will be diffuse and the distinction between direct radiation and diffuse radiation may be lost (Loomis and Williams 1969). Most plants seem to have evolved with leaves maintaining a vertical distance twice that of their width. In breeding for dwarf varieties of cultivated species (e.g., sorghum and maize), this

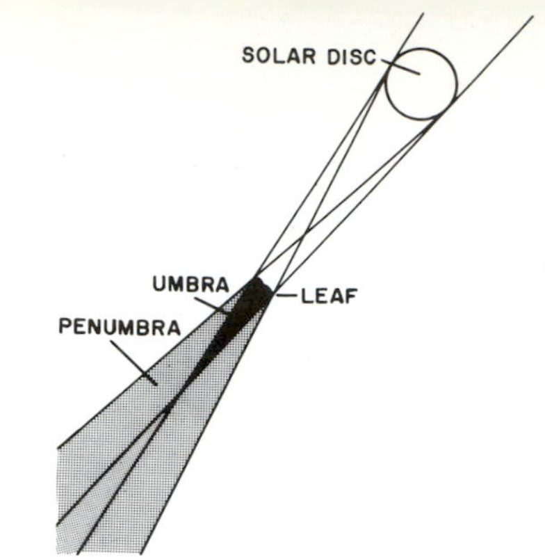

Fig. 2.12. Diffusion of shadow edges from a leaf in direct sunlight illustrating the irradiance changes in the canopy.

distance factor has changed so that wide leaves are closer together in relation to leaf width. Loomis and Williams suggest that the leaf arrangements in dwarfs might be improved by reducing leaf width, by having fewer leaves, and/or by arranging leaves in a whorled pattern.

Strategies for Maximizing Solar Energy Utilization

Yield is the accumulation of dry matter over time. How efficiently the crop utilizes solar radiation and how long it can efficiently maintain utilization results in the final dry matter yield of the crop.

LEAF AREA DURATION

To correlate dry matter yield with LAI, Watson (1947) integrated the LAI with time and called it *leaf area duration* (LAD), which takes into account both the duration and extent of photosynthetic tissue of the crop canopy. In annual crops the leaf area emerging from the seed is small but, under favorable conditions, increases at an exponential rate. Calculating the area under a curve of the LAI over time (Fig. 2.13B) gives leaf area duration; LAD units, ex-

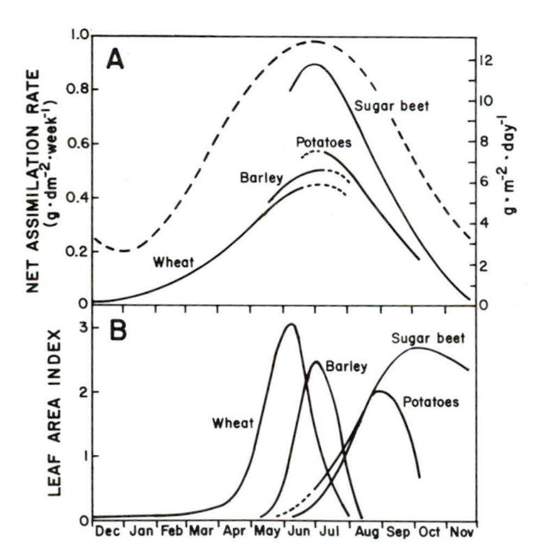

Fig. 2.13. Smoothed curves showing the average change with time in (*A*) the NAR and (*B*) the LAI of different species. The area under each LAI curve represents LAD (Watson 1947, by permission). The dashed line *A* is an approximation of the relative radiation levels in England during the year. Note that the NAR is highly correlated with radiation level.

TABLE 2.3. Dry weight at harvest in relation to leaf area duration (LAD) and net assimilation rate (NAR) of different crops

	Dry Matter at Harvest		NAR	
Crop[a]	(cwt • acre^{-1})	LAD (wk)	(cwt • acre^{-1} • wk^{-1})	(g • dm^{-2} • wk^{-1})
Barley	58	17	3.4	0.43
Potato	61	21	2.9	0.36
Wheat	76	25	3.0	0.38
Sugar beet	96[b]	33	2.9	0.36

Note: Cwt = hundredweight = 100 lb.
Source: Watson 1947, by permission.
[a]Same crops as shown in Figure 2.13.
[b]Harvested at the end of October.

pressed as time (e.g., LAI days or weeks), are the average LAI multiplied by the time from beginning to end of leaf area (e.g., 360 LAI days).

Usually the LAD is closely correlated with yield because interception of solar radiation over longer periods of time generally means greater total dry matter production (Table 2.3). Large differences in total biomass yields are often as much or more the result of duration of photosynthesis as of the photosynthetic rate. For example, in southern pine, 12 months of photosynthesis a year produce acceptable biomass production (pulpwood) even though the leaf photosynthetic rate is low compared with many crop species. However, the LAD does not take into account the amount of solar radiation available for crop photosynthesis, the attenuation of irradiance within the canopy, or the efficiency of leaves in utilizing available radiation (Fig. 2.13A). Photosynthesis by nonlaminar parts and shading from nonlaminar tissue (i.e., maize tassel) can also influence utilization of sunlight by the crop canopy, regardless of the LAD.

Leaf area duration is fairly easy to measure and, since it is related to dry matter yield, can give an indication of crop productivity. It can even give a good measure of grain yield in wheat if measured from ear emergence to maturity, even though head photosynthesis is a major contributor to yield. Because most of the carbohydrate in wheat grain is derived from photosynthesis after ear emergence and because ear photosynthetic duration is related to the LAD, the LAD should be correlated to yield. Studies cited by Evans et al. (1975) showed that the LAD could account for about half of the variation in grain yields, even with great differences in climate, agronomic practices, and cultivars (Fig. 2.14).

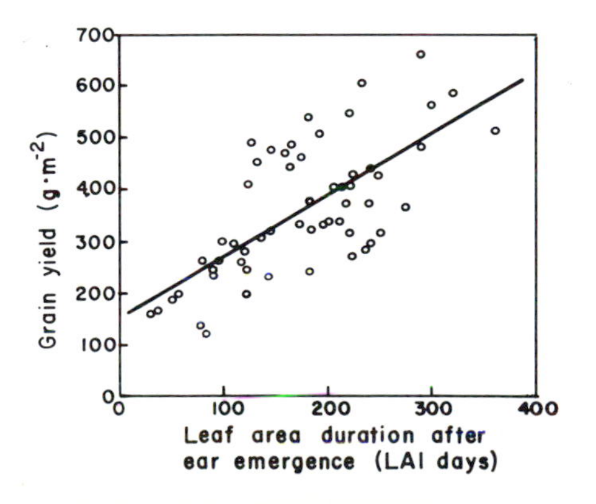

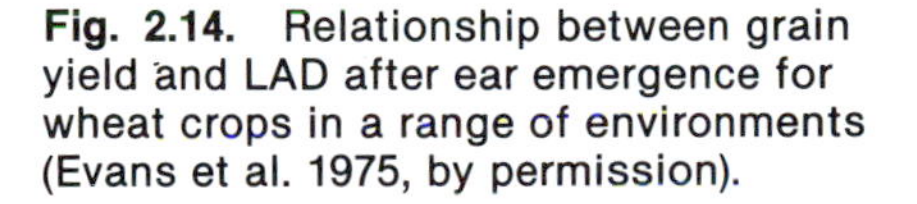
Fig. 2.14. Relationship between grain yield and LAD after ear emergence for wheat crops in a range of environments (Evans et al. 1975, by permission).

As efficient as the LAD is in predicting yield, it is only an estimate of light utilization over time. Actual measurements of irradiance and the interception of irradiance integrated over time would correlate much better with yield than would the LAD.

THE SOLAR ENERGY–TEMPERATURE INTERACTION

Total dry matter yield of a crop is the integral of net CO_2 assimilation over the total growing season. Some environmental factors change predictably during the growing season. Often the yield of a crop is enhanced by taking advantage of these changes.

In temperate regions, radiant energy levels and the temperatures of both soil and air are the two major environmental variables with predictable changes during the growing season. Since surface temperatures at any one location are largely influenced by the amount of radiant energy received, these two factors tend to fluctuate together. The earth, however, maintains some residual energy, taking time to heat up or cool down according to changes in incoming solar energy. This creates a lag time between the lowest solar energy level and the lowest temperature at a location or between the highest solar energy level and the highest temperature level (Fig. 2.15A). Because of this lag time the radiant energy levels at a particular temperature are higher in the spring than in the fall. Since crop growth rate is closely correlated with the interception of solar energy (Fig. 2.13A), the occurrence of the critical LAI during the period of highest solar energy gives the potential for highest yields.

Crops vary as to the range of temperatures in which they will grow. Those crops that grow under cooler conditions (i.e., grow at minimum cardinal temperatures of 0–5°C, like wheat) have the advantage of being able to produce the critical LAI early enough to coincide with the time of maximum solar energy (Fig. 2.15B). Warm-season crops (i.e., grow at minimum cardinal temperatures of 5–15°C, like maize) have to wait until soil temperatures are high enough to promote growth; they cannot produce leaf area fast enough to reach the critical LAI at maximum solar energy (Fig. 2.15B). This solar energy–temperature interaction becomes more acute at higher latitudes.

The challenge to crop physiologists and plant breeders is to develop plants that will grow sufficient leaf area before the radiant energy peak and maintain an active leaf area over the major portion of the solar energy peak. Winter and spring barley have different rates of leaf area development (Fig. 2.16A). Winter barley develops sooner because it does not have to be planted in the spring and thus flowers and dies before spring barley. Both reach maximum leaf area during the period of maximum solar energy levels. The area under the irradiance curve in each graph represents the amount of solar energy intercepted, which is proportional to potential yield.

For earlier leaf area development of a crop seeded in the spring, genotypes usually must be found that will grow and develop at lower temperatures and

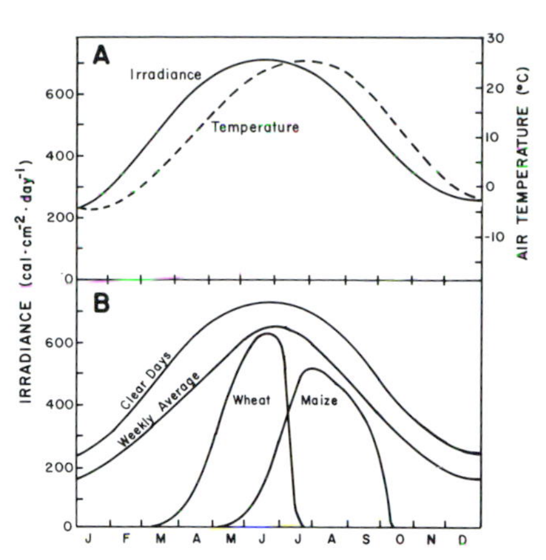

Fig. 2.15. Seasonal irradiance and its relationship to (*A*) seasonal temperatures and (*B*) interception of irradiance by a wheat and a maize crop. These curves are based on a continental climate at about 42°N latitude.

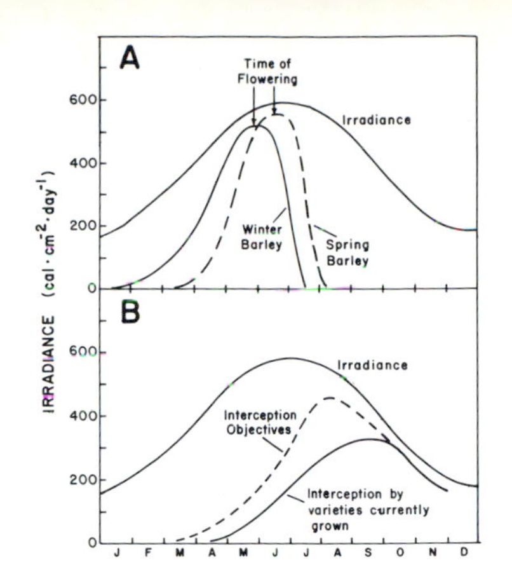

Fig. 2.16. Approximate seasonal irradiance in England and its relationship to (*A*) interception of solar radiation by winter and spring barley crops and (*B*) interception of solar radiation by sugar beet crops using current varieties and management, and the interception pattern for each sought after by plant breeders. (From Ivins 1973)

have frost tolerance. For many crops, plant breeders in temperate regions are selecting for increased cold tolerance so they can be planted earlier and thus reach the critical LAI earlier in the season. Ivins (1973) described increasing sugar beet yield by planting earlier, getting more leaf area earlier in the growing season, and intercepting more solar energy (Fig. 2.16B).

LIFE PERIOD OPTIMAL FOR SEED YIELD

In theory the longer the stage of maximum net assimilation per unit of soil surface, the higher the total dry matter production and the more fruits or other parts the plant can produce. Some data support this view. For example, in the temperate zone the best late-maturing varieties of a crop normally outyield the best early-maturing varieties.

In this respect the tropics might seem to have an advantage over the temperate zone; the duration of the vegetation period is not limited by temperature so varieties with very long vegetative periods, producing high yields, could be used. Best (1962) suggests that this premise is incorrect, stating that many other factors come into play, most of which should probably be considered under the heading of plant senility. In the examples in Figure 2.17, seed yield was studied as a function of the vegetation period. Using a rice variety and a soybean variety sensitive to photoperiod, variations in the duration of the vegetation were obtained by changing photoperiod with weak supplementary irradiation, while maintaining a standard amount of daylight per 24 hours.

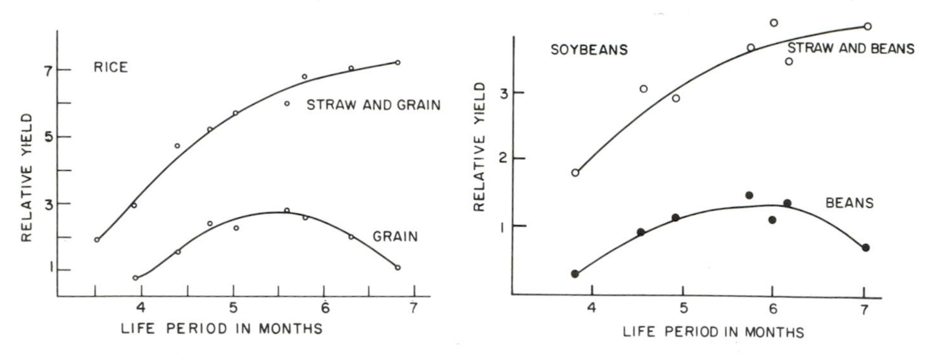

Fig. 2.17. Effect of duration of the vegetative period on yield of soybean and rice, with and without straw. Duration of vegetative periods was regulated by photoperiodic treatments, with the same daily amount of photosynthetic light in all treatments (Best 1962, by permission).

Best then plotted the yield of grain in these figures against the length of the vegetation period. The graphs show that there were certain optimum vegetation periods for seed production, which were not correlated with maximum dry matter production. Best cited other information to suggest that in the tropics light and duration of particular photoperiods may not favor high rates of photosynthetic productivity. Therefore it may be suggested that high rates of grain dry matter production over relatively short periods may best be obtained in the temperate zone, and crops producing vegetative products (e.g., sugarcane, cassava, forages) that have steadier rates of dry matter production may be better produced in the tropics. Best also points out that temperature has a distinct influence on the distribution of dry matter in the crop; cool temperatures favor carbohydrate transport to the grain.

Plant Density

Efficient interception of radiant energy incident to the crop surface requires adequate leaf area, uniformly distributed to give complete ground cover. This is achievable by manipulating stand density and its distribution over the land surface.

Measurements of the LAI and CGR in crop communities reveal much about how crops achieve high yields. However, they are difficult to measure and so, for practical reasons, crop managers use *plant densities* (plants per unit of ground area) and final yield.

PLANT AND ENVIRONMENTAL FACTORS AFFECTING OPTIMUM PLANT DENSITY

Selections of the most suitable stand density must be based on the following plant and environmental factors.

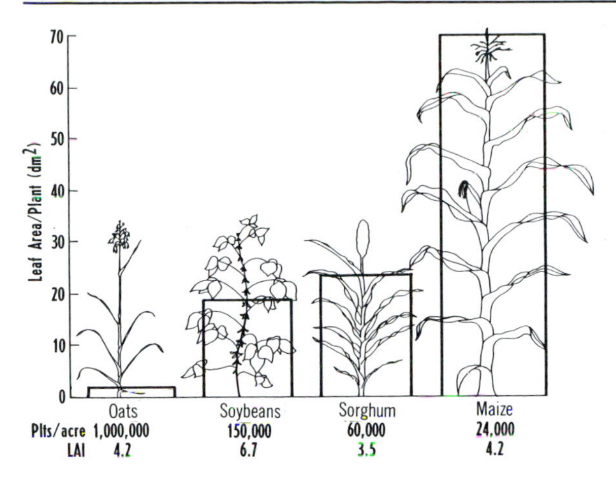

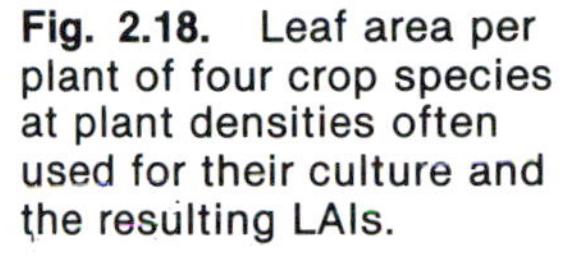

Fig. 2.18. Leaf area per plant of four crop species at plant densities often used for their culture and the resulting LAIs.

1. Plant size (which primarily reflects leaf area per plant). Four species—oats, soybeans, sorghum, and maize—are compared in Figure 2.18 at commonly used plant densities that result in LAIs of 4.2, 6.7, 3.5, and 4.2, respectively. The leaf area per plant determines the number of plants needed to develop a critical LAI. Maize hybrids adapted to northern latitudes have one to three fewer leaves than those in the southern United States and require higher stand densities for maximum yield. Leaf inclination would modify the critical LAI, and the stand density would have to be adjusted accordingly.

2. Tillering and/or branching. Branching, an effective way of increasing the leaf area per plant, decreases the sensitivity of yield to plant density. In sorghum plants, which *tiller* (branch from nodes close to or below the soil surface), the number of heads per acre increased only slightly when plant density was changed from 13,000 to 52,000 plants per acre (32,000 to 128,000

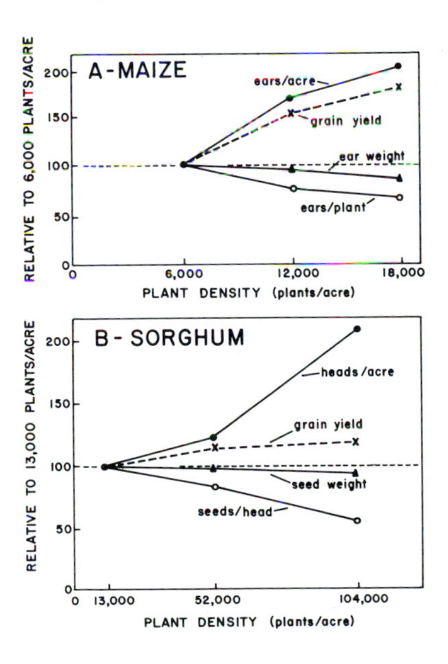

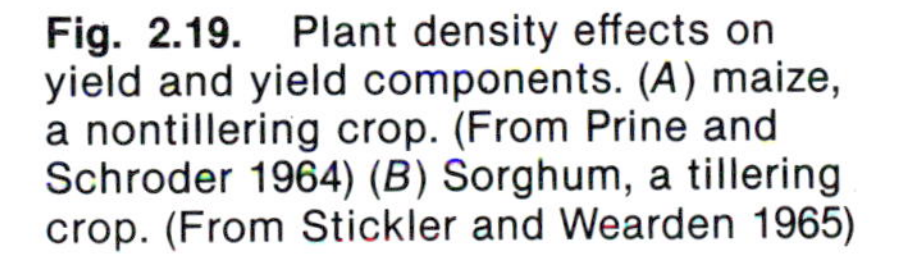

Fig. 2.19. Plant density effects on yield and yield components. (*A*) maize, a nontillering crop. (From Prine and Schroder 1964) (*B*) Sorghum, a tillering crop. (From Stickler and Wearden 1965)

per ha) (Fig. 2.19), indicating over three tillers per plant at 13,000 plants per acre. When plant density was doubled from 52,000 to 104,000 plants per acre the heads per acre also doubled, indicating that little tillering occurred at 52,000 plants per acre. Increased plant density did not increase grain yield because as heads per acre increased, the seeds per head decreased proportionately. Modern maize varieties do not tiller much, even at low plant densities, and usually produce only one ear per plant. Therefore maize grain yield is much more sensitive to plant density than is sorghum because both the LAI and the number of ears per acre increase or decrease with density. Maize does not have the flexibility of most crop species, which can increase leaf area and number of reproductive units by branching at low crop densities.

3. Lodging. Increased density causes plants and stems to become smaller, weaker, and often taller. Thus, strong-stemmed cultivars are required or plant density has to be decreased to reduce *lodging* (the leaning or falling over of plants). Lodging decreases harvestable yield by putting the seeds too close to the ground for equipment to harvest them and decreases absolute yield by ruining leaf display.

4. Reduction in fruit set. As density increases, potential flowers and fruits do not set or are aborted. This reduces the possible seed yield by decreasing the total possible amount of assimilate that the seed could retain.

The environment also influences the optimum plant density for yield. The primary environmental factors include (1) irradiance, (2) moisture, and (3) soil fertility. Limitation of these environmental factors lowers the optimum plant density for maximum production. Weeds compete with crop plants for these environmental factors, which decreases the optimum plant density.

PLANT DENSITY AND YIELD

Holliday (1960 a,b) summarized a large volume of literature and emphasized the two density-yield interactions that occur when crop plant density is increased. These interactions depend on whether the yield is a product of the plant's growth in the reproductive phase (seed yield) or a product of growth in the vegetative phase. The key consideration is whether the economic yield is a plant component (e.g., seed weight) or the entire plant (biological yield).

To illustrate yield from the reproductive phase, wheat grain yields in England were cited. These data described a parabolic response curve (Fig. 2.20), typically a flat-topped one with decreases in yield on both sides of an optimum. When seed yield is the desired product, there is an optimum plant density beyond which plant density can become too high because available photosynthate is partitioned more to vegetative growth or to maintenance respiration than to seed growth.

The curve for seed yield in Figure 2.19 can be fitted to the following quadratic equation:

$$Y = a + bx - cx^2$$

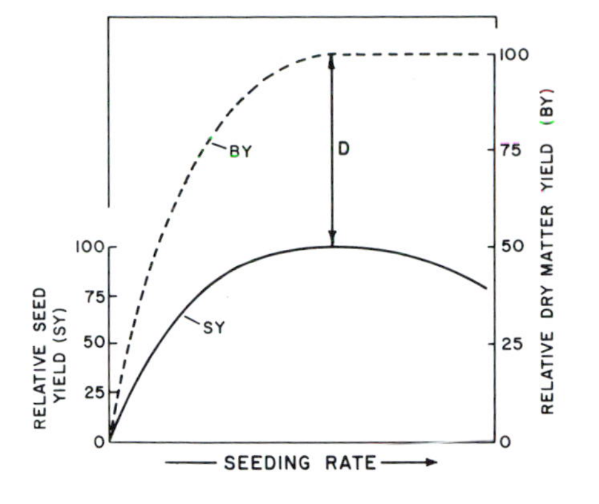

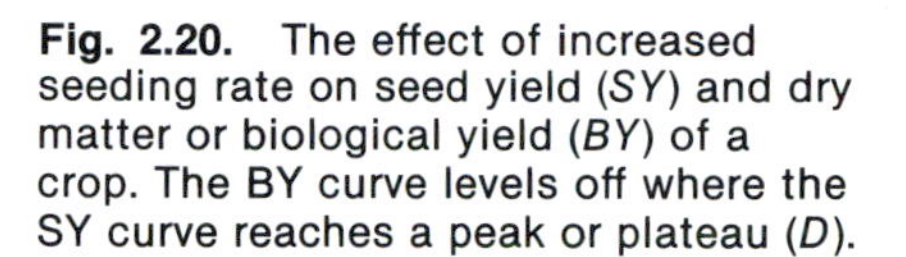
Fig. 2.20. The effect of increased seeding rate on seed yield (*SY*) and dry matter or biological yield (*BY*) of a crop. The BY curve levels off where the SY curve reaches a peak or plateau (*D*).

where Y = yield per unit area, x = plant density (plants/area), and a, b, and c are regression constants.

When yield is the product of growth of vegetative material, the yield response to increasing plant density is asymptotic, similar to the critical LAI. In this case a dense stand for maximum radiation interception must be achieved as rapidly as possible; but if the stand is too dense, the only loss is from greater seeding expense. This partially explains why recommended seeding rates for forages are often so high. Although there is no loss from being over the critical plant density, there is also no gain because only 100% of solar radiation can be intercepted. Uniform stands of forages are often difficult to establish, which motivates super-high seeding rates. The curve for biological yield in Figure 2.20 can be defined by the following expression for a rectangular hyperbole:

$$Y = Ax \; 1/(1 + Abx)$$

where Y = yield of dry matter per unit, A = apparent maximum yield per plant, x = number of plants per unit area, and b = linear regression coefficient.

In this expression the term $1/(1 + Abx)$ represents the manner in which the maximum individual plant yield (A) is reduced by the increasing competition resulting from greater plant density. It may be termed the *competition factor.*

While plant density and yield have been determined in many studies, the three parameters, plant density, the yield of dry matter, and the yield of grain, have not often been measured together. Six such studies, however, are cited by Donald (1963); in each instance the peak of the grain curve occurred approximately at the density at which the *biological yield* (dry matter yield) leveled off (Fig. 2.20). Therefore grain yield had an optimum LAI at the critical LAI for biological yield. At this plant density any gain in total yield per acre due to the addition of extra plants is offset by the decrease in the weight per plant. No doubt these relationships represent conditions of either limiting radiation or limiting nutrients and would not hold, for example, under conditions in which water supply becomes exhausted before grain is formed.

Duncan (1958) presented an interesting discussion of the relationship between plant density and yield of maize, with special emphasis on the interaction of number of plants and yield per plant. He formed the hypothesis, confirming it with results from many field experiments, that the logarithm of average yield for individual plants had a negative linear relationship to plant density (Fig. 2.21). He concluded that one could grow a maize variety at two widely divergent plant densities (6,000 and 25,000 plants per acre, or 15,000 to 62,000 per ha) and calculate the density at which the maximum yield could be expected from that variety.

The yield per plant decreases as the number of plants increase; this relationship can be determined by plotting the yield per plant on semilog paper. Since it is a straight line plot, one only needs to know the yield per plant at two plant densities. The yield per acre is the yield per plant multiplied by the plant density, so the yield at any plant density can be calculated and the results plotted into the usual graph form (Fig. 2.22). When the logarithm plot is

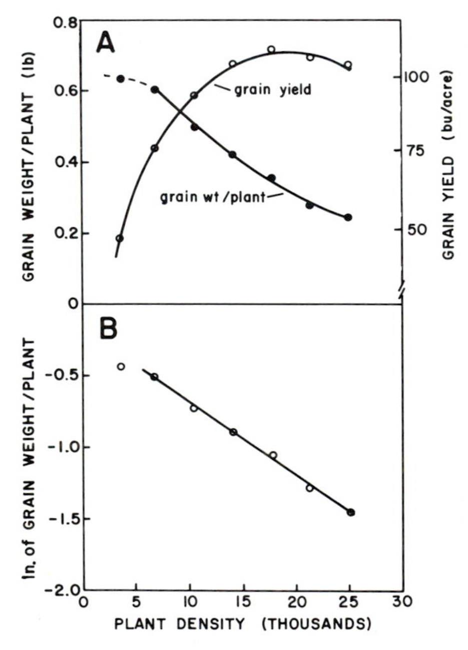

Fig. 2.21. Example of the relationship between grain yield per plant and grain yield per acre at increasing plant densities. (*A*) The arithmetic relationship; (*B*) the logarithmic relationship. (From Duncan 1958)

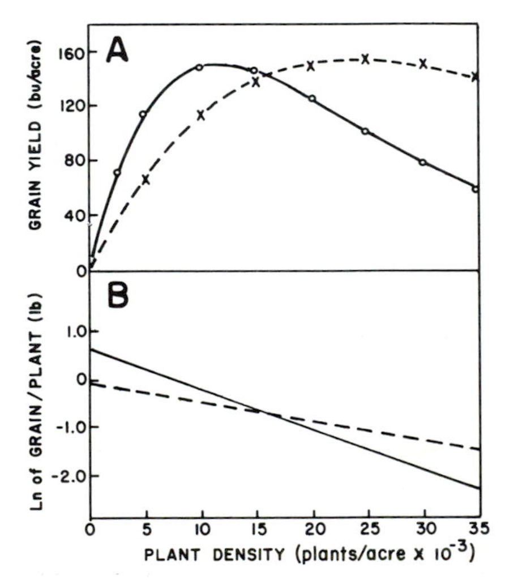

Fig. 2.22. Two examples of relationships between grain yield per plant and plant density. The differences between the two examples could represent different varieties or different environments (e.g., fertility, moisture, soil types) (Duncan 1958, by permission).

compared with the arithmetic plot in Figure 2.22, it can be seen that the flatter the logarithm slope, the higher the plant density required for maximum yield.

Willey and Heath (1969) commented that while Duncan's approach is sound, it would seem safer in practice to include a third intermediate density, so that the point calculated for maximum yield is not too far from an experimental treatment.

PLANT RESPONSES TO STAND DENSITY CHANGES

Donald (1963) presented an explanation of plant responses to stand density changes. Both he and Duncan (1969) relied extensively on the work of Hozumi et al. (1955), who studied the yield of adjacent plants.

Donald suggested that the greater seed weight and number of seeds per inflorescence at intermediate densities are due to the timing of *interplant* (between plants) and *intraplant* (within a plant) competition. At the widest spacing (lowest plant density), both types of competition are absent during early stages of growth. Flower primordia are formed in large numbers. As growth proceeds, there is little interplant competition and even less intraplant competition until after flowering and seed setting. The large load of inflorescences leads to competition for assimilate among inflorescences and seeds on the same plant, that is, intraplant competition. This loss of efficiency at the widest spacing reflects greater intraplant competition, resulting in fewer seeds per inflorescence and reduced seed size compared with denser stands. Thus intraplant competition may be intense at low densities.

In moderately dense stands interplant competition apparently becomes operative at the time of flower initiation or formation. The number of floral primordia laid down by each plant is considerably reduced; this reduced load lies more closely within the capacity of the plant as interplant competition intensifies. Seeds per inflorescence and seeds per unit area achieve maximum values. Still higher plant densities should reduce seed number, causing reduced seed yield, because the interplant competition is already intense at the time of flower primordia formation.

PLANT DISTRIBUTION—ROW SPACING

Throughout this chapter has been the assumption that plants were uniformly placed in the field, producing a uniform canopy of leaves that uniformly intercepted solar radiation. In managed crops, however, this is not the case. Seeds are placed in the soil with a mechanical planter, usually in discrete rows. The wider the row, the more seeds must be planted per length of row to achieve a particular plant density.

The objective for higher yields is to intercept as much solar radiation as possible, and equidistant planting would give the earliest and the maximum light interception (Fig. 2.23). As rows are widened and spacing becomes less uniform, interplant competition occurs earlier. Plants in rows that are farther

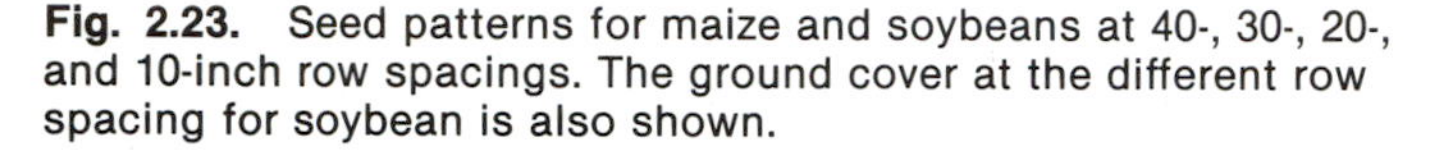

Fig. 2.23. Seed patterns for maize and soybeans at 40-, 30-, 20-, and 10-inch row spacings. The ground cover at the different row spacing for soybean is also shown.

apart must be closer together within rows to achieve a particular plant density. The major factor determining the distance between plants is plant density; the same factors that affect optimum plant density influence optimum row spacing. Crop plants with high leaf area per plant grown at lower plant densities (e.g., maize) respond less to reductions in row spacing than do smaller crop plants grown at higher plant densities (e.g., soybean).

Shibles and Weber (1966) conducted a classic comparison of a large soybean variety ('Hawkeye') at different plant densities and row spacings (Table 2.4). Their results showed that a particular combination of earlier leaf area development, maximum interception of solar energy, conditions favoring re-

TABLE 2.4. Row width and plant density effects on leaf area index (LAI), solar energy interception, branching, yield, and percent dry weight of beans in soybean fields of different plant densities and row spacing

Planting Factors[a]	Plants per Foot of Row	L_{95}[b]	Days to L_{95}	Branches per Plant	% Yield[c]	% Bean Wt. of Total Wt.
Plant density						
25,000	1	3.3	69	9.0	108	31
50,000	2	3.1	61	4.0	123	27
100,000	4	3.6	57	0.5	115	27
200,000	8	4.0	52	0.1	100	21
Row spacing[d]						
5″	1	3.6	53	1.2	126	
10″	2	3.6	55	1.1	133	
20″	4	3.6	57	0.5	115	
40″	8	4.2	66	0.6	100	

Source: Shibles and Weber 1966.
[a]Plants per acre in 20-in. rows.
[b]L_{95} = LAI at which 95% solar energy was first intercepted.
[c]Compared with 1000 plants per acre in 40 in.-row spacing.
[d]At 100,000 plants per acre.

duced lodging, and efficient translocation of dry matter to the seed was needed to get the highest yields. They attained a 33% increase in yield by reducing row spacing from 40 in. to 10 in. (102 to 25 cm) at 100,000 plants per acre (247,000 per ha).

Plant stature affects optimum plant density. Soybean seeded late in the season is usually short statured due to early flowering induced by photoperiod. Plant densities have to be increased and row widths narrowed to obtain maximum yield potential, compared with taller statured soybean.

Under unfavorable environments, narrowing the rows of most crop plants will not increase yield. Taylor (1980) tested the hypothesis that during years of low water supply soybean grown in wide rows would yield as much as or more than soybean grown in narrow rows. In a season of high water supply, seed yield in narrow (25-cm) rows yielded 17% more than in 100-cm rows. During 2 yr of lower seasonal water supply, there was no difference in seed yield among 25-, 50-, 75-, and 100-cm row spacings. In dry years, severe water deficits occurred in the narrow rows first, resulting in plants smaller in both height and LAI.

Weeds compete with crop plants for environmental factors, so good weed control is important for high yields. Weed control is difficult in rows too narrow to cultivate. Weber (1962) observed that in soybean the yields were reduced in 6- to 7-in. rows compared with 21- to 28-in. (53- to 71-cm) rows because of poorer weed control. Narrow row culture calls for higher plant densities that ensure faster canopy development to compete successfully against weeds (Shibles and Green 1979).

Using narrow rows appears to be one of a series of steps that has led to higher crop yields for producers. However, to obtain a high yield response from narrow row widths, the producer must have already adopted other managerial tools leading to high yields (e.g., using adapted varieties, fertilization, weed control, insect control, timely cultural practices, uniform plant distribution within the row, and opimum plant densities). When shifting to narrow rows the producer must decide on the variety to plant; the seeding rate (plant density); how to deal with potential lodging problems, the expense of buying narrow row equipment (planting and harvesting), and the higher investment needed for seed and fertilizer; and how to get good early-season weed control.

Plant breeders and crop physiologists are attempting to identify genotypes adapted to high plant densities and narrow rows (Mock and Pearce 1975; Cooper 1980).

Summary

Total dry matter yield is a result of crop canopy efficiency in intercepting and utilizing the solar radiation available during the growing season. The primary plant organs intercepting solar radiation are the leaves. For maximum crop growth rates, enough leaves must be present in the canopy to intercept most of the solar radiation incident on the crop canopy. When this occurs the

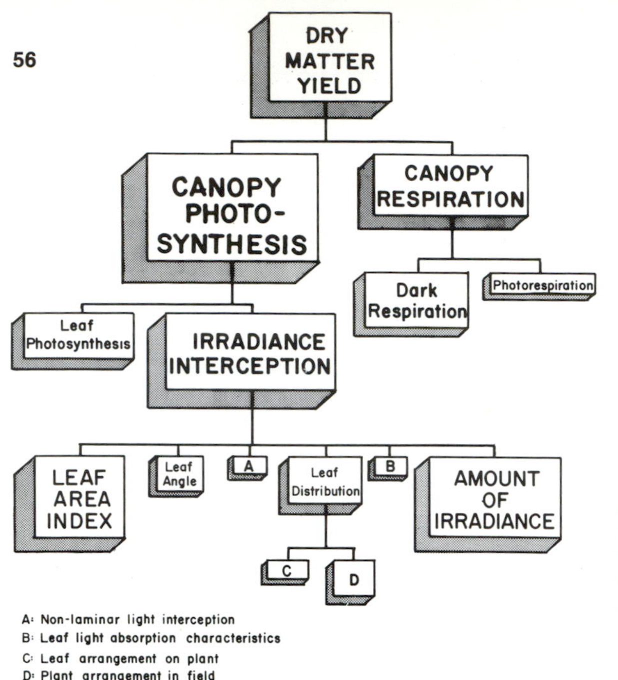

Fig. 2.24. Flow chart showing the different factors affecting total dry matter yield. The size of the boxes is an estimate of relative importance.

level of crop photosynthetic efficiency (or the CGR) is determined by the photosynthetic efficiency of leaves (or the NAR). The efficiency of the NAR can be influenced by the amount of solar radiation, the ability of leaves to photosynthesize, the LAI, how evenly the solar radiation level is divided among leaf surfaces, and the amount of plant respiration (Fig. 2.24).

Crop plants do not maintain a critical LAI over the total growing season. Annuals start leaf area accumulation from seedlings, in which radiation interception by the crop canopy is almost zero. But the LAI increases and eventually intercepts most of the solar radiation. After total ground cover is achieved, total dry matter production is a factor of how long the crop can maintain an active, green leaf canopy.

Strategies for maximizing solar radiation utilization and crop yields include

1. Planting early for earlier leaf area development. Often, varieties with greater resistance to frost and cool temperatures must be developed.
2. Planting at a seeding rate that will develop an optimal LAI at the maximum leaf area development.
3. Planting at a time that provides total ground cover during the period of maximum solar radiation levels.
4. Planting plants uniformly or nearly uniformly over the land to reduce early interplant competition and increase the rate of solar radiation interception.
5. Fertilizing to increase the rate of growth and photosynthetic efficiency of leaf surfaces.

6. Extending the time of maximum radiation interception by active leaf surfaces (or the LAD).

Many plant and environmental factors can modify the ability of a crop canopy to utilize solar radiation. Much of what we know about these factors is discussed in the following chapters.

References

Best, R. 1962. Neth. J. Agric. Sci. 10:347–53.
Brougham, R. W. 1956. Aust. J. Agric. Res. 7:377–87.
Cooper, R. 1980. In Solid Seeded Soybeans—Systems for Success, American Soybean Association.
de Wit, C. T. 1965. Versl. Landbouwkd. Onderz. Ned. 663.
Donald, C. M. 1963. Adv. Agron. 15:1–118.
Duncan, W. G. 1958. Agron. J. 50:82–84.
______. 1969. In Physiological Aspects of Crop Yield, ed. J. D. Eastin et al. Madison, Wis.: American Society of Agronomy.
______. 1971. Crop Sci. 11:482–85.
Duncan, W. G., R. S. Loomis, W. A. Williams, and R. A. Hanau. 1967. Hilgardia 38:181–205.
Evans, L. T., I. F. Wardlaw, and R. A. Fisher. 1975. In Crop Physiology, ed. L. T. Evans. London: Cambridge University Press.
Holliday, R. 1960a. Field Crop Abstr. 13:159–67.
______. 1960b. Field Crop Abstr. 13:247–54.
Hozumi, K., H. Koyama, and T. Kira. 1955. J. Inst. Polytech. Osaka City Univ. Ser. D 6:121–30.
Ivins, J. D. 1973. Phil. Trans. Roy. Soc. Lond. [B] 267:81–91.
Kasanga, H., and M. Monsi. 1954. Jpn. J. Bot. 14:304–24.
Kawashima, R. 1969. Proc. Crop Sci. Soc. Jpn. 38:718–42.
King, R. W., and L. T. Evans. 1967. Aust. J. Biol. Sci. 20:623–31.
Loomis, R. S., and W. A. Williams. 1963. Crop Sci. 3:67–72.
______. 1969. In Physiological Aspects of Crop Yield, ed. J. D. Eastin et al. Madison, Wis.: American Society of Agronomy.
Mock, J. J., and R. B. Pearce. 1975. Euphytica 24:613–23.
Monsi, M., and T. Saeki. 1953. Jpn. J. Bot. 14:22–52.
Pearce, R. B., R. H. Brown, and R. E. Blaser. 1965. Crop Sci. 5:553–56.
Pendleton, J. W., G. E. Smith, S. R. Winter, and T. J. Johnson. 1968. Agron. J. 60:422–24.
Prine, G. M., and V. N. Schroder. 1964. Crop Sci. 4:361–62.
Ross, J. K. 1970. In Prediction and Measurement of Photosynthetic Productivity, ed. I. Setlik. Wageningen, Netherlands: IBP/PP.
Shibles, R. M., and D. E. Green. 1979. Proc. Ninth Soybean Seed Res. Conf., Washington, D.C.: American Seed Trade Association.
Shibles, R. M., and C. R. Weber. 1965. Crop Sci. 5:575–77.
______. 1966. Crop Sci. 6:55–59.
Stern, W. R., and C. M. Donald. 1962. Aust. J. Agric. Res. 13:615–23.
Stickler, F. C., and S. Wearden. 1965. Agron. J. 57:564–67.
Taylor, H. M. 1980. Agron. J. 72:573–77.
Trenbath, B. R., and J. F. Angus. 1975. Field Crop Abst. 28:231–44.
Warren Wilson, J. 1959. In The Measurement of Grassland Productivity, ed. J. D. Ivins. London: Butterworth.
Watson, D. J. 1947. Ann. Bot. n.s. 11:41–76.
______. 1958. Ann. Bot. n.s. 22:37–55.
Weber, C. R. 1962. Iowa State Univ. Pam. 290.
Willey, R. W., and S. B. Heath. 1969. Adv. Agron. 21:281–322.
Wolf, D. D., and R. E. Blaser. 1971. Crop Sci. 11:55–58.

3 Transport and Partitioning

TO UTILIZE solar radiation efficiently and to store the resulting photosynthate (*assimilate*), a plant needs a transport system to move assimilate from areas of synthesis to areas of utilization. At germination, stored assimilate in the seed is mobilized and moved to newly activated meristems for leaf, stem, and root development, and soon the seedling becomes autotrophic. Assimilate produced by green tissue is translocated throughout the plant for growth, development, storage, and cell maintenance. The division of assimilate among these processes, termed *partitioning,* affects both productivity and survival of the plant.

Chapters 1 and 2 have dealt primarily with how dry matter is produced. This chapter discusses how assimilate moves in plants and how it is partitioned among different organs of the plant. Usually only certain organs, not the whole plant, are harvested—often only the seeds, leaves, stems, flowers, or roots. Even these products may be utilized for only a certain chemical component (e.g., oil, protein, or starch from seeds or sucrose from stems or roots). For such crops, yield is the amount of oil, sugar, seed, leaf, stem, flower, or root that is produced per unit of land area. Thus, the partitioning of assimilate and inorganic nutrients can affect both the efficiency of dry matter production and the portion of dry matter in the harvested plant part.

Phloem Transport

In plant growth and development, materials are moved from the *source* (where they enter the plant or are synthesized) to the *sink* (where they are utilized). Interorgan translocation in the plant is primarily through the vascular system, the xylem and phloem. Movement in the xylem tissue is essentially a one-way *acropetal* (upward) movement from the roots via the transpiration stream. In contrast, substances in the phloem have bidirectional movement; movement may be acropetal or *basipetal* (downward). Assimilate produced in leaves moves to sinks, while substances absorbed by roots move upward. In both xylem and phloem there are lateral connections, *plasmodesmata,* which allow some lateral movement.

The bulk of translocated substances, other than water, is the result of photosynthesis or remobilization of assimilate in storage. This is indicated by the fact that 90% of the total solids in the phloem consists of carbohydrates, mostly *nonreducing sugars* (sugars without an exposed aldehyde or ketone group, e.g., sucrose and raffinose), which occur in phloem sap at the rather high concentrations of 10 to 25%. The predominant sugar translocated in the phloem of most crop species is sucrose; in some species it is the only one. The phloem sap also contains nitrogenous substances, especially amino acids, amides, and urides, at concentrations of 0.03 to 0.4%. Extremely low quantities of many other compounds are also translocated in the phloem, including many growth regulators, nucleotides, some inorganic nutrients, and systemic pesticides. However, many compounds, such as reducing sugars, contact herbicides, proteins, most polysaccharides, calcium, iron, and most micronutrients, do not normally move in phloem.

The most widely proposed translocation mechanism is the mass flow or pressure flow hypothesis originally suggested by Munch (1930), which postulates that assimilate moves in a mass flow along a hydrostatic pressure gradient. The active (metabolic) loading and unloading of assimilate in the source and sink regions, respectively, are responsible for differences in osmotic potential in the sieve tubes in these regions (Giaquinta 1980). At the source, where sugars are produced, the phloem increases in sugar concentration. This reduces the water potential in sieve tubes, which causes water to move into sieve tubes from surrounding tissue. This, in turn, increases the hydrostatic pressure, causing mass flow of water and assimilate to areas of less pressure. At sinks the sugar concentration is reduced by sink utilization. This removes sugars from the sieve tubes, which increases the water potential, and water moves in from the sieve tubes, which reduces the hydrostatic pressure in the tubes and thus results in a hydrostatic pressure gradient from source to sink. A presentation of the pressure flow hypothesis has recently been presented by Milburn (1975).

TRANSLOCATION RATES

The rate at which a compound is moved in the phloem can be affected by the rate of acceptance by sinks (phloem unloading), the chemical nature of the compound as it affects movement in phloem tissue, and the rate at which the source is moving the compound into sieve tube elements (phloem loading). One way of measuring the translocation rate of assimilate is to allow leaves to photosynthesize $^{14}CO_2$ and measure the rate of ^{14}C movement from the leaf. Velocity of front molecules with ^{14}C have been measured at over 500 cm • hr^{-1}. However, when the bulk of assimilate is measured, velocities usually range 30–150 cm • hr^{-1} (Salisbury and Ross 1978).

For yield, velocity is less important than *specific mass transfer* (SMT) (Canny 1973), which is the weight of assimilate moved per cross-sectional area of phloem per unit of time. SMTs measured for several species have been surprisingly similar, ranging 3–5 g • cm^{-2} • hr^{-1} (Canny 1960). These observations suggest that the cross-sectional phloem area might limit the translocation rate.

Phloem size seems to develop according to the size of the source or sink it is serving. For example, the cross-sectional area of phloem within the peduncle of modern wheat is greater than that of wheat ancestors and is correlated to greater translocation rates (Evans et al. 1970). The fact that larger leaves have a proportionally larger cross-sectional phloem area than do smaller leaves (Segovia and Brown 1978) is specific for leaves of the same species and generally true for leaves among species.

Although the cross-sectional phloem area is fairly uniform among plants, there seems to be more phloem tissue than is needed for adequate translocation. In experiments in which the cross-sectional phloem area of peduncles was reduced by incision, the grain growth rate was not reduced in either wheat (Wardlaw and Moncur 1976) or sorghum (Muchow and Wilson 1976). In addition, when the cross-sectional phloem area of wheat roots was reduced the specific mass transfer (based on cross-sectional phloem area) increased more than 10 times. Considering these results, it seems unlikely that the volume of phloem tissue limits the flow from source to sink in most crops.

Different translocation rates occur among species, especially between the plants exhibiting C_4-type and C_3-type photosynthesis. Leaves of C_4 species have higher CO_2 exchange rates, a larger ratio of cross-sectional phloem area to leaf area, and greater translocation rates (Gallaher et al. 1975). Leaves of C_4 species also export a larger percentage of their assimilate within a few hours than do C_3 species (Fig. 3.1). This improved export of assimilate by leaves of

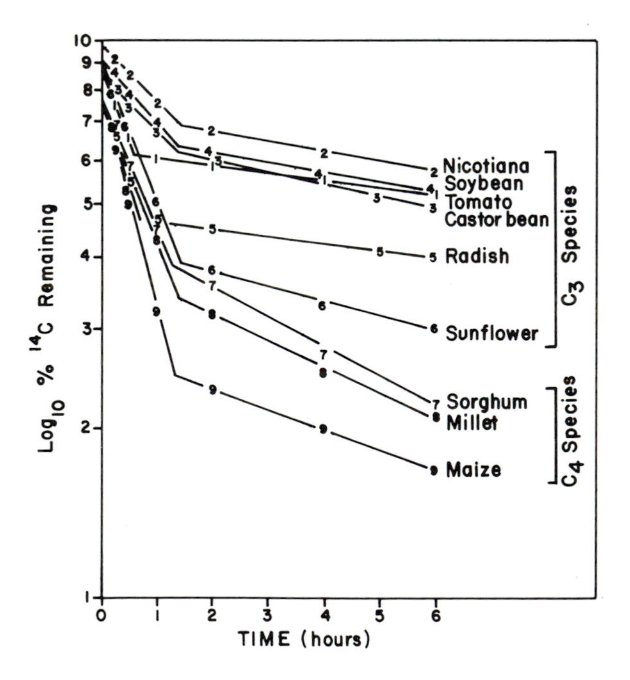

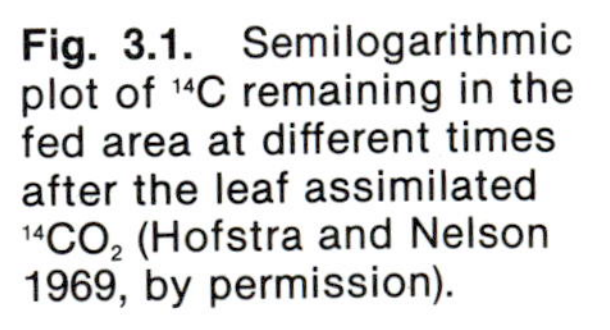

Fig. 3.1. Semilogarithmic plot of ^{14}C remaining in the fed area at different times after the leaf assimilated $^{14}CO_2$ (Hofstra and Nelson 1969, by permission).

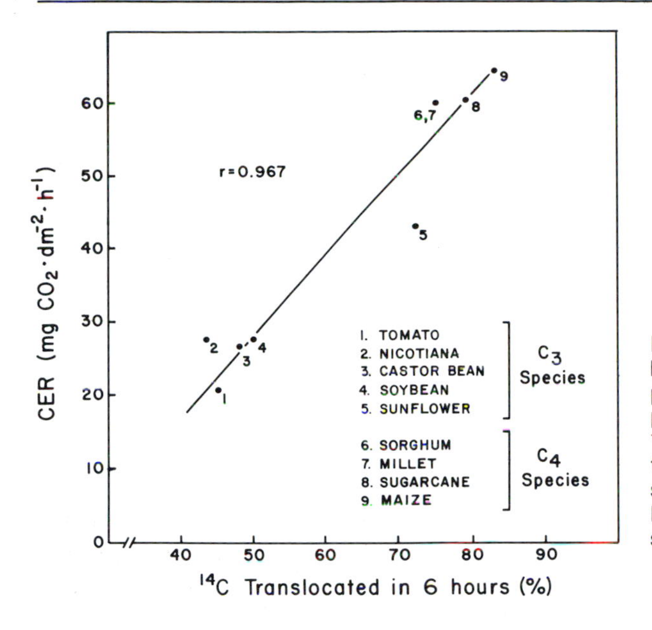

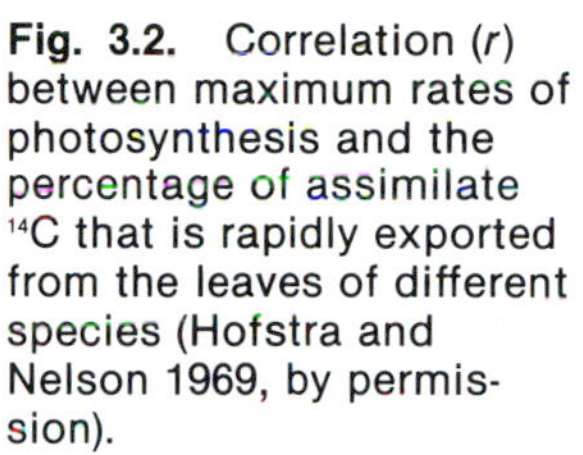
Fig. 3.2. Correlation (*r*) between maximum rates of photosynthesis and the percentage of assimilate ^{14}C that is rapidly exported from the leaves of different species (Hofstra and Nelson 1969, by permission).

C_4 species may be due to their specialized anatomy, in which vascular sheath cells have chloroplasts (Kranz anatomy), or the result of a greater cross-sectional phloem area (Hofstra and Nelson 1969; Moss and Rasmussen 1969). However, there is evidence to indicate that improved export might be related more to higher CO_2 exchange rates than to leaf anatomy (Fig. 3.2).

PHLOEM LOADING AND UNLOADING

Phloem loading (transfer of photosynthate from the mesophyll cells of the leaf to the phloem sieve tube elements) and *phloem unloading* (transfer of photosynthate from phloem sieve tube elements to the cells of a sink) can be rate limiting and can affect translocation. During phloem loading the mesophyll cells are typically at a lower osmotic potential (higher water potential) than the sieve tube elements; thus phloem loading requires an energy input to move sugars into an area of higher concentration. Phloem loading generates the increased osmotic potential in the sieve tube elements, supplying the driving force for mass flow of assimilate. It consists of movement of sugars from symplast (mesophyll cells) into apoplast (cell walls) and then into symplast (phloem cells). When sugars move into sieve elements, the movement may be aided by adjacent companion cells (Giaquinta 1980).

The greater rate of movement in C_4 species may be due to the vascular sheath cells, which surround the veins in the leaf and have chloroplasts (see Fig 1.17). Under illumination, chloroplasts can help provide photosynthetic energy (adenosine triphosphate, or ATP) needed for loading. It has also been suggested that under high leaf sucrose levels the bundle sheath cells might have a higher osmotic potential than adjacent sieve tubes to facilitate loading through a sugar concentration gradient (Troughton and Currie 1977).

At the other end of the translocation process, phloem unloading can also limit the rate at which a sink receives assimilate. Studies on unloading are scarce, so description is difficult (Giaquinta 1980). Some studies have shown that unloading is similar to loading in that the sugars move from the phloem symplast to the apoplast and then are transferred to the symplast of sink cells. However, there are indications that unloading may occur by a direct symplast transfer from phloem cells to sink cells (McNeil 1976). Current indications are that unloading occurs by different mechanisms in different tissues and may vary with the developmental status of the sink (Giaquinta 1980).

ASSIMILATE PARTITIONING

Partitioning of assimilate is generally to the sinks closest to the source. For example, upper leaves export principally to the shoot apex, lower leaves to roots, and middle leaves to both (Wardlaw 1968). Since phloem sieve connections are on one side of the stem, the leaves on one side may be more efficient at exporting assimilate to sinks on the same side. This has been shown for many crops (Wardlaw 1968); for example, the upper, expanding leaves of soybean will import more assimilate from the second leaf below them, which is on the same side of the stem, than from the closest leaf, which is on the other side of the stem (Thrower 1962). Cross-linking of sieve tube elements occurs in most species, but some seem more efficient than others. Grasses have extensive cross-linking at nodes, which essentially eliminates a preferred route by assimilate from any leaf to any particular sink (Gifford and Evans 1981).

Sink-Source Relationships and Partitioning

The movement of assimilate from source to sink is currently believed to occur something like the following (Fig. 3.3). The photosynthetic source cell

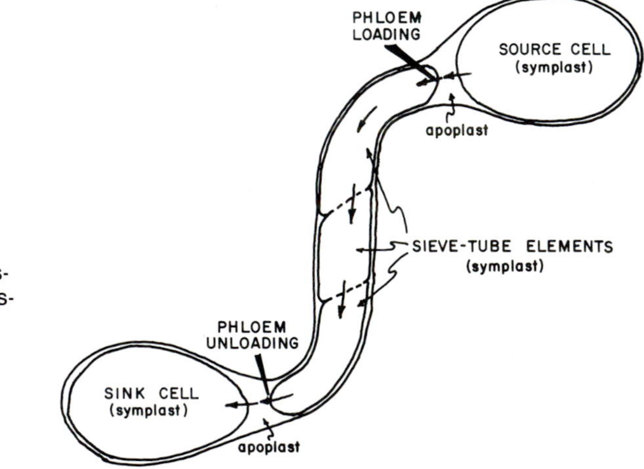

Fig. 3.3. Simplified illustration of assimilate transport from source to sink.

produces the sugars, which can move symplastically to the sieve tubes. Phloem loading increases the sugar concentration of sieve tubes above that of the apoplast.

At the sink, carbohydrates are being absorbed and either actively partitioned into cell constituents (e.g., starches) or changed to other carbohydrates that have little effect on hydrostatic pressure of the phloem. Phloem unloading lowers the concentration of sugars in sieve tubes. The buildup of sugars at the source and the removal of sugars at the sink establish a hydrostatic pressure gradient, which moves water and sugars from sources to sinks.

Where are the limitations for movement of assimilate from sources to sinks? According to the mass flow hypothesis, anything increasing photosynthesis increases hydrostatic pressure and translocation rate. However, this is true only if sinks have the ability to utilize more assimilate. If they are unable to utilize the increased production there would be a steady buildup of sugars in the system, causing a feedback inhibition resulting in reduced photosynthesis (Mondal et al. 1978). Presumably, photosynthetic rate would be reduced to the rate at which sinks could accept assimilate. For leaf photosynthesis to be at maximum potential rates, sinks must be able to utilize all the assimilate produced. Under these conditions partitioning would be controlled by sink strength, that is, sink availability and the rate at which available sinks can utilize assimilate (Gifford and Evans 1981).

Factors that control sink strength also control the partitioning in crop plants. The effect of hormones on enzymatic activity and the elasticity of sink cells can have a dramatic effect on partitioning. Indoleacetic acid (IAA), cytokinins, ethylene, and gibberellic acid, when applied to cut stem surfaces, cause assimilate to accumulate in the region of application (Gifford and Evans 1981). In bean seedlings, the main control over the distribution of sucrose between root and shoot sinks can be attributed to auxin and cytokinin (plant growth regulators, see Chap. 7) concentrations in various sinks (Gersani et al. 1980). Hormonal influences on initiation, development, and abortion of flowers and seeds have a significant effect on the source-sink relationships in crops.

Although there is some evidence that hormones may have a direct effect on translocation rates, most results show an indirect influence through affecting sink demand (Gifford and Evans 1981).

Assimilate Partitioning during the Vegetative Phase

Leaves and other green tissues are the original sources of assimilate. Some remains in the green tissue for cell maintenance and, if translocation is slow, can be converted to starch or some other form of storage. The rest is exported (translocated) to vegetative sinks, which are composed of growth, maintenance, and storage functions.

During vegetative growth, roots, stems, and leaves are competitive sinks for assimilate. The proportions of assimilate partitioned to these three organs can influence plant growth and productivity. The investment of assimilate into greater leaf area development results in greater light interception. However, the leaves also require water and nutrients, so investment in root growth is necessary. Some crop plants, such as most grasses, have essentially no stem growth during vegetative development and favor partitioning to leaves and roots.

Some meristems are in more favorable positions to intercept assimilate. For example, the intercalary meristems of leaves are in a better position to intercept translocated assimilate than are the peripheral root and shoot meristems (Evans and Wardlaw 1976).

Young developing leaves need imported assimilate to provide the energy and carbon skeletons for growth and development until they produce enough assimilate to handle their own requirements. Thrower (1962) and Webb and Gorham (1964) have shown that the leaves of soybean and squash are largely self-sufficient when 50% of their final area is developed (Fig. 3.4). After full expansion and under good environmental conditions for photosynthesis, leaves may export 60 to 80% of their assimilate to other areas of the plant (Hofstra and Nelson 1969). As the leaf gets older and begins senescence, it may fail to support its own energy requirements because of age or shading or both. Under these conditions the leaf does not export or import assimilate. Instead, cell maintenance requirements (respiration) are often greatly reduced, which allows the leaf merely to survive. Before death, many of the inorganic and organic compounds in the leaf are remobilized and translocated to other parts of the plant.

The early growth of branches and tillers requires importing assimilate

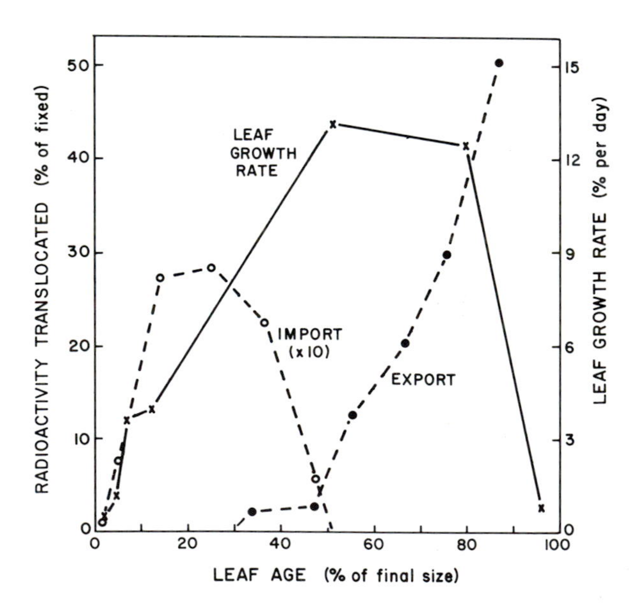

Fig. 3.4. Photosynthate import and export of a soybean leaf with age (Thrower 1962).

from the main stem or other branches until they become autotrophic. In oats, this usually occurs between the two- and four-leaf stage (Labanauskas and Dungan 1956). Whether a branch or tiller becomes completely independent of the rest of the plant is variable among species. In timothy, the tillers behave as separate units once they become autotrophic (St. Pierre and Wright 1972). Little interaction between timothy tillers occurs even under stress conditions, and roots are supplied only by the tillers to which they are attached. When under stress the autotrophic tillers of some species, such as ryegrass (Marshall and Sagar 1968) and oats (Labanauskas and Dungan 1956), will again start transporting assimilate from the main culm. How partitioning of assimilate among tillers affects total yield is influenced by how much the additional leaf area of the tiller contributes to the total dry weight of the plant and how much the tiller contributes to harvestable yield; for example, tillers of maize do not usually produce grain.

Assimilate Partitioning during the Reproductive Phase

Reproductive growth is often the primary part of the plant harvested for yield. Crops, whose flowers, fruits, and seeds (and their products) are the economic yield, have been selected over time to partition large amounts of their total dry matter into reproductive parts. In such plants a large photosynthetic surface and supporting structure are required prior to fruiting. After flowering the reproductive sink becomes extremely strong, which limits the assimilate partitioned for additional leaf, stem, and root growth. In determinate species, leaf and stem growth cease at flowering (Fig. 3.5), while indeterminate species may have vegetative and reproductive growth occurring simultaneously. Thus indeterminate species are variable in the relative strength of their vegetative and reproductive sinks. If there is much vegetative growth during reproductive development, reproductive yield may be reduced.

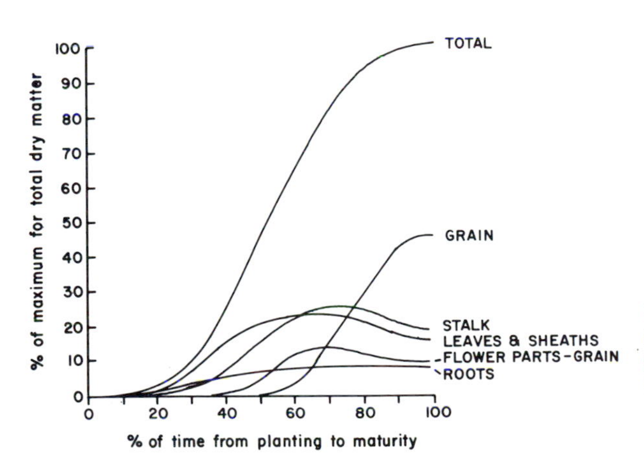

Fig. 3.5. Dry matter accumulation of a determinate grain crop and its component parts.

In determinate grain crops early growth is vegetative, allowing the plant to intercept more light energy for photosynthesis as it increases in size and allowing for adequate water and nutrient absorption to support leaf growth. The number of leaves is established at the initiation of inflorescence and is affected by temperature and photoperiod (see Chap. 12). Shortly after seed initiation, seeds become the dominant sink of annual plants. Therefore, during seed filling, the major part of assimilate, both current and stored, is used for increasing seed weight.

HARVEST INDEX

Two useful terms used to describe partitioning of dry matter by the plant are *biological yield* and *economic yield*. The term biological yield was proposed by Nichiporovich (1960) to represent the total dry matter accumulation of a plant's system. Economic yield and agricultural yield have been used to refer to the volume or weight of those plant organs that constitute the product of economic or agricultural value. The proportion of biological yield represented by economic yield has been called the *harvest index*, the coefficient of effectiveness, or the migration coefficient. All these terms characterize the movement of dry matter to the harvested part of the plant. The harvest index, the most widely used term, is defined as follows:

$$\text{harvest index} = \frac{\text{economic yield}}{\text{biological yield}} \times 100$$

(It must be remembered that the biological yield total often does not include root weights because of the difficulty of obtaining those values.)

Crop yield can be increased either by increasing the total dry matter produced in the field or by increasing the proportion of economic yield (the harvest index) or both. There is potential for increasing yields by both

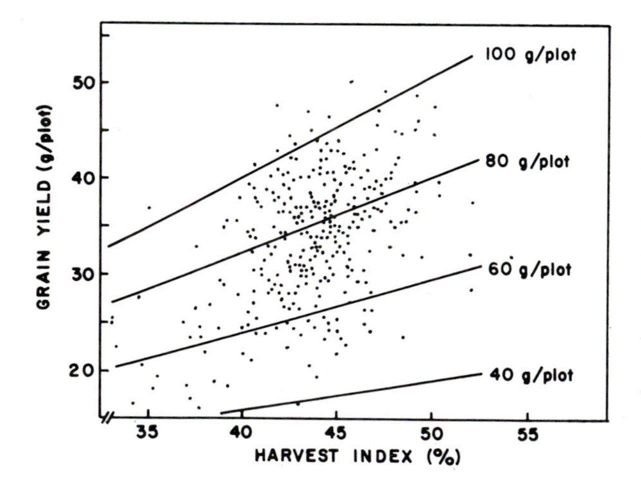

Fig. 3.6. Grain yield of a genetic population of oats in relation to the harvest index and total plant dry weight (biological yield). All plants were grown in the same environment, which indicates the variability in the ability of oat plants to partition their assimilate to grain (Takeda and Frey 1976, by permission).

methods. In oats (Takeda and Frey 1976) a large genetic population showed variability in both biological yield and harvest index (Fig. 3.6). Oat lines with high biological yield and a harvest index of 40 to 50% showed the highest grain yield. Crosbie and Mock (1981) showed that increases in both biological yield and harvest index were responsible for increased grain yield of three maize populations.

In some grain crops, the increase in seed yield has been primarily due to increases in harvest index. In other words, the plants are not producing any more total dry matter but are rather partitioning more of their dry matter into seed yield. Donald and Hamblin (1976) reported that increased grain yields in small grains were primarily due to increases in the harvest index. Research on peanuts (Duncan et al. 1978) showed the same phenomenon (Fig. 3.7). 'Dixie

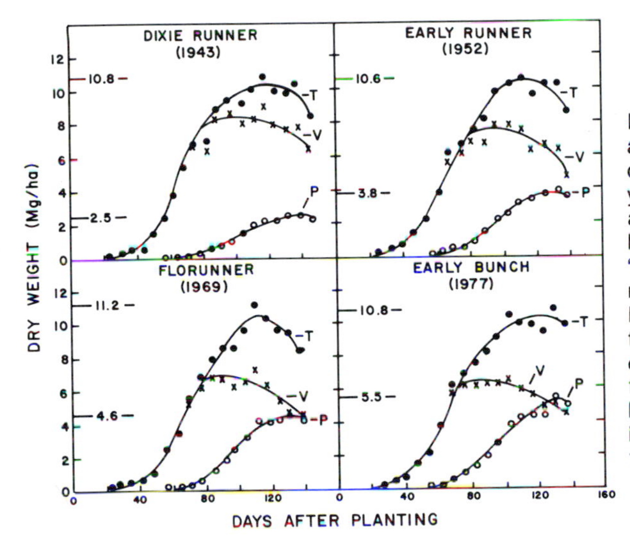

Fig. 3.7. Simulated (*lines*) and experimental (*points*) data for fruit, or economic yield, (*P*); top weight (*V*), and total dry weight, or biological yield, (*T*) of 'Dixie Runner,' 'Early Runner,' 'Florunner,' and 'Early Bunch.' Note that *T* for the four varieties remains fairly constant, while *P* increases from 2.5 Mg/ha to 5.5 Mg/ha as the varieties were improved (Duncan et al. 1978, by permission).

Runner,' developed in 1943, had a harvest index of 23 and a biological yield of 10.8 Mg (metric ton)/ha. In 1952, 'Early Runner' showed a 50% increase in seed yield over 'Dixie Runner,' primarily due to an increased harvest index of 36. In 1969, 'Florunner' showed a 20% increase in seed yield over 'Early Runner,' due to a harvest index of 41; and in 1977 'Early Bunch' was introduced with a 10% increase in seed yield over 'Florunner,' due to a harvest index of about 51. Yet the total dry matter yields of all these varieties are essentially the same. The question now for peanut breeders is whether they should select for plants with a harvest index even higher than 51, if possible, or instead breed for plants that will yield more dry matter and still have a harvest index of 51. Obviously, there must be a limit to the amount of total dry matter that can efficiently be converted to seeds.

YIELD COMPONENTS

Grain yield is a product of a number of subfractions called *yield components* and can be expressed as follows:

$$Y = N_r N_g W_g$$

where Y = grain yield, N_r = the number of reproductive units (e.g., heads, ears, panicles) per unit of ground area, N_g = the number of grains per reproductive unit, and W_g = the average weight per grain.

Yield components are affected by management, genotype, and environment, which often helps explain why a reduction in yield occurred (Table 3.1).

The genotype can influence emergence capabilities and sets the potential for tillering, flower number, number of flowers that develop into grain, amount of assimilate produced, and assimilate partitioning. The environment affects the ability of the plant to express its genetic potential. Managerial factors include the number of seeds planted and the ability of the crop manager to produce a growth environment favoring maximum yield. Water, nutrients, temperature, light, and other environmental factors at levels other than optimum can reduce one or more yield components. Examples of water stress effects on yield and yield components are illustrated in Chapter 4 in Figures 4.17, 4.18, and 4.19.

REMOBILIZATION

Once produced, assimilate is transported to many areas in the plant. It can be transformed into many compounds, some of them structural compounds, such as cellulose and hemicellulose, that provide for the physical structure of the plant and usually remain where they are synthesized. Plant cells do not have enzyme systems to degrade structural compounds, but many storage compounds that can be changed back into forms that can be translocated to other parts of the plant are also produced. These storage compounds are significant in maintaining growth and development constancy despite photosynthetic fluctuations. Storage compounds are mostly composed of carbohydrates but often include significant amounts of lipids and proteins. The movement of compounds from an area where they were once deposited to an area where they can be reutilized is referred to as *remobilization*. During certain phases of development more assimilate is being produced than is used in growth and development, and this excess can be directed to storage compounds. At a later phase, for example, fruiting, when photosynthesis is not able to furnish the assimilate requirements of plant sinks, storage compounds can be remobilized and moved to active sites, such as seed development. Remobilization occurs with both organic and inorganic compounds. During leaf senescence carbohydrates, nitrogenous compounds, phosphorus, sulfur, and other mobile elements are remobilized and translocated to current plant sinks.

TABLE 3.1. Grain yield components, the factors affecting them, and the stages of development in which they are affected

Symbol	Description	Primary Factors Affecting Yield Component	Stage of Development and Amount of Effect				
			Early vegetative	Floral initiation	Pollination	Grain filling	
						Early	Late
N_r	Reproductive units/ground area	Plants/ground area Seeding rate Emergence Tillering	Large	Small	...	...	...
N_g	Grain/reproductive unit	No. flowers initiated No. flowers pollinated No. pollinated flowers that fill	*	Large	Large	Small	...
W_g	Weight/grain	Assimilate availability	*	*	*	Large	Large

Note: *Large* indicates a developmental stage in which management, environment, and genotype have primary effects on the yield component. *Small* indicates a stage in which there is less effect. An *ellipsis* indicates a stage in which there is little or no effect. An *asterisk* indicates interactions among yield components. If there are many reproductive units and much grain per reproductive unit, even optimum conditions during grain filling will mean lighter kernels since the assimilate must be partitioned over a large number of kernels. However, if environmental conditions reduced the grain per reproductive unit, heavier kernels result, due to fewer kernels to fill with similar amounts of assimilate.

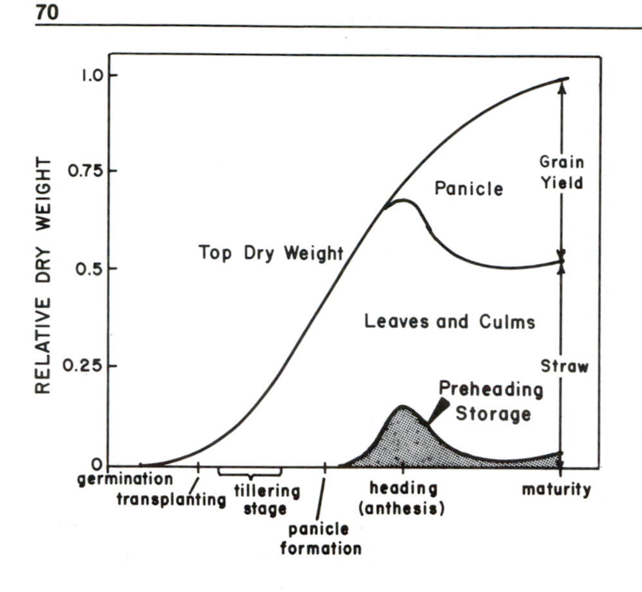

Fig. 3.8. Changes in the amount of temporarily stored carbohydrates (preheading storage) and dry weight of various parts, according to growth stages in rice (Murata and Matsushima 1975, by permission).

An example of remobilization has been shown in rice (Fig. 3.8). During the heading and flowering stage of the plant, the assimilate produced by photosynthesis is more than is required by these processes. The extra assimilate is moved to the stem and stored primarily as starch. However, as the plant goes into grain fill, starch is converted to sugars and translocated to the filling grain.

ASSIMILATE PARTITIONING DURING GRAIN FILL

Photosynthate deposited in grain can come from three major sources: current leaf photosynthesis, current photosynthesis from nonleaf parts, and remobilization of assimilate deposited in other plant organs. How much each of these factors contributes to final grain yield is affected by species and environment.

Partitioning has been extensively studied in small grain crops. Work in wheat and barley has shown that photosynthesis of the flag leaf, stem, and head, which are the closest sources to the grain, is the primary contributor to the grain. Lower leaves supply the needs of lower stem and roots (Lupton 1966; Wardlaw 1968). The strength of the grain as a sink and the relative availability and strength of sources affect the assimilate partitioning. If the top leaves are removed, the lower leaves will supply assimilate to the grain; if the lower leaves are removed, the flag leaf will transport assimilate to roots (Marshall and Wardlaw 1973).

It would be helpful to know just how much each source contributes to grain yield and the variability involved. Early work on shading the head of wheat or barley showed a 20 to 30% reduction in grain weight (Porter et al. 1959). Using shading and measuring photosynthesis, these investigators calcu-

TABLE 3.2. Contribution of preanthesis photosynthesis to the grain yield of wheat and barley under wet and dry growing conditions

Authors	Species	Growing Conditions	Grain Yield ($g \cdot m^{-2}$)	Grain Yield due to Remobilization ($g \cdot m^{-2}$)	Contributions from Preanthesis Photosynthesis (%)
Austin et al. (1980)	Barley	Wet	673	74	11
		Dry	302	133	44
Bidinger et al. (1977)	Barley	Wet	530	65	12
		Dry	384	67	17
	Wheat	Wet	509	64	13
		Dry	294	79	27

lated the contribution of different photosynthetic sources to final grain yield. They estimated that the contribution of preanthesis photosynthesis (remobilized assimilate) was 25%, current leaf and stem photosynthesis was 45%, and head photosynthesis was 30%. These percentages were confirmed by recent studies using more sophisticated procedures (Table 3.2). Drought stress during grain filling reduces grain yield through reduced photosynthesis. Thus the sink demand for grain filling uses more remobilized stored assimilate, which results in a much higher proportional contribution by remobilization (Table 3.2). Although remobilization is an important component of grain yield, photosynthesis during the grain filling period is normally the most important source of weight for grain yield. This is because most assimilate is used for vegetative or flower production before grain filling, whereas during grain filling most assimilate is used for that process (Fig. 3.9).

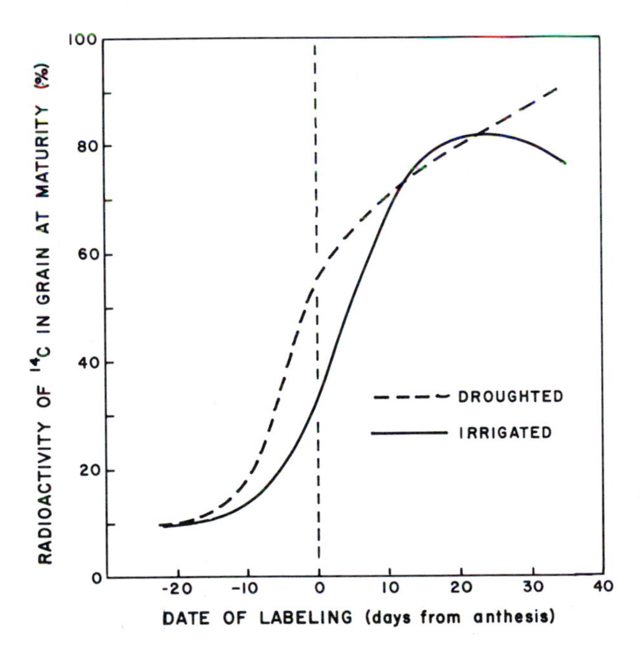

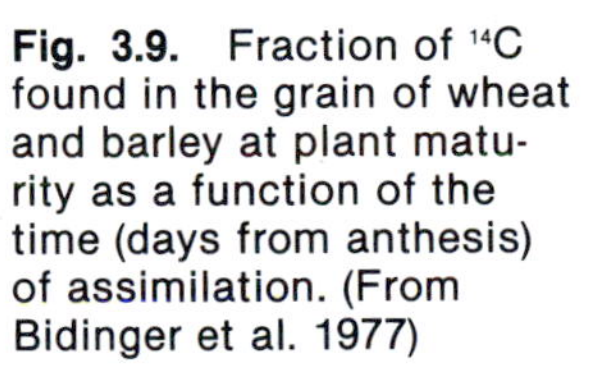

Fig. 3.9. Fraction of ^{14}C found in the grain of wheat and barley at plant maturity as a function of the time (days from anthesis) of assimilation. (From Bidinger et al. 1977)

Since the heads of small grains are located at the top of the canopy in the best light conditions for photosynthesis, and since the assimilate produced is next to the grain, head photosynthesis would be expected to contribute heavily to grain yield. Primitive wheat types have lower yields, and thus lower sink demand, than modern wheats. They rely primarily on head photosynthesis for grain yield, partitioning very little from leaves. Wheat developed with greater yield has an increased partitioning of photosynthate from upper leaves (Evans and Dunstone 1970). Increasing photosynthesis of small grain heads could also increase yields; one way is to add *awns* (thin extensions of the glume or lemma), which have been shown to double the photosynthesis rate of heads (McDonough and Gauch 1959). Several elaborate studies have shown that in semiarid and arid environments isolines of wheat with awns will outyield isolines without awns by as much as 12% (Suneson et al. 1948; McDonough and Gauch 1959). Illustrating the same effect, awn removal can decrease yield as much as 21% (Saghir et al. 1968). The primary effect of awns on yield components is to increase kernel weight (Suneson et al. 1948). Lupton (1966) showed that the amount of assimilate from glumes and flag leaves partitioned to the seeds increased as grain filling progressed and that glumes, which are closest to the seeds, partitioned a larger percentage of assimilate to seeds than did flag leaves (Table 3.3). Awns have shown no yield advantage in humid climates, possibly because of an increased susceptibility to disease or lodging (McKenzie 1972).

TABLE 3.3. Amount of ^{14}C left in the applied organ or translocated to grain 7 days after $^{14}CO_2$ is applied to the glume or the flag leaf of wheat

	^{14}C Left in Applied Organ (%)		^{14}C in Grain (%)	
Days after Anthesis	Glume	Flag leaf	Glume	Flag leaf
7	49	37	46	45
21	23	19	72	72
28	14	20	84	75

Source: From Lupton 1966.

In maize, in which the ear is located in the middle of the stem, almost all the assimilate produced is from leaves or sheaths. During grain filling, the upper leaves distribute about 85% of their assimilate to the ear (Table 3.4). Lower leaves contribute to root growth and stem and leaf maintenance as well as to ear weight. In contrast to small grains, all leaves of maize may contribute some assimilate to grain yield.

Remobilization of stalk reserves in modern maize is not much different from that in primitive maize (Valle 1981). Stalk reserves are positively correlated with stalk strength; lower reserves mean weaker stems, favoring stalk rot disease organisms. Apparently, modern maize cultivars have not been selected for strong remobilization, so the plants maintain resistance to lodging.

TABLE 3.4. Percentage ^{14}C distributed to different parts of the maize plant from a leaf that had assimilated $^{14}CO_2$ 4 days earlier

Plant Part Measured[a]	^{14}C Found in Plant Part (%)[b]							
	At anthesis				After anthesis			
	$^{14}CO_2$ added to leaf in stem[c]				$^{14}CO_2$ added to leaf in stem			
	1	2	3	4	1	2	3	4
Tassel	4	N	N	N	N	N	N	N
Stem 1	22	N	N	N	10	N	N	N
Stem 2	19	15	N	N	5	11	N	N
Ear	40	62	3	3	83	85	65	45
Stem 3	8	12	33	8	2	3	23	8
Stem 4	4	6	29	65	N	N	11	40
Roots	2	5	35	24	N	N	1	8

Source: From Eastin 1969.

[a]Measurements were taken at pollination (anthesis) and during grain filling (after anthesis).

[b]Leaves and nodes are numbered sequentially from the top down. Stems include leaves, sheaths, and stem.

[c]Stem 1 is node 3 up, stem 2 is node 3 to ear node (usually node 7), stem 3 is ear node to node 10, and stem 4 is node 10 to stem base. N = negligible measurement.

In soybean, in which almost every node provides seed growth and development, the pattern of translocation from each leaf is similar. The greatest amount of assimilate remains with the pods at the node of the applied leaf, with the rest transported to upper and lower nodes (Fig. 3.10). Lower leaves retain more assimilate than upper nodes, which may be due to the lower light levels reducing the amount of assimilate produced (Shibles et al. 1975).

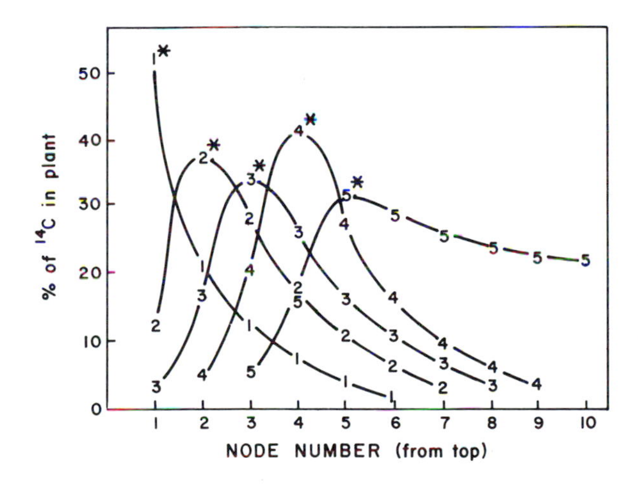

Fig. 3.10. Distribution of ^{14}C added to a leaf (*asterisk*) of different nodes of a soybean plant during bean filling. All leaves but five were removed from the plant, which changed the distribution of the fifth node. (From Belikov and Pirskii 1966)

Summary

Partitioning of assimilate in the plant takes place in the phloem where assimilate translocates from photosynthetic sources to sinks. Source cells synthesize sugars, which move to the apoplast around the phloem. These sugars are actively taken up by the phloem (phloem loading), which increases the hydrostatic pressure in that area of phloem by increasing water uptake. Assimilate moves to sinks, where the removal of sugars in the apoplast around the phloem causes sugars and water to move out of the phloem (phloem unloading), keeping a low hydrostatic pressure in that area of the phloem.

In terms of yield, assimilate partitioning is important in both the vegetative and reproductive growth phases. Partitioning during the vegetative phase will determine the final leaf area, root development, and branching. Investment of assimilate into plant growth during the vegetative period determines the productivity at later stages of development, including seed number just prior to anthesis. Partitioning during reproductive development is important for flower, fruit, and seed crops. Assimilate can be distributed from current leaf photosynthesis, nonlaminar photosynthesis, and remobilization of stored assimilate. The proportion of assimilate coming from each source depends on genotype and environment.

For high yield, the crop should quickly produce enough leaf area index to intercept most of the light for maximum dry matter production, after which it should maintain high light interception and should partition assimilate in the largest quantities possible to the organs of economic value without affecting quality or harvestability.

References

Austin, R. B., C. L. Morgan, M. A. Ford, and R. D. Blackwell. 1980. Ann. Bot. 44:309–19.
Belikov, I. F., and L. I. Pirskii. 1966. Sov. Plant Physiol. 13:361–64.
Bidinger, F., R. B. Musgrave, and R. A. Fischer. 1977. Nature 270:431–33.
Canny, M. J. 1960. Biol. Rev. 35:507–32.
______. 1973. Phloem Translocation. London and New York: Cambridge University Press.
Crosbie, T. M., and J. J. Mock. 1981. Crop Sci. 21:255–59.
Donald, C. M., and J. Hamblin. 1976. Adv. Agron. 28:361–405.
Duncan, W. G., D. E. McCloud, R. L. McGraw, and K. J. Boote. 1978. Crop Sci. 18:1015–20.
Eastin, J. A. 1969. Proc. 24th Annu. Corn Sorghum Res. Conf. Washington, D.C.: American Seed Association.
Evans, L. T., and R. C. Dunstone. 1970. Aust. J. Biol. Sci. 23:725–41.
Evans, L. T., and I. F. Wardlaw. 1976. Adv. Agron. 28:301–59.
Evans, L. T., R. L. Dunstone, H. M. Rawson, and R. F. Williams. 1970. Aust. J. Biol. Sci. 23:743–52.
Gallaher, R. N., D. A. Ashley, and R. H. Brown. 1975. Crop Sci. 15:55–59.
Gersani, M., S. H. Lips, and T. Sachs. 1980. J. Exp. Bot. 31:177–84.
Giaquinta, R. T. 1980. Biochem. Plants 3:271–320.
Gifford, R. M., and L. T. Evans. 1981. Annu. Rev. Plant Physiol. 32:485–509.
Hofstra, G., and C. P. Nelson. 1969. Planta 88:103–12.
Labanauskas, C. K., and G. H. Dungan. 1956. Agron. J. 48:265–68.
Lupton, F. G. H. 1966. Ann. Appl. Biol. 57:355–64.
McDonough, W. T., and H. G. Gauch. 1959. Maryland Agric. Exp. Stn. Bull. A103, pp. 1–16.
McKenzie, H. 1972. Can. J. Plant Sci. 52:81–87.
McNeil, D. L. 1976. Aust. J. Plant Physiol. 3:311–24.

Marshall, C., and G. R. Sagar. 1968. J. Exp. Bot. 19:785–94.
Marshall, C., and I. F. Wardlaw. 1973. Aust. J. Biol. Sci. 26:1–13.
Milburn, J. A. 1975. In Transport in Plants, ed. M. H. Zimmerman and J. A. Milburn. Berlin and New York: Springer-Verlag.
Mondal, M. H., W. A. Brun, and M. L. Brenner. 1978. Plant Physiol. 61:394–97.
Moorby, J., M. Ebert, and L. T. Evans. 1963. J. Exp. Bot. 14:210–20.
Moss, D. N., and H. P. Rasmussen. 1969. Plant Physiol. 44:1063–68.
Muchow, R. C., and G. L. Wilson. 1976. Aust. J. Agric. Res. 27:489–500.
Munch, E. 1930. Die Staffbewegungen in Der Pflanze [Translocation in Plants]. Jena: Fisher.
Murata, Y., and S. Matsushima. 1975. In Crop Physiology, ed. L. T. Evans. London: Cambridge University Press.
Nichiporovich, A. A. 1960. Field Crop Abstr. 13:169–75.
Porter, H. K., N. Pal, and R. V. Martin. 1950. Ann. Bot. n.s. 15:55–67.
Saghir, A. R., A. R. Khan, and W. Worzella. 1968. Agron. J. 60:95–97.
St. Pierre, J. C., and M. J. Wright. 1972. Crop Sci. 12:191–94.
Salisbury. F. B., and C. W. Ross. 1978. Plant Physiology. Belmont, Calif.: Wadsworth.
Segovia, A. J., and R. H. Brown. 1978. Crop Sci. 18:90–93.
Shibles, R. M., I. C. Anderson, and A. H. Bigson. 1975. Agron. J. 60:95–97.
Suneson, C. A., B. B. Bayles, and C. C. Fifield. 1948. USDA Circ. 783, pp. 1–8.
Takeda, K., and K. J. Frey. 1976. Crop Sci. 16:817–21.
Thrower, S. L. 1962. Aust. J. Biol. Sci. 15:629–49.
Troughton, J. H., and B. G. Currie. 1977. Plant Physiol. 59:808–20.
Valle, M. R. R. 1981. Ph.D. diss., University of Florida, Gainesville.
Wardlaw, I. F. 1968. Bot. Rev. 34:79–105.
Wardlaw, I. F., and L. Moncur. 1976. Planta 128:43–100.
Webb, J. A., and P. R. Gorham. 1964. Plant Physiol. 39:663–72.

4 Water Relations

A RAPIDLY GROWING HERBACEOUS CROP is composed primarily of water. Water content varies between 70 and 90%, depending on age, species, particular tissue, and environment. It is indispensable for numerous plant functions:

1. Solvent and medium for chemical reactions
2. Medium for organic and inorganic solute transport
3. Medium that gives turgor to plant cells. Turgor promotes cell enlargement, plant structure, and foliar display.
4. Hydration and neutralization of charges on colloidal molecules. For enzymes, water of hydration helps to maintain structure and facilitates catalytic functions.
5. Raw material for photosynthesis, hydrolytic processes, and other chemical reactions in the plant
6. Water evaporation (transpiration) for cooling plant surfaces

Under field conditions, the roots permeate a relatively moist soil while the stems and leaves grow into a relatively dry atmosphere. This causes a continuous flow of water from soil through the plant to the atmosphere along a gradient of decreasing energy potential. On a daily basis, this flow amounts to 1 to 10 times the amount of water held in plant tissues, 10 to 100 times the amount used in expansion of new cells, and 100 to 1,000 times the amount used in photosynthesis (Jarvis 1975). The primary movement of water, therefore, is from the soil to the leaf to replace transpiration loss.

Because of the high demand for and importance of water, a plant requires a consistent water source for growth and development. Anytime water becomes limiting, growth is reduced and usually also crop yield. The amount of yield reduction is affected by genotype, water deficit severity, and the stage of development.

Water Potential

The system that describes the behavior of water and water movement in soils and plants is based on a potential energy relationship. Water has the capacity to do work; it will move from an area of high potential energy to an area of low potential energy. The potential energy in an aqueous system is expressed by comparing it with the potential energy of pure water. Since the water in plants and soils is usually not chemically pure due to solutes and is physically constrained by forces such as polar attractions, gravity, and pressure, the potential energy is less than that of pure water. In the plant and soil, potential energy of water is called the *water potential,* symbolized by the Greek letter *psi* (ψ_w) and expressed as force per unit of area. The unit measurement is usually the *bar* or *pascal* (Pa): 1 bar = 10^5 Pa = 10^6 dynes • cm^{-2} = 0.99 atmospheres, or 10^2 J • kg^{-1}. Pure water has a water potential equal to 0 bar. The water potential in plants and soils is usually less than 0 bar, which means that it is of negative value. The more negative the value, the lower the water potential.

The water potential of plants and soils is the sum of a number of component potentials:

$$\psi_w = \psi_m + \psi_s + \psi_p + \psi_z \tag{4.1}$$

where ψ_m = *matrix potential,* the force with which water is held to plant and soil constituents by forces of adsorption and capillarity. It can only be removed by force and so has a negative value.

ψ_s = *solute potential* (*osmotic potential*), the potential energy of water as influenced by solute concentration. Solutes lower the potential energy of water and result in a solution with a negative ψ_w.

ψ_p = *pressure potential* (*turgor pressure*), the force caused by hydrostatic pressure. Since it is a force, it usually has a positive value. It is generally of minor importance in soils but of primary importance in plant cells, as will be explained later.

ψ_z = *gravitational potential,* which is always present but usually insignificant in short plants, compared with the other three potentials. It can be significant in tall trees.

SOIL WATER AVAILABILITY

Plant roots grow into moist soil and extract water until a critical water potential in the soil is reached. Water that can be extracted from the soil by plant roots, termed *available water,* is the difference between the amount of water in the soil at *field capacity* (water held in the soil against the gravitational force) and the amount of water in the soil at *permanent wilting percentage* (the percentage of soil moisture at which a plant will wilt and not recover in an atmosphere of 100% relative humidity).

The availability of soil moisture is affected by the *colloidal property* (i.e., the surface area of soil particles). A clay loam soil holds about 20% of its weight as available moisture, whereas a coarser textured soil, such as fine

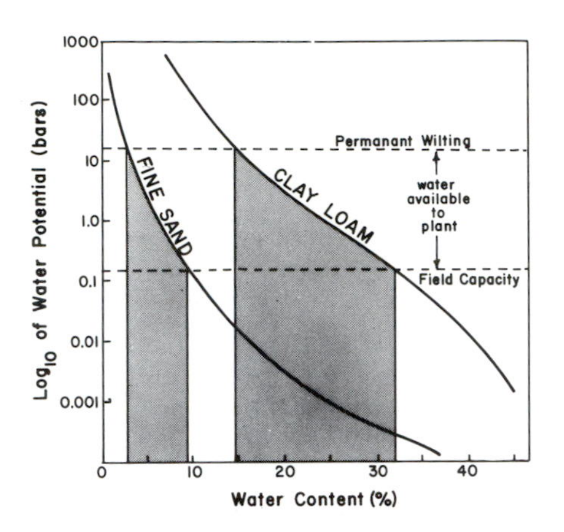

Fig. 4.1. Water potential for fine sand and clay loam soils at different water contents. Note the 20% available soil moisture in the clay loam soil, only 7% in the fine sand.

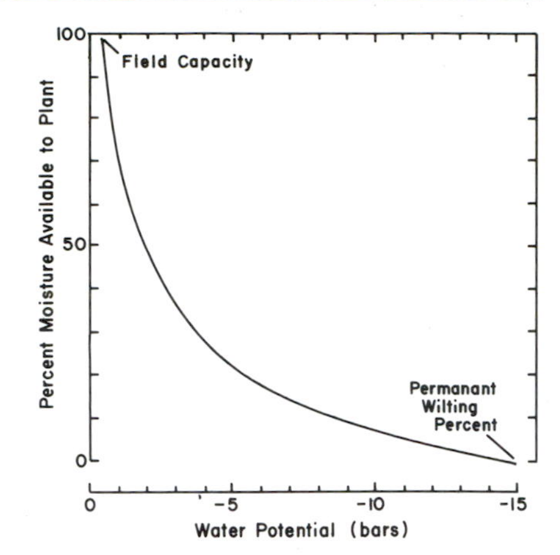

Fig. 4.2. Moisture percentage available to the plant in a loam soil at different soil water potentials. In this soil 50, 75, and 90% of the available water is held in the soil at water potentials of −2, −5, and −10 bars, respectively.

sand, only holds about 7% (Fig. 4.1). On a soil volume basis, the clay loam at field capacity will hold about 17 cm available water per meter of soil depth whereas the fine sand will hold less than 8 cm. A fine-textured loam soil at field capacity can supply about 25 cm (10 in.) of water to a plant that has roots extending 1.5 m (5 ft) into the soil.

Soil water potential (ψ_{soil}) in agricultural soils is primarily affected by matrix potential (ψ_m) and secondarily by solute potential (ψ_s). Soil water potential can be related to field capacity and permanent wilting percentage. At field capacity the ψ_{soil} is about -0.1 to -0.3 bars. Permanent wilting percentage varies among crop plant species (-15 to -50 bars) but is often arbitrarily set at -15 bars. The water potential at the permanent wilting point is of minor importance, since over 70% of the available water has been removed from a soil at -5 bars (Fig. 4.2) and very little water is available from -15 to -30 bars (Verasan and Phillips 1978).

WATER UPTAKE AND MOVEMENT

Air usually has an extremely low water potential (Fig. 4.3), compared with plants or soils. Since a living leaf usually has a water potential greater than -15 bars, there is a steep energy gradient and continual movement of water as vapor from the leaf to the air. When there is no loss of water from the plant to the air (e.g., at night), the ψ_{plant} would come close to equilibrium with the ψ_{soil}. When the stomata are open, the water loss from the leaf is continual,

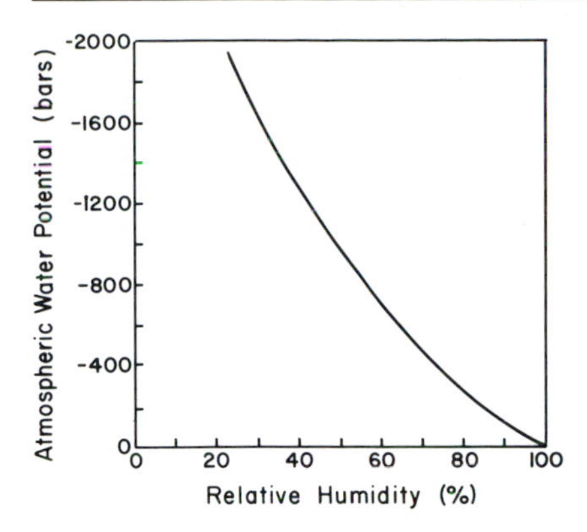

Fig. 4.3. Water potential of the atmosphere plotted against relative humidity at 25°C.

which lowers the ψ_{leaf} below that of the $\psi_{petiole}$. Since water moves from a high ψ_w to a low ψ_w, water will flow from the petiole to the leaf. This water flux reduces the $\psi_{petiole}$, which was at equilibrium with the ψ_{stem}, so water moves from the stem to the petiole. This energy gradient is continuous down to the root and soil. In other words, the system develops a ψ_w gradient from the soil to the air. Rates of water absorption and movement through the plant can be affected by the amount of soil moisture, root to soil contact, plant and soil resistances to water flux, and the ψ_w gradient (Begg and Turner 1976).

Figure 4.4 is a schematic illustration of changes in the ψ_w of a soil-plant system with a 5-day drying period. When adding water to the soil and allowing the soil to lose gravitational water, the ψ_{soil} is approximately -0.3 bars. At night the stomata are closed so water moves into the plant and creates a ψ_{soil} and ψ_{plant} equilibrium. During the day, the stomata are open and transpiration occurs. As water is lost from the leaf the ψ_{leaf} becomes reduced, initiating a ψ_w gradient. This provides the energy differential to cause water movement from the soil to replace the water lost by transpiration. At night stomata close and transpiration is reduced to near zero. However, water will continue to flow in the system until the ψ_{plant} and ψ_{soil} again approach equilibrium. As water is removed from soil by plants, the ψ_{soil} becomes lower and the ψ_{leaves} become relatively lower, establishing a ψ_w gradient for continual uptake. The ψ_{leaves} on day 4 (Fig. 4.4) drop to -15 bars and remain there, indicating that transpiration is being reduced by stomatal closure, which is usually accompanied by temporary leaf wilting. On day 5 the water potential of leaves, roots, and soil go below a water potential of -15 bars. This usually indicates that the water available to the plant is not enough to prevent wilting and recovery will not occur unless water is added to the soil. Soil moisture dropping to permanent wilting percentage in 5 days (Fig. 4.4) indicates a very limited soil volume and a root system in close contact with the total soil volume. Under most field conditions, the soil volume per plant is larger, allowing for a much slower reduction of soil water content. In the field the water content is not uniform

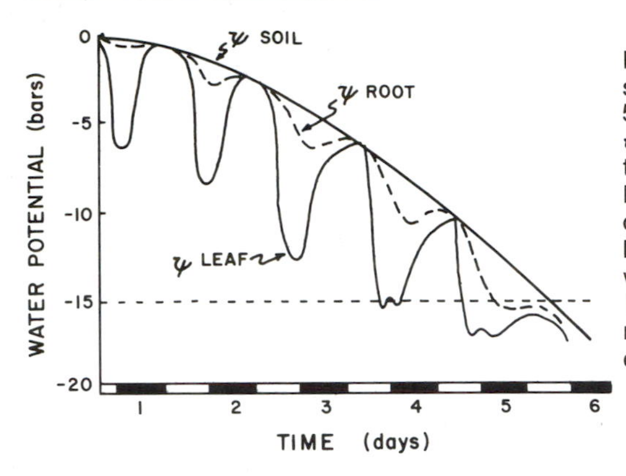

Fig. 4.4. Schematic representation of soil and plant water potentials (ψ) for a 5-day drying period. During the day the ψ declines as water is lost through transpiration. This builds a ψ gradient. During the evening transpiration declines and the ψ in the plant rises. The horizontal dashed line indicates where wilting occurs. (From Slatyer, *Plant-Water Relationships,* © 1967, by permission of Academic Press, Inc. [London] Ltd.)

throughout the soil profile; while roots extract moisture from one area, they are extending into new areas of the soil that may have a high ψ_w. In this manner, the plant is often able to stay at a higher ψ_w than the average ψ_{soil}. However, as the volume of moist soil diminishes, the plant will need large ψ_w gradients for roots to absorb enough moisture to satisfy transpirational loss. This is a gradual process in the field and may last for weeks in medium- to low-textured soils, which allows the plants to become acclimated to lower water potentials. Under limited soil volumes, the change in ψ_w is rapid and the plant has less chance to adjust to a lower ψ_w.

Evapotranspiration

The total amount of water lost from the field by both soil evaporation and plant transpiration is called *evapotranspiration* (ET).

Evaporation is an energy-dependent process involving a change in state from a liquid to a vapor phase. The rate of transpiration is a function of the vapor pressure gradient, the resistance to flow, and the ability of the plant and soil to transport water to the transpiration sites. Transpiration provides the major driving force for plant water absorption against the gravitational pull and frictional resistances in the water pathway through the plant (Jarvis 1975). The uptake rate is controlled primarily by the transpiration rate; root pressure, the active absorption of water, plays a minor role in absorption and is only evident when transpiration is low or ceases (Kramer 1959).

ENVIRONMENTAL FACTORS AFFECTING EVAPOTRANSPIRATION

Water loss to the atmosphere is determined by both environmental and plant factors. The environmental effect on ET is called *atmospheric demand* or *evaporatory demand*. The greater the atmospheric demand, the faster water can be evaporated from a free water surface. The following factors influence atmospheric demand:

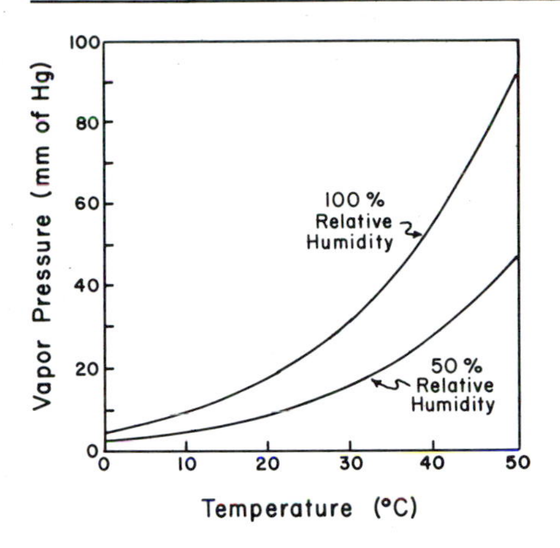

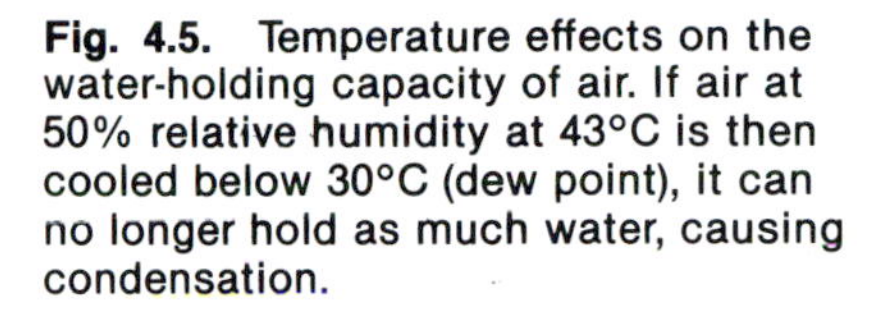
Fig. 4.5. Temperature effects on the water-holding capacity of air. If air at 50% relative humidity at 43°C is then cooled below 30°C (dew point), it can no longer hold as much water, causing condensation.

1. Solar radiation. Of solar radiation absorbed by the leaf, 1 to 5% is used for photosynthesis and 75 to 85% is used to heat the leaf and for transpiration. Increased solar radiation increases atmospheric demand.

2. Temperature. Increasing the temperature increases the capacity of air to hold water (Fig. 4.5), which means a greater atmospheric demand.

3. Relative humidity. The greater the water content of air, the higher the ψ_{air}, which means that atmospheric demand decreases with increasing relative humidity (Figs. 4.3, 4.5).

4. Wind. Transpiration occurs when water diffuses through the stomata. A diffusion gradient barrier builds up around stomata when the air is still (Fig. 4.6). This means that water diffusing from the wet leaf interiors is almost matched by water buildup outside the leaf, which reduces the diffusion gradient and thus reduces transpiration. When turbulence (wind) removes the moisture next to the leaf, the difference in water potential inside and immediately outside the stomatal opening is increased and net water diffusion from the leaf is increased (Fig. 4.7).

Climatologists measure atmospheric demand by determining the amount

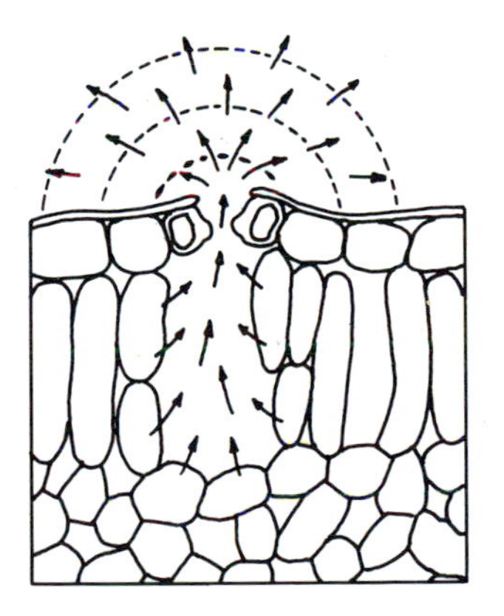
Fig. 4.6. Water diffusing through a stomatal opening with no turbulence at the leaf exterior, so a diffusion gradient builds up that reduces transpiration.

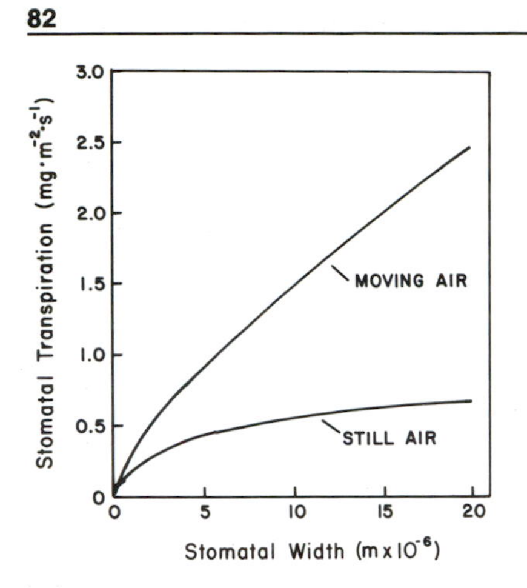

Fig. 4.7. Wind and its effect on transpiration at different structural openings. At narrow stomatal openings the difference between moving and still air is not as great as at wide openings. (From Bange 1953)

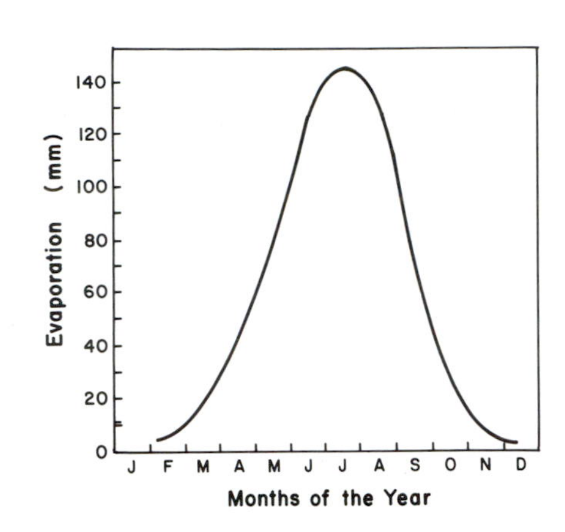

Fig. 4.8. Open-pan evaporation during different months of the year in a temperate region.

of water evaporation from an open pan. The highest atmospheric demand occurs at the time of year when solar radiation and temperatures are greatest (Fig. 4.8).

PLANT FACTORS AFFECTING EVAPOTRANSPIRATION

Plant factors, as well as atmospheric demand, modify the ET rate by affecting the resistance to water movement from soil to air:

1. Stomatal closure. Most transpiration occurs through stomata because of the relative impermeability of the cuticle, and little transpiration occurs when stomata are closed. As stomata open wider, more water is lost (Fig. 4.7), but the loss increase is less for each unit increase in stomatal width. Many factors influence stomatal opening and closing, the major ones under field conditions being light and moisture level. In most crop plants light causes stomata to open. A low moisture level in the leaf (low ψ_{leaf}) causes guard cells to lose their turgor, resulting in stomatal closure.

2. Stomatal number and size. Most leaves of productive crops have many stomata on both sides of their leaves (see Table 1.1). Stomatal number and size, which are affected by both genotype and environment, have much less effect on total transpiration than stomatal opening and closing.

3. Leaf amount. The more leaf surface area, the greater the ET. Figure 4.9 illustrates that when LAI increases in a field, the amount of ET, compared with open-pan evaporation, increases. However, there is less increase in water

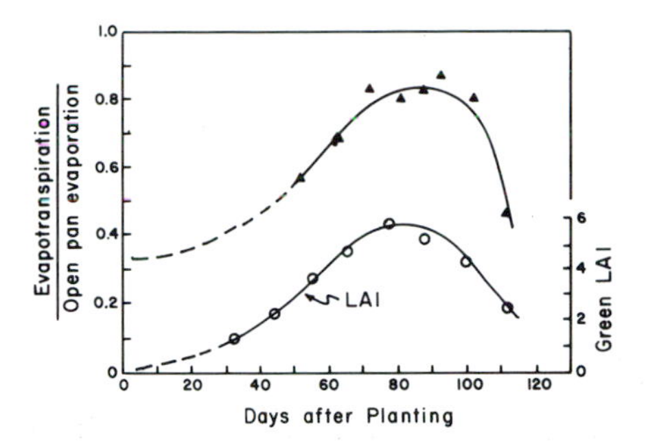

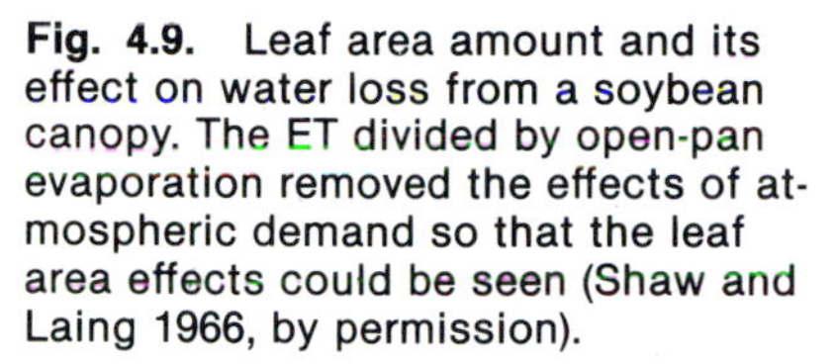

Fig. 4.9. Leaf area amount and its effect on water loss from a soybean canopy. The ET divided by open-pan evaporation removed the effects of atmospheric demand so that the leaf area effects could be seen (Shaw and Laing 1966, by permission).

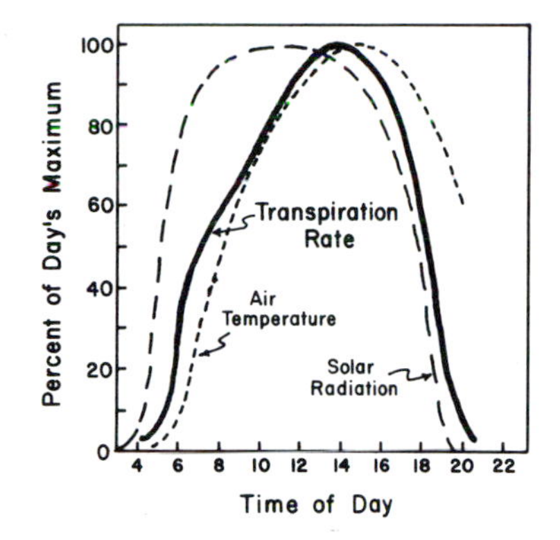

Fig. 4.10. Daily relationships among solar radiation, temperature, and ET (Briggs and Shantz 1916)

loss for each unit increase in the LAI. There are some indications that ET will not increase with increases in the LAI over that required to intercept 80% of solar radiation (Stern 1965).

4. Leaf rolling or folding. Many plants have mechanisms in leaves that favor reduced transpiration when water becomes limiting. Some grass species, like maize, reduce their exposed leaf area by leaf rolling, while many others, like bluegrass, reduce exposed leaf area by leaf folding. Broadleaves have other mechanisms to reduce water loss; for example, soybean has a tendency to roll the leaves over so the silvery pubescence (hair) on the exposed lower surface can reflect more light.

5. Root depth and proliferation. The availability and extraction of soil moisture by the crop is highly dependent on rooting depth and proliferation. Deeper rooting increases water availability, and root proliferation (roots per unit soil volume) increases water extraction from a unit volume of soil before permanent wilting occurs.

Knowing how environment and the plant influence ET helps to explain the daily pattern of ET in the field. Stomata open in response to light, and ET increases with increases in solar radiation and air temperature. If atmospheric demand does not become greater than the ability of the plant to supply water to leaves, the greatest ET will occur during the afternoon when air temperatures are highest (Fig. 4.10). Daily ET starts to decrease in the late afternoon, primarily because of less light energy and decreasing temperature.

Under high soil water, ET would usually increase with increased atmospheric demand. Limiting soil moisture, however, causes changes in the relationships among atmospheric demand, soil moisture, stomatal closure, and rate of water flow through the plant (Fig. 4.11). As moisture level in the soil is reduced, the ET level of a high atmospheric demand day (clear, dry) is reduced to a level similar to that of a lower atmospheric demand day (partly cloudy, humid). This is probably caused by stomatal closure or by increased resistance to translocation during the afternoon of high, but not low, atmospheric demand days. In other words, when moisture is limiting in the soil in a high atmospheric demand day, the leaves are losing water more rapidly than the roots or the translocation system can supply it. This results in a ψ_{leaf} low enough to cause stomatal closure and/or results in slower water absorption and movement, due to increased resistance within the soil and plant. On the lower demand day, water uptake by roots can keep up with the water lost by leaves and so water loss continues unencumbered until a low ψ_{soil} is reached. This illustrates the interaction among atmospheric demand, soil factors, and plant factors that influences the ET rate in the field.

POTENTIAL EVAPOTRANSPIRATION

Potential ET is a combination of evaporation and transpiration with the soil surface completely covered with vegetation and abundant moisture. It can be estimated by open-pan evaporation (Fig. 4.8). Most crops do not stay at the potential ET throughout their life cycles because there are times when they do not have a full plant canopy and/or the soil is unable to supply the water to replace transpiration. Annual crops start out with very little leaf area and increase it through the growing season. Since crop plants grow fastest under warm temperatures and high solar radiation and since atmospheric demand is greatest with these conditions, high leaf area occurs during the peak of potential ET. This usually causes the peak water demand to occur in midsummer (Fig. 4.8).

When potential evapotranspiration is compared with precipitation, it is apparent why moisture deficits often occur during the period of most rapid growth rate (Fig. 4.12). For high yields, the crop must be supplied with water during this period. This can be done either by sufficient storage of soil moisture to supply crops during the deficit period or by irrigation. In many farming areas, the most productive soils are those with a high water-storage capacity, which allows crops to keep producing during periods when precipitation is less than ET.

Moisture Stress

Water often limits crop growth and development. The plant's response to water stress is relative to its metabolic activity, morphology, stage of growth, and yield potential. The sequence of the response to a drying cycle follows.

Cellular growth is the plant function most sensitive to water deficits (Table 4.1). The daytime ψ_w of meristematic tissues often causes the ψ_p to decrease

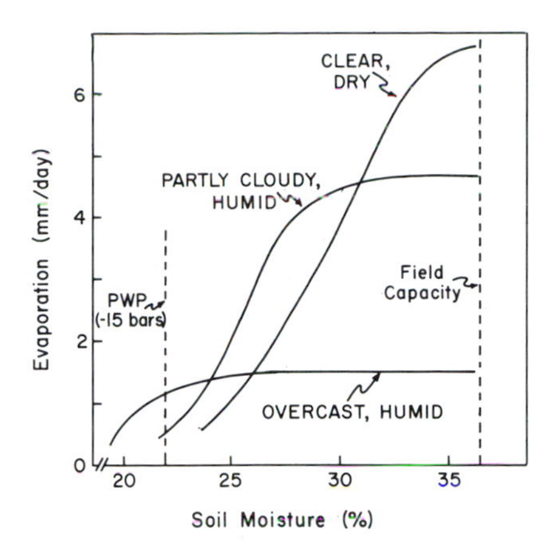

Fig. 4.11. ET in a crop in relation to atmospheric demand and water availability in the soil (Denmead and Shaw 1962, by permission)

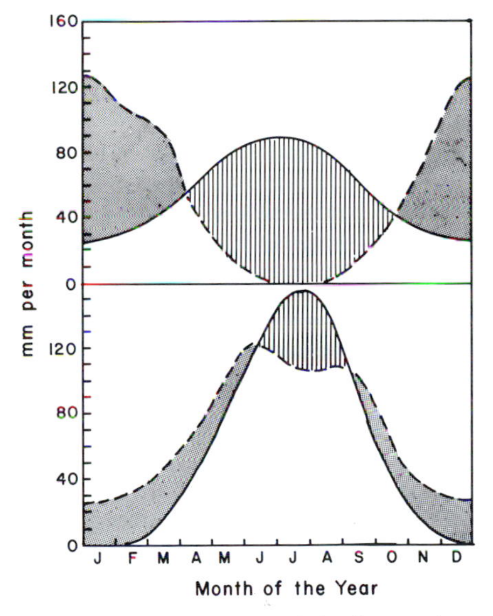

Fig. 4.12. Average precipitation and potential ET of (*above*) a Mediterranean climate and (*below*) a continental climate. (Dashed line represents precipitation; solid line represents ET; striped area indicates precipitation less than ET; shaded area indicates precipitation in excess of ET.)

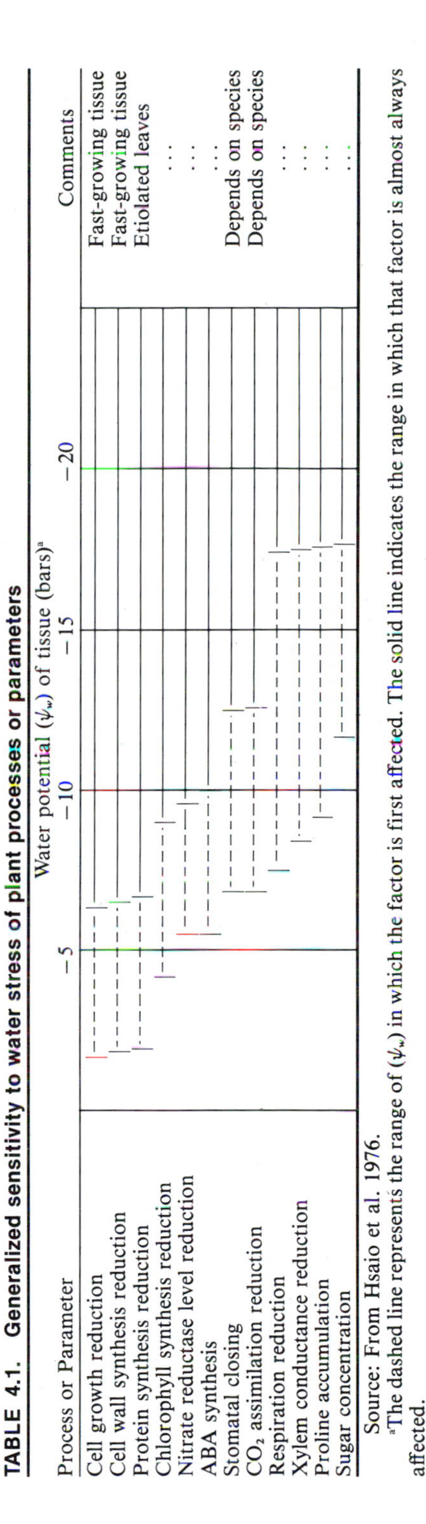

TABLE 4.1. Generalized sensitivity to water stress of plant processes or parameters

Process or Parameter	Water potential (ψ_w) of tissue (bars)[a] (−5, −10, −15, −20)	Comments
Cell growth reduction		Fast-growing tissue
Cell wall synthesis reduction		Fast-growing tissue
Protein synthesis reduction		Etiolated leaves
Chlorophyll synthesis reduction		. . .
Nitrate reductase level reduction		. . .
ABA synthesis		. . .
Stomatal closing		Depends on species
CO_2 assimilation reduction		Depends on species
Respiration reduction		. . .
Xylem conductance reduction		. . .
Proline accumulation		. . .
Sugar concentration		. . .

Source: From Hsaio et al. 1976.

[a]The dashed line represents the range of (ψ_w) in which the factor is first affected. The solid line indicates the range in which that factor is almost always affected.

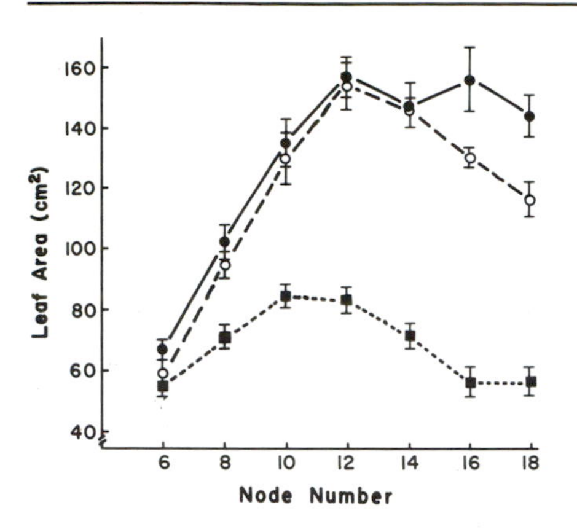

Fig. 4.13. Final area of leaf at each node of field beans for wet (●), medium (○), and dry (■) treatments (Elston et al. 1976).

below that needed for cell enlargement. This in turn causes a reduction in protein synthesis, cell wall synthesis, and cell enlargement, which may account for the observation that many species have their greatest growth at night when the ψ_w is greatest (Boyer 1968). The effect of stress during the vegetative stage is the development of smaller leaves (Fig. 4.13), which can reduce the LAI at maturity and result in less light interception by the crop. Chlorophyll synthesis is inhibited at greater water deficits. With moisture stress most enzymes show reduced activity (e.g., nitrate reductase), but some hydrolytic enzymes show increased activity (e.g., amylase) (Hsaio et al. 1976). The breakdown of reserved polymer molecules decreases the ψ_s, resulting in increased ψ_p and counteracting the effects of a water deficit.

With reduced water potentials, plant hormones also change in concentration. For example, abscisic acid (ABA) increases in leaves and fruits. ABA accumulation induces stomatal closure, which results in reduced CO_2 assimilation; older leaves and fruits often abscise if accumulation is high (Hsiao 1973). Not all plants show an ABA increase with moisture stress. Cytokinins and ethylene can counter the ABA effect and often increase when ABA increases (Tal and Imber 1971). This may account for the more rapid fruit ripening under water stress conditions.

Under moderate to severe stress conditions, the amino acid *proline* increases in concentration more than any other amino acid. Proline seems to aid in drought tolerance, acting as a storage pool for nitrogen and/or as a solute molecule reducing the ψ_s of the cytoplasm (Stewart 1982). At extreme levels of stress (water potential greater than -15 bars), respiration, CO_2 assimilation, assimilate translocation, and xylem transport rapidly diminish to lower levels while the activity of hydrolytic enzymes increases.

Stressed plants growing in a soil water level at permanent wilting will usually recover when irrigated if wilting is of short duration. However, old leaves may abscise, new leaves will be reduced in size, and several days may be required for leaf photosynthesis to reach prestress levels (Begg and Turner 1976).

OSMOTIC ADJUSTMENT

Much of the research on water stress effects on plants has been performed on tissue excised from plants or on plants grown in pots with restricted soil volumes. There is increasing evidence that plants grown in pots respond differently to water deficits than those grown under field conditions; plants grown with small soil volumes experience water stress more rapidly than under usual field conditions. The root density is likely to be high throughout the soil volume, water extraction from the whole soil profile is uniform, and the drying cycle is relatively fast (Fig. 4.4).

The roots of field-grown plants usually grow in large soil volumes. High root densities occur in the upper soil profiles where water is extracted rapidly, but as water becomes limiting in the upper soil profile, roots expand into lower soil profiles where water is more abundant. Thus in field-grown plants the development of stress during a drying cycle is much more gradual, the possibility for overnight recovery of ψ_w is much greater, and the plant has time to adapt to the developing stress (Begg and Turner 1976).

Plants grown in growth chambers show rapid reduction in leaf expansion, starting at a leaf ψ_w of -2 to -4 bars and photosynthesis at -6 to -12 bars (Fig. 4.14), while field data show rapid leaf expansion rates at -8 to -10 bars (McCree and Davis 1974; Watts 1974). Water movement in plants is determined by ψ_w, but physiological processes affected by water availability are better predicted by ψ_w components. Solute potential (ψ_s) and turgor pressure

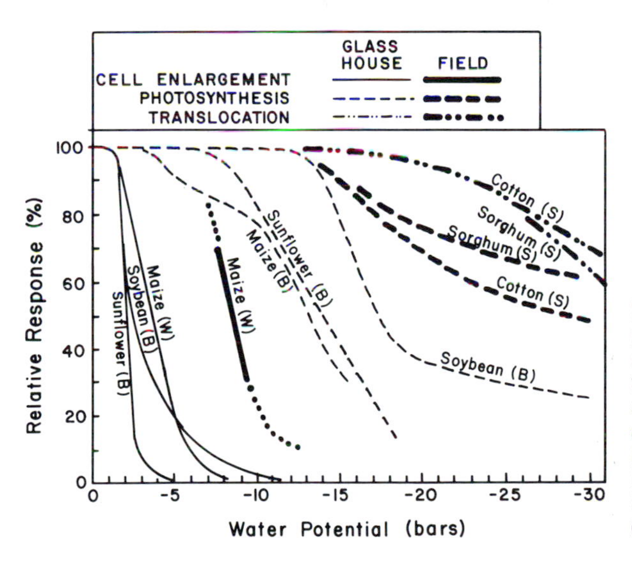

Fig. 4.14. Effect of water potential on cell enlargement, photosynthesis, and translocation for several species under glasshouse and field environments. Data compiled from (*B*) Boyer (1970), (*S*) Sung and Krieg (1979), and (*W*) Watts (1974).

(ψ_p) must be known in order to evaluate the effect of water stress on physiological processes. The primary factor affecting growth or cell expansion is ψ_p, which can vary considerably at any ψ_w because it is the positive value equal to the negative values of ψ_s and ψ_m. For example, when using equation (4.1) a cell could have the following ψ_w:

$$\underset{(-6\ \text{bars})}{\psi_w} = \underset{(-7\ \text{bars})}{\psi_s} + \underset{(-2\ \text{bars})}{\psi_m} + \underset{(+3\ \text{bars})}{\psi_p}$$

However, if (due to starch decomposition or potassium movement) the solute level should increase in the cell, water would diffuse into the cell, increasing the ψ_p even though the ψ_w may be decreasing. This is called *osmotic adjustment.*

$$\underset{(-11\ \text{bars})}{\psi_w} = \underset{(-15\ \text{bars})}{\psi_s} + \underset{(-2\ \text{bars})}{\psi_m} + \underset{(+6\ \text{bars})}{\psi_p}$$

Research on the expansion of maize and sorghum leaves by Acevedo et al. (1979) illustrates osmotic adjustment. They measured ψ_p and ψ_s of expanding leaves during a 24-hr period and found that leaf expansion occurred rapidly late in the day due to increased ψ_p even though the ψ_w was below -6 bars (Fig. 4.15). The increase in the ψ_p resulted from a drop in the ψ_s due to sugar accumulation in expanding cells. Shaded leaves did not accumulate sugars and had reduced leaf expansion, and little leaf expansion occurred at night because of low temperature. Even though leaf expansion occurred under low ψ_w, irrigated plants still had greater expansion during the morning period than did leaves of nonirrigated plants.

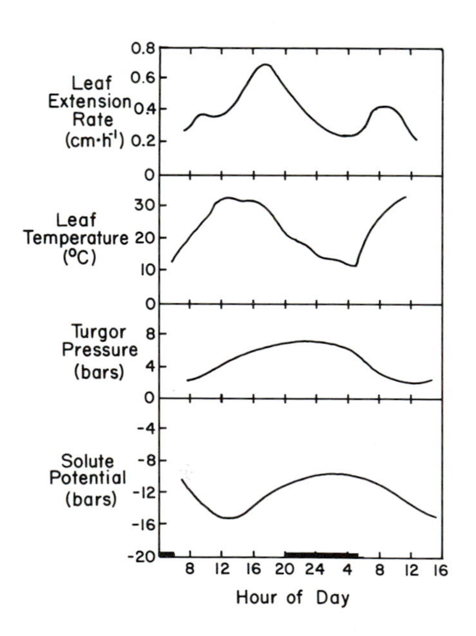

Fig. 4.15. Diurnal course for leaf extension (elongation) rates, leaf temperatures, turgor pressures, and solute potentials for unirrigated maize 42 days after planting (Acevedo et al. 1979, by permission).

Plants grown in small pots, which have restricted root systems and rapid development of water deficits, do not seem to be able to create the osmotic adjustment found in plants grown in the field. It cannot be assumed, however, that all crop species can generate osmotic adjustment under field conditions. More research must be done before this phenomenon is well understood.

STOMATAL RESPONSE TO MOISTURE STRESS

Stomatal opening results from an increase in turgor pressure of guard cells in relationship to surrounding cells. This turgor is in response to environmental stimuli, sometimes an influx of potassium ions, affecting osmotic adjustment (Humble and Hsiao 1970). Light, low CO_2 concentrations, adequate water, and low levels of ABA are the factors needed to stimulate the influx of potassium ions into guard cells (Humble and Raschke 1971). Therefore water stress, which can reduce stomatal opening, may be mediated through ABA.

Stomatal response can be different in plants grown in small pots and those grown under field conditions. Figure 4.14 shows a differential response of CO_2 uptake to leaf ψ_w. The measurements for maize and sunflower were made on plants with reduced soil volumes. Stomatal closure had to be the factor reducing photosynthesis because transpiration (stomatal resistance) was reduced to the same degree as CO_2 intake. In potted plants, stomata start closing at a leaf ψ_w of about -8 bars. However, work by Sung and Krieg (1979) showed that in sorghum and cotton, CO_2 uptake started a decline at a leaf ψ_w of -30 bars. It is apparent that stomata under field conditions, in which the drying cycle occurs over a period of weeks instead of days, are often able to stay open at a much lower ψ_{leaf}.

In some species under field conditions, the stage of development influences the stomatal opening. Research on maize and sorghum by Ackerson and Krieg (1977) showed that in the vegetative stage a low ψ_w would cause stomatal closure under sunlight. Stomatal closure was never complete under sunlight because the highest leaf resistance was around 10 s • cm^{-1} at a ψ_w of -20 bars, whereas in the dark the resistance increased to 30 s • cm^{-1}. During the reproductive stage, both maize and sorghum showed no change of leaf resistance with changes in leaf ψ_w. Thus stomatal control showed a complete insensitivity to moisture stress during the reproductive stage of growth. Under these conditions, it is difficult to determine what does limit water loss from plants. Plants may be able to create internal resistances that limit the ability to transpire (Ackerson and Krieg 1977).

Stomatal behavior is different in different environments, at different stages of development, at different leaf positions on the plant, and in different crop species. Further study is needed to better understand the factors affecting plant responses to different water regimes.

WATER STRESS EFFECTS ON YIELD

The effects of water stress on yield are manifold. During vegetative devel-

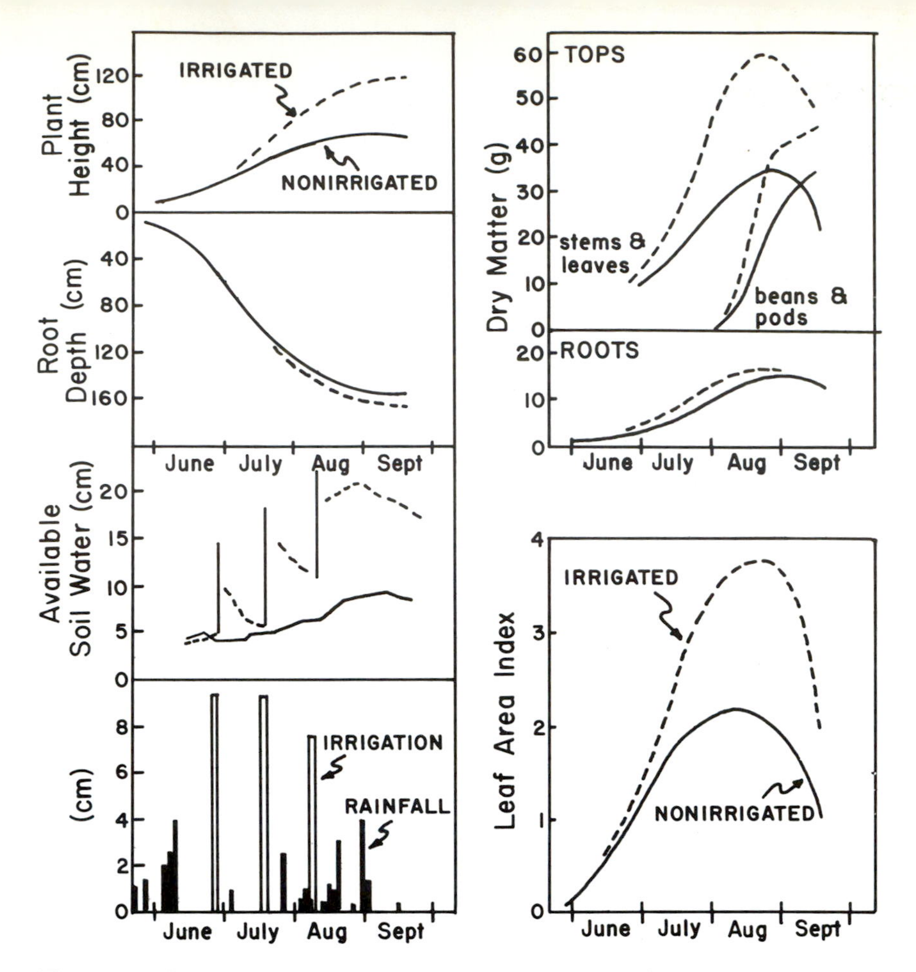

Fig. 4.16. Changes in length, dry weight, and leaf area of irrigated and unirrigated soybean. Note the much larger effect of drought on tops than on roots (Mayaki et al. 1976, by permission).

opment even minor stresses can reduce the rate of leaf expansion and LAI at later stages of development (Fig. 4.13). Severe water stress can cause stomatal closure, which reduces CO_2 uptake and dry matter production. A continuation of stress can cause such a severe reduction in photosynthesis rate that it may take several days after irrigation to resume the original rate. A number of soybean growth parameters, for example, were found to be greatly affected by water stress (Mayaki et al. 1976) (Fig. 4.16). Root elongation and dry weight were not affected as much as leaf area, stem elongation, and dry weight of tops. Roots were expanding into areas where available water was not depleted, resulting in less reduction of cell elongation. Seed yield was not affected as drastically as vegetative yield, possibly reflecting the greater water availability during seed fill and remobilization of assimilate stored in vegetative parts. The most dramatic effect of the early vegetative moisture deficit was reduction in leaf area.

For seed yield the timing of water stress may be as important as the degree of stress. To a determinate species such as maize, a severe 4-day stress at certain stages of the reproductive cycle may be critical (Fig. 4.17A). Pollination (silking) and the 2 wk following was the period most sensitive to water stress; the number of kernels per ear was the yield component most drastically affected (Fig. 4.17B). That the plants could produce more photosynthate than the kernels would accept is indicated by the increase in stalk weight (Fig. 4.17C). Three wk after pollination, water stress no longer affected kernel number but did decrease kernel weight, indicating that the period of kernel loss had passed and that moisture stress reduced leaf photosynthesis and/or translocation. A similar pattern also exists for wheat, another determinate species (Fig. 4.18).

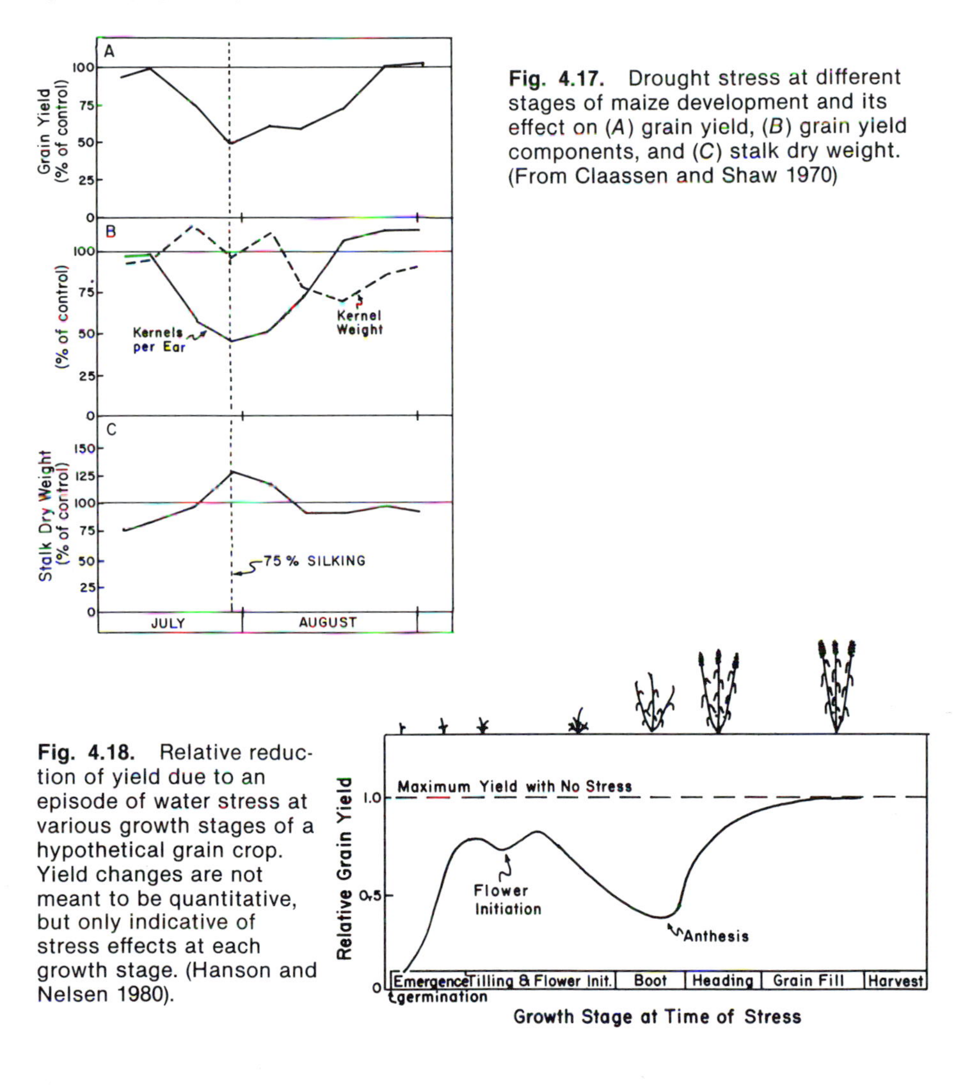

Fig. 4.17. Drought stress at different stages of maize development and its effect on (*A*) grain yield, (*B*) grain yield components, and (*C*) stalk dry weight. (From Claassen and Shaw 1970)

Fig. 4.18. Relative reduction of yield due to an episode of water stress at various growth stages of a hypothetical grain crop. Yield changes are not meant to be quantitative, but only indicative of stress effects at each growth stage. (Hanson and Nelsen 1980).

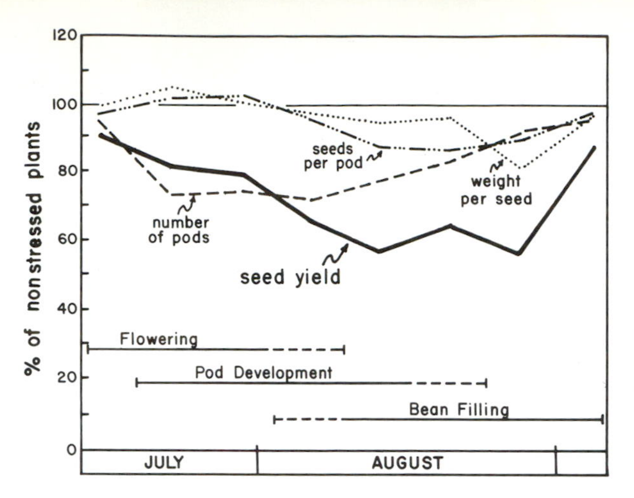

Fig. 4.19. Drought stress at different stages of soybean development and its effect on seed yield and yield components. (From Shaw and Laing 1966)

A relatively short but severe stress may have no influence on grain yield if imposed during the vegetative stage of development (Fig. 4.17A). Longer periods of less severe stress might have a greater influence on yield, indicated by soybean in Figure 4.16.

Indeterminate species that have the potential to flower over a longer period of time may not be as sensitive to water stress (Fig. 4.19). Short-term severe water stress during early flowering of soybean caused little reduction in seed yield; even though water stress caused flower abortion, the plant had time to generate more flowers after stress was removed (Shaw and Laing 1966). Flowers produced late in the flowering period, however, were less likely to produce mature pods by harvest. The yield component most influenced by water stress at flowering was the number of pods per plant. The stages most sensitive to water stress were late pod development and midbean filling. At late pod development water stress still caused pod abortion, poorer pod development (fewer seeds per pod), and reduced photosynthesis (reduced weight per seed). In later stages of seed filling, although there was some effect on pods per plant and seeds per pod, the greatest influence was on the weight per seed.

The effect of drought during pollination in areas where determinate maize and indeterminate soybeans are grown usually causes more severe seed yield losses in the maize.

Water Use Efficiency

The *water use efficiency* (WUE) of field crops for this discussion is defined as follows:

$$\text{WUE} = \frac{\text{dry matter production (DM)}}{\text{evapotranspiration}}$$

and is expressed as g DM • kg^{-1} water, kg DM • $(ha/cm)^{-1}$ water, or lb DM • $(acre/in.)^{-1}$ water. WUE measurements have been made on plants in containers, on individual plants in the field, and on crop communities. They can be used for economic yield as well as total dry matter. A related term, *water requirement,* is the reciprocal of WUE.

$$\text{water requirement} = \frac{\text{evapotranspiration}}{\text{dry matter production}}$$

Water requirement is usually expressed in weights of equal magnitude, such as g water • (g • DM)$^{-1}$.

WUE is not the same as drought resistance. WUE refers to yield in relation to the water used to produce the yield. Most of the research on WUE has been oriented toward attaining high WUE while maintaining high productivity. In drought resistance research the emphasis is often placed on survival during periods of high atmospheric demand and low water availability. In many cases, the ability to withstand severe moisture stress is negatively correlated with productivity (Reitz 1974). Many species that can tolerate severe water deficits do not make efficient use of water in the absence of stress (Levitt 1972). Some species, well adapted to severe water deficits, have moderate efficiency even in the presence of stress. The succulents are one such group. Their crassulacean acid metabolism (CAM), stomata that are closed during the day and open at night during water deficits, and leaf structure that allows minimum water loss when stomata are closed reduce transpiration more than photosynthesis and result in a higher WUE than most other species (Neales 1970).

Crop yields have increased considerably over the past 40 years. These yields have been achieved without much increase in seasonal ET. For this reason the WUE has increased along with increases in yields. Any management factors that reduce the limitations to growth without significantly increasing ET will increase WUE. Such factors as fertilizer application, control of weeds and other crop pests, water conservation, improved tillage techniques, timely planting, and improved crop varieties have substantially increased both yield and WUE.

Large differences in WUE occur when species are categorized by CO_2 fixation pathways. It is now accepted that the WUE of C_4 species is generally higher than that of C_3 species (Downes 1969; Bjorkman 1971; Brown and Simmons 1979). Earlier field data for WUE, when regrouped into C_3 and C_4 species, illustrate a 2-fold increase for C_4 species when calculated for either grasses or dicots (Table 4.2). Differences between C_3 and C_4 species increase as the temperature rises from 20 to 35°C (Bjorkman 1971).

TABLE 4.2. Water use efficiency [g DM • (kg H_2O)$^{-1}$] for C_4 and C_3 species

Species	Grasses	Dicots
	(g • kg^{-1})	
C_3	1.49	1.59
C_4	3.14	3.44

Note: Species as measured by Shantz and Piemeisel (1927). Downes (1969) regrouped them into C_3 and C_4 species.

The factors contributing to the higher WUE of C_4 species include higher photosynthesis and growth rates under high light and temperature (Bjorkman 1971; Downton 1971) and higher stomatal resistance (Begg and Turner 1976). The higher WUE of C_4 species is thus a result of higher photosynthetic rates under high light and temperature and lower transpiration rates under low light (Downton 1971). So WUE can be increased by growing C_4 crops in high solar energy regions or seasons and growing C_3 crops only in temperate humid regions or seasons (Begg and Turner 1976).

The WUE values for both C_3 and C_4 species are low compared with CAM plants. One CAM species, pineapple (*Ananas comosus*), has shown a WUE of 20 g • DM • kg water^{-1} (Joshi et al. 1965). Use of crop species with CAM is limited because the CO_2 fixation and overall productivity of CAM plants is low (Osmond 1978).

In most crop species, field ET is influenced more by atmospheric demand, amount of ground cover, and water availability than by the specific crop species. Table 4.3 illustrates that well-watered crops differed in average daily ET from 4.2 to 5.7 mm • d^{-1} (Jensen 1973). The primary factors that influenced ET for different species, with water availability remaining high, were time of year (atmospheric demand) and rate of canopy development. The consumptive use coefficient (k), which is calculated as follows,

$$\text{consumptive use coefficient } (k) = \frac{\text{actual evapotranspiration}}{\text{potential evapotranspiration}}$$

ranges from 0.65 to 0.87. It is primarily affected by the amount of ground covered by the crop canopy integrated over the growth period. Wheat has a low k because it is grown in the relatively cooler spring season and slowly develops leaf area from seeds. Alfalfa has a high k because in the spring it develops leaf area rapidly from reserve carbohydrates; although harvested during the year, it recovers leaf area rapidly from reserve carbohydrates of the root and crown and maintains a ground cover for a longer time during the growing season. Sorghum and soybean have intermediate k values because they are grown in the warm spring and summer seasons but have relatively slow leaf area development from seeds.

The WUE values in Table 4.3 illustrate that the C_4 species have an advantage, but not as great as that shown in Table 4.2. This may indicate that over a full season soil evaporation and atmospheric demand somewhat reduce the advantage one species has over another in WUE efficiency.

Improved crop management and plant breeding have led to substantial gains in WUE. Most of these gains are derived from increased leaf area production (which increases transpiration, reduces soil evaporation, and increases light interception for increased photosynthesis), greater water availability due to deeper roots and/or better water extraction, and (for economic yield) an increase in harvest index.

TABLE 4.3. Water use and dry matter productivity of seven crop species under well-watered conditions

Crop	CO_2 Fixation Pathway	Growth Period (days)	Consumptive[a] Use Coefficient (k)	ET		DM		Water[b] Requirement ($g\ H_2O \cdot g\ DM^{-1}$)	Water Use[c] Efficiency ($g\ DM \cdot kg\ H_2O^{-1}$)
				Seasonal (mm)	Daily average (mm)	Seasonal ($kg \cdot ha^{-1}$)	Daily average ($kg \cdot ha^{-1}$)		
Maize	C_4	135	0.75	658	4.9	17,000	126	388	2.58
Sorghum	C_4	110	0.78	583	5.3	14,500	132	402	2.49
Potato	C_3	128	0.65	532	4.2	10,000	78	532	1.88
Sugar beet	C_3	190	0.72	876	4.6	14,500	76	606	1.65
Wheat	C_3	112	0.66	473	4.2	7,700	69	613	1.63
Soybean	c_3	113	0.78	599	5.3	8,500	75	704	1.42
Alfalfa	C_3	195	0.87	1112	5.7	11,200	57	993	1.01

Source: From Jensen 1973.

Note: The environmental conditions are those for Kimberly, Ida.

[a]Consumptive use coefficient = actual ET/potential ET (depends primarily on the rate of full canopy development in relation to the length of season).

[b]Water requirement = ET/DM.

[c]Water use efficiency = DM/ET.

Summary

Between 70 and 90% of an actively growing herbaceous crop is water, which is indispensable for most plant functions. The roots take up water from a moist soil and move it to plant tops where it is transpired into a dry atmosphere. Thus crop plants require a consistent source of water for consistent growth and development. The system used to describe the behavior of water movement in soils and plants is based on water potential (ψ), which is the sum of component potentials: matrix potential, solute potential (osmotic potential), pressure potential (turgor pressure), and gravitational potential. Osmotic potential and turgor pressure exert the greatest influence on how plants respond to moisture stress. High turgor pressure is required for cell elongation; some plants can maintain high turgor pressure even at fairly low water potentials by increasing the osmotic potential through increased solute levels in cells, a process called osmotic adjustment. The plant's ability to adjust osmotically is greatly influenced by its growth environment.

The amount of ET from a crop canopy is a function of the water potential gradient from within the soil out to the air and the resistance to flow through the plant or from soil surfaces. Solar radiation, temperature, relative humidity, and wind are the primary environmental factors affecting ET. Stomatal closure, stomatal number and size, leaf amount, and leaf characteristics are the plant factors determining the resistance of water movement from soil to air. Potential ET is ET with a full crop canopy and abundant moisture (for minimum resistance). It is an indication of environmental effects on atmospheric demand and has both annual and diurnal fluctuations. It is estimated using open-pan evaporation.

Water deficits reduce vegetative development and yield through reduced leaf expansion and reduced leaf photosynthesis, resulting in reduced canopy photosynthesis. These reductions are primarily affected by the degree of stress. For seed yield, the timing of stress may be as important as the degree. Water stress during floral initiation, pollination, or seed development may greatly reduce the number of seeds that develop. If water stress is alleviated during grain filling, the potential seed yield may be below that of potential photosynthate production.

WUE is the yield produced per unit of water used. Since crop yields have increased considerably in the last 40 years with little increase in seasonal ET, the WUE has been increased by reducing the limitations to crop growth. WUE is particularly important in areas where water is normally a major limitation to crop yields.

References

Acevedo, E., E. Fereres, T. C. Hsiao, and D. W. Henderson. 1979. Plant Physiol. 64:476–80.
Ackerson, R. C., and D. R. Krieg. 1977. Plant Physiol. 60:850–53.
Bange, G. G. J. 1953. Acta Bot. Neerl. 2:255–97.

Begg, J. E., and N. C. Turner. 1976. Adv. Agron. 28:161–217.
Bjorkman, O. 1971. In Photosynthesis and Photorespiration, ed. M. D. Hatch et al. New York: Wiley.
Boyer, J. S. 1968. Plant Physiol. 43:1056–62.
______. 1970. Plant Physiol. 46:233–35.
Briggs, I. J., and H. L. Shantz. 1916. J. Agric. Res. 5:583–651.
Brown, R. H., and R. E. Simmons. 1979. Crop Sci. 19:375–79.
Claassen, M. M., and R. H. Shaw. 1970. Agron J. 62:652–55.
Denmead, O. T., and R. H. Shaw. 1962. Agron. J. 54:385–90.
Downes, R. W. 1969. Planta 88:261–73.
Downton, W. J. S. 1971. In Photosynthesis and Photorespiration, ed. M. D. Hatch et al. New York: Wiley.
Elston, J. A., J. Karamanos, A. H. Kassam, and R. M. Wadsworth. 1976. Philos. Trans. R. Soc. Lond. [B] 273:581–91.
Hanson, A. D., and C. E. Nelsen. 1980. In The Biology of Crop Productivity, ed. P. S. Carlson. New York: Academic Press.
Hsiao, T. C. 1973. Annu. Rev. Plant Physiol. 24:519–70.
Hsiao, T. C., E. Acevedo, E. Fereres, and D. W. Henderson. 1976. Philos. Trans. R. Soc. Lond. [B] 273:479–500.
Humble, G. D., and T. C. Hsiao. 1970. Plant Physiol. 46:483–87.
Humble, G. D., and K. Raschke. 1971. Plant Physiol. 48:447–53.
Jarvis, P. G. 1975. In Heat and Mass Transfer in the Biosphere, ed. D. A. de Vries and N. H. Afgan. Washington, D.C.: Halsted.
Jensen, M. E. 1973. Consumptive Use of Water and Irrigation Water Requirements. New York: American Society of Civil Engineers.
Joshi, M. C., J. S. Boyer, and P. J. Kramer. 1965. Bot. Gaz. 126:174–79.
Kramer, P. J. 1959. Adv. Agron. 11:51–70.
Levitt, J. 1972. Responses of Plants to Environmental Stresses. New York: Academic Press.
McCree, K. J., and S. D. Davis. 1974. Crop Sci. 14:751–55.
Mayaki, W. C., I. D. Teare, and L. R. Stone. 1976. Crop Sci. 16:92–94.
Neales, T. F. 1970. Nature [Lond.] 228:880–82.
Osmond, C. B. 1978. Annu. Rev. Plant Physiol. 29:379–414.
Reitz, L. P. 1974. Agric. Meteorol. 14:3–11.
Shantz, H. L., and L. N. Piemeisel. 1927. J. Agric. Res. [Washington, D.C.] 34:1093–1190.
Shaw, R. H., and D. R. Laing. 1966. In Plant Environment and Efficient Water Use, ed. W. H. Pierre et al. Madison, Wis.: American Society of Agronomy.
Slatyer, R. O. 1967. Plant-Water Relationships. London: Academic Press.
Stern, W. R. 1965. Aust. J. Agric. Res. 16:921–27.
Stewart, C. R. 1982. In Physiology and Biochemistry of Drought Resistance in Plants, ed. L. G. Paleg and D. Aspinall. New York: Academic Press.
Sung, F. J. M., and D. R. Krieg. 1979. Plant Physiol. 64:852–56.
Tal, M., and D. Imber. 1971. Plant Physiol. 47:849–50.
Verasan, V., and R. E. Phillips. 1978. Agron. J. 70:613–18.
Watts, W. R. 1974. J. Exp. Bot. 25:1085–96.

5 Mineral Nutrition

SCIENTIFIC ADVANCES in plant nutrition and fertilization have revolutionized crop production. As much as 50% of the high yields of maize and other cereal grains, not to mention the improvement in crop quality and nutritional value, can be attributed to commercial fertilizers. Low crop yields in many countries are often primarily due to lack of plant nutrients.

The science of plant nutrition began about 150 years ago with the classic experiments of Liebig, Lawes and Gilbert, de Saussure, Boussingault, and others. It probably remains one of the best hopes for solving the world food crisis.

Higher plants are unique organisms in that they can synthesize all the substances they require, including amino acids, hormones, and vitamins, if given 13 essential mineral elements along with carbon dioxide and water. Green plants are *autotrophic,* that is, they can synthesize all the necessary growth constituents from the basic elements. Green plants are also *photolithotrophic,* which means that essential components of growth are synthesized in the presence of light from inorganic or *lithic* (soil) elements.

Essential Elements

Sixteen elements (Table 5.1) were classified as essential for all crops, and sodium (Na), silicon (Si), and cobalt (Co) essential to some plants (Epstein 1972). Molybdenum (Mo) was declared essential in 1939 only after techniques were sufficiently refined to reduce the Mo content in the nutrient solution to less than 10 ppm, the deficiency threshold (Arnon and Stout 1939). Chlorine (Cl) was the last element declared essential (Boyer et al. 1954). Further refinement in assay techniques may demonstrate the requirement of still other elements.

Two criteria are used to establish essentiality of an element, both subject to certain limitations and qualifications. (1) An element is declared essential if the plant fails to grow and complete its life cycle in a medium devoid of the element, compared with normal growth and reproduction in a medium containing the element. Indirect or secondary beneficial effects do not qualify for essentiality. (2) An element is declared essential if it is a constituent of a necessary metabolite, such as sulfur (S) in the amino acid methionine.

Essentiality has been established using the technique of hydroponics be-

TABLE 5.1. Concentrations of nutrient elements in plant material at levels considered adequate

Element	Atomic Weight	Concentration in Dry Matter $\mu mol \cdot g^{-1}$	Concentration in Dry Matter Amount	Relative Number of Atoms with Respect to Mo
			(ppm)	
Molybdenum	95.95	0.001	0.1	1
Copper	63.54	0.10	6	100
Zinc	65.38	0.30	20	300
Manganese	54.94	1.0	50	1,000
Iron	55.85	2.0	100	2,000
Boron	10.82	2.0	20	2,000
Chlorine	35.48	3.0	100	3,000
			(%)	
Sulfur	32.07	30	0.1	30,000
Phosphorus	30.98	60	0.2	60,000
Magnesium	24.32	80	0.2	80,000
Calcium	40.08	125	0.5	125,000
Potassium	39.10	250	1.0	250,000
Nitrogen	14.01	1,000	1.5	1,000,000
Oxygen	16.00	30,000	45	30,000,000
Carbon	12.01	35,000	45	35,000,000
Hydrogen	1.01	60,000	6	60,000,000

Source: Bonner and Varner 1965.

cause of the ease of withdrawing an element by using chemically pure salts and deionized water. However, essentiality of an element has been much easier to prove than nonessentiality; the procedure may not be sufficiently sensitive to show that an element is not essential.

Requirement for Other Elements

Certain mineral elements are essential only for some species. Lower forms of plant life require fewer elements than higher plants.

Silicon was declared essential for rice, based on the observation that growth was not normal without it in the culture medium (Yoshida et al. 1959; Okuda and Takahashi 1964). Silicon was shown to be essential for sugarcane, based on yield responses to silicate applied on a muck soil. The addition of 15 $t \cdot ha^{-1}$ of silicate materials increased cane and sugar yields by approximately 70% in new crop and 125% in ratoon crop (Elawad et al. 1982). While Si is apparently not essential for maize and several other members of the grass family, these species still accumulated large amounts, up to 1 to 4% of total dry matter (Salisbury and Ross 1978).

Halogeton, a weed on saline soils of some 5,000,000 ha of western range, had a requirement for Na as a microelement (Williams 1960), as did *Atriplex vesicera,* a pasture plant in Australia (Brownell 1965). Evidently it is required as a microelement in certain species having the C_4 photosynthetic pathway (Brownell and Crossland 1975). In sugar beet and cotton, Na apparently substituted for much or most of the K requirement, apparently due to its role in ion balance (Gauch 1972). Na is, as well, a major element requirement for animals.

Certain plant species adapted to high-selenium (Se) soils not only tolerated it but apparently required it as a nutrient (Shrift 1969). Animals also require Se.

Symbiotic and free-living nitrogen (N)-fixing organisms required Co (Gauch 1972). Evidently, Co was required in these lower plant forms, as it is in animals, for the formation of vitamin B_{12} (Salisbury and Ross 1978).

The qualitative and quantitative nutrient requirements of lower forms of plant life are often different from those of higher plants. For example, calcium (Ca) and magnesium (Mg) are micronutrients for fungi and macronutrients for higher plants. Boron apparently is not required by bacteria and fungi.

Sources of Plant Nutrients

Natural organic and inorganic substances are the primary sources of plant nutrients in agricultural and natural ecosystems. Supplementation of the natural fertility with commercial fertilizers is a modern agricultural practice. However, a segment of modern society rejects the concept, claiming that commercial fertilizers contain toxic chemicals harmful to humans, animals, and the environment and, further, that fertilizer nutrients should come from "natural or organic" materials. The fact that nutrients enter the plant as ions, whether the fertilizer source is organic, such as manure, or inorganic, such as commercial fertilizers, is evidently disregarded. The rigid philosophy of organic gardening/farming ignores the fact that higher plants are autotrophic and do not require supplementation with organic materials.

All the chemical elements in plants come from the soil, water, and atmosphere, collectively termed the *biosphere.* Over 75% of the solid phase (soil) consists of Si, oxygen (O), and aluminum (Al), which are not plant nutrients per se. The atmosphere is 79% N and is the sole source of carbon (C) as carbon dioxide (CO_2), although the concentration is only about 0.034%. Water of the soil solution contains cations and anions in concentrations characteristic of the particular soil, but generally in minute concentrations. However, saline soils have high levels of Na, carbonates, and Cl.

Nutrients of the biosphere are continuously recharged by recycling; otherwise they would eventually become exhausted. Plant nutrient movement is a two-way street; nutrients enter plants as elements or ions and eventually return to the environment as elements, via biodegradation by microorganisms (Fig. 5.1). Carbon and phosphorus (P) may be precipitated as marine deposits of Ca and Mg carbonates or phosphates and are consequently lost to the cycle. In Florida, phosphate materials deposited over many millenia are mined industrially. Currently, the CO_2 content of the atmosphere is being enriched at a rate of about 2 ppm/yr, largely the result of burning of fossil fuels. The ambient level now is about 340 ppm, depending on proximity to industrial

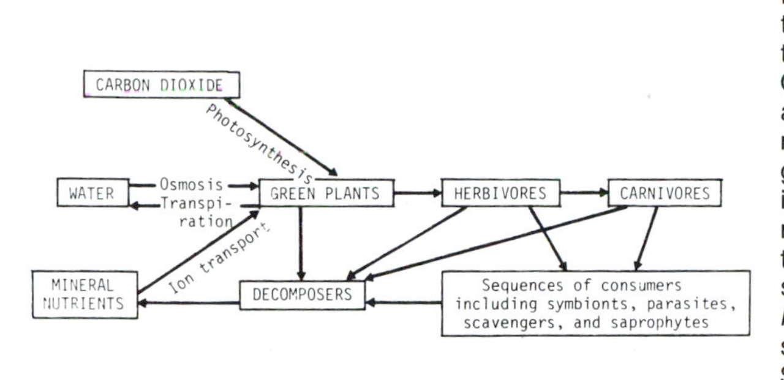

Fig. 5.1. Flow of nutrients in the biosphere; the return of C to the free CO_2 pool from respiration and burning of organic materials. Release of oxygen in photosynthesis and its uptake in aerobic respiration are not included in the diagram. (From Epstein, *Mineral Nutrition of Plants,* © 1972, by permission of John Wiley and Sons, Inc.)

centers, compared with 290 ppm in preindustrial times. Municipal and industrial wastes are enriching the nutrient supplies in air and water, particularly S in the atmosphere and N and P in water. Human and industrial waste products also contribute heavily to micronutrient recycling (Fig. 5.2). The use of sewage slurry on agricultural lands is usually restricted by statute because of the high content of heavy metals such as lead (Pb), cadmium (Cd), zinc (Zn), nickel (Ni), and manganese (Mn), which can accumulate to levels toxic to plants and consumers of plant products.

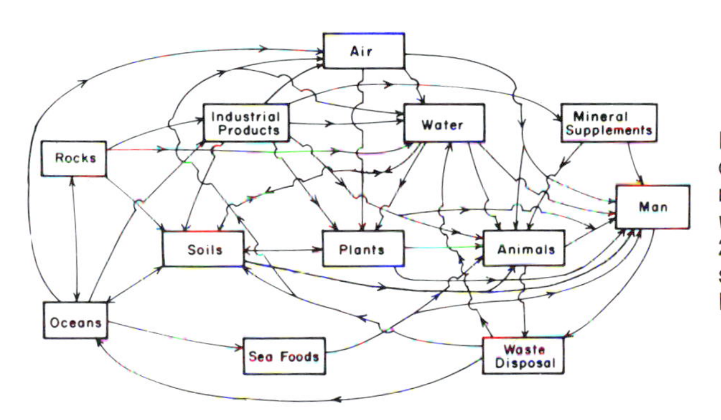

Fig. 5.2. Cycling of microelements in the environment. (From Allaway, *Advances in Agronomy* 20[1968]:235–74, by permission of Academic Press, Inc.)

Soil Nutrients

Generally soil nutrients, irrespective of soil origin or type, are derived by weathering of inorganic minerals (*parent material*) and by biodegradation of organic matter.

Soils from global, local, or even small areas, such as experimental plots, may vary morphologically, physically, chemically, biologically, and therefore in nutrient-supplying power. For example, soils high in montmorillonitic clay and/or organic matter have a high cation exchange capacity (CEC) of 25 meg • 100 g^{-1}. These soils hold vast quantities of nutrients as exchangeable ions,

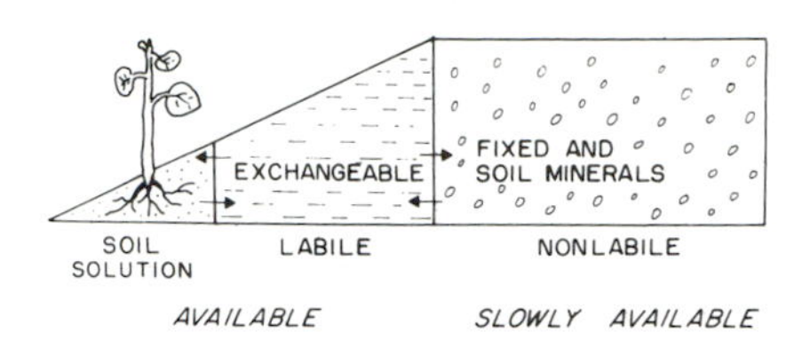

Fig. 5.3. Schematic representation of plant nutrient availability in the soil. Concentration in the soil solution is low but is in equilibrium with other fractions and therefore renewable.

which are partially available for plant growth (Fig. 5.3). Sandy soils, such as those in Florida, have a CEC of 5 meg • 100 g^{-1} or less and therefore have a low nutrient-holding and -supplying power. Mollisols, formed under tall, native grasses, such as those in the Corn Belt, are high in organic matter, CEC, and many nutrients. These soils are usually high in potassium (K), Ca, and Mg (i.e., a high base saturation) and generally are among the most fertile in the world.

Laterite soils (Oxisols, Ultisols, and Alfisols) of tropical Africa, South America, and Southeast Asia are acid or Al saturated and have an unusually high capacity to fix P, due to the high levels of soluble iron (Fe), manganese (Mn), and Al, as follows:

$$Fe^{3+} \text{ or } Al^{3+} + H_2O + H_2PO_4^- \longrightarrow Al(H_2O)_3(OH)_2H_2PO_4$$

Soluble Al can be toxic to plants and is a common problem in these soils (Arnon 1974). Desert soils are generally alkaline and high in Ca, Mg, and K. They may have toxic quantities of Na, Cl, (SO_4), and carbonates.

Nutrient Availability

Generally nutrient availability, more than the absolute quantity, determines the plant nutrient status. Soil pH is the primary factor influencing solubility and hence availability of plant nutrients (Fig. 5.4). Most nutrients were more available between the pH values of 6.0 and 7.0 (Truog 1961). Calcium, Mg, K, and Mo are more available in alkaline soils, and Zn, Mn, and B less available. Iron, Mn, and Al may be soluble to the extent of toxicity in strongly acid soils (Fig. 5.4). Liming Podzol soils (formed under forest and leached) frequently induces B deficiency, especially in alfalfa, which requires high lime (Table 5.2). Heavy N fertilization, common on grass crops such as maize or wheat, increases acidity and may induce Al toxicity, low base saturation, and Ca, K, and Mg deficiencies. Nitrification of N fertilizers was the principal cause of acid agricultural soils (Pierre et al. 1970), and Ca, Mg, and K deficiencies are commonly associated with high N fertilization.

Another primary cause of unavailability of nutrients is microbial immobilization. Immobilization of N by microbes is a common result after large additions of residues with a high C-N ratio (e.g., maize stalks or sawdust).

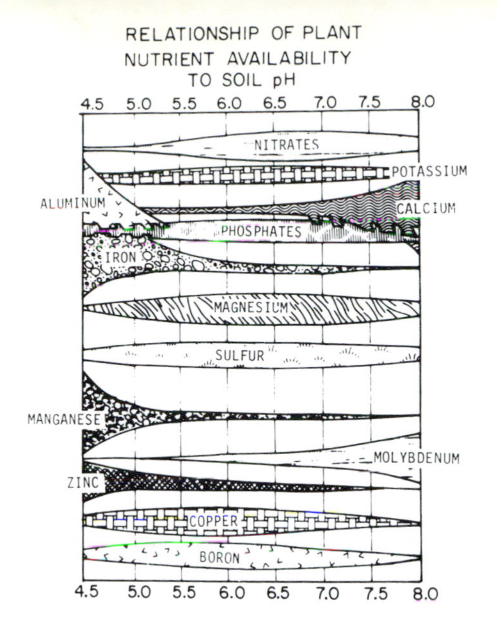

Fig. 5.4. Effect of soil pH on the availability of mineral nutrients. At low pH levels, toxic amounts of Fe, Mn, and Al may be present, but P is unavailable due to conversion to insoluble phosphates with Fe and Al. At high pH levels, P reacts with Ca and becomes insoluble. (Courtesy of University of Kentucky Extension Service)

TABLE 5.2. Classification of plants according to sensitivity to pH

	Soil pH	
4.5–5.5	5.5–6.5	6.5–7.5
azalea	barley	alfalfa
bent grass	bean	apple
blueberry	brussels sprout	asparagus
cranberry	carrot	sugar beet
dandelion	maize	broccoli
fescue	fescue	cabbage
potato	oat	cauliflower
poverty grass	pea	celery
red top	rye	soybean
rhubarb	strawberry	sweet clover
sorrel	timothy	
sweet potato	tobacco	
	tomato	
	wheat	

Note: Acid soil (from humid regions) = 4–6, neutral soil = 7, alkaline soil (from arid regions) = 8–10.

Similarly other nutrients may be immobilized by microorganisms, for example, copper (Cu) in organic soils. Soil sterilization, on the other hand, kills microorganisms and can result in the release of micronutrients such as Mn to the extent of toxicity.

Because of unfavorable pH or for other reasons, micronutrients are sometimes not available in sufficient quantities to support healthy plants. Since elements such as Fe and Mn are converted to insoluble salts in the soil, the common method of application has been aerial application to leaves. Modern methods consist of soil or foliar application of *chelates* (organic compounds containing the microelement). Ethylenediaminetetraacetic acid (EDTA) is a common chelating agent for Fe, Zn, and other microelements. In calcareous

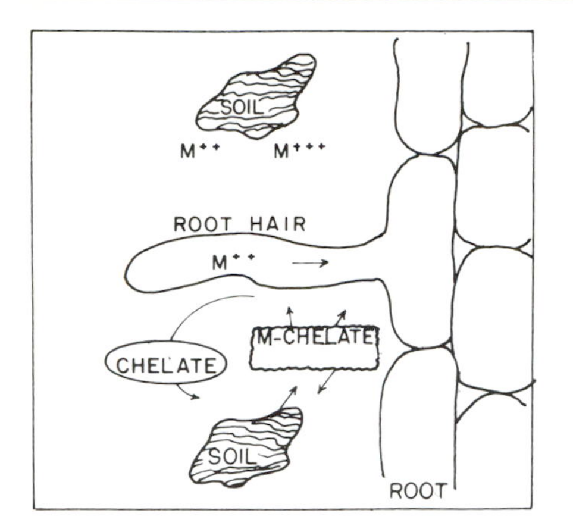

Fig. 5.5. Mobilization of mineral ions in the soil (*M*) by chelates. (From Mengel and Kirkby 1979)

soils, ethylenediaminedi-o-hydroxyphenylacetic acid (EDDHA) is superior to EDTA. FE- or Zn-EDDHA, for example, is less reactive with soil Ca. Micronutrients form chelates naturally in the soil and with organic molecules in the plant, resulting in greater solubility and availability (Fig. 5.5). For example, chlorophyll is a Mg-chelate; hemoglobin and cytochrome C are Fe-chelates.

Quantitative Requirements

Variations in the quantity of the various essential nutrients required for normal plant growth are large. Quantitative requirements depend on the crop, the yield level, and the particular nutrient. For example, on an atomic basis approximately 60 million times as much H as Mo is required (Table 5.1), on a mole basis 60 thousand times as much. Carbon, H, and O are required in tons per hectare by crop plants; N, P, K, S, Ca, and Mg in tens to hundreds of kilograms per hectare, and micronutrients in grams per hectare. Due to the low requirements, fertilization with micronutrients is generally not necessary on most cropland, although it is a common practice on muck and marine soils.

The status of nutrients in plant tissue and the corresponding plant growth (Fig. 5.6) can be described as (1) deficient, (2) transitional, (3) adequate, and (4) toxic. The critical tissue concentration is defined as that just below the level that gives optimum growth; the minimum tissue level is that giving near maximum growth (Epstein 1972). This phenomenon is the basis for tissue testing for nutrient status as a guide for fertilizer recommendations. The critical level for the essential elements has been determined for numerous crops, but absolute values should be regarded only as a guide since genetics, environment, and sampling procedure may appreciably alter this level. In the deficiency zone adding increments of a nutrient results in increased dry matter production, while in the adequate zone adding increments of a nutrient results in increased

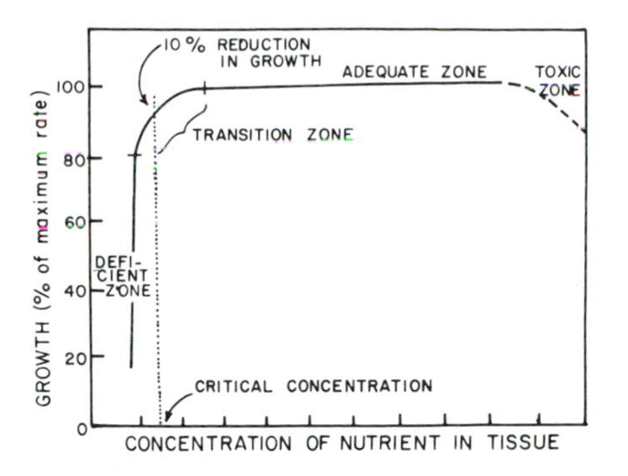

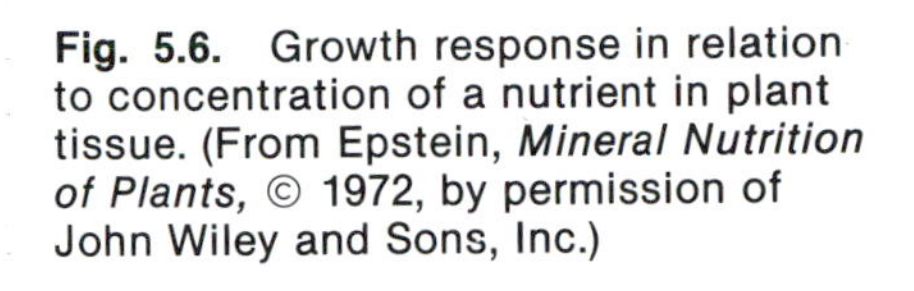
Fig. 5.6. Growth response in relation to concentration of a nutrient in plant tissue. (From Epstein, *Mineral Nutrition of Plants,* © 1972, by permission of John Wiley and Sons, Inc.)

element content in the plant tissue but little or no yield increase. This part of the response curve is termed *luxury consumption.* In the transitional zone, adding increments of a nutrient increases both yield and nutrient concentration.

Fertilization with some nutrients (e.g., K) results in more luxury consumption than with other nutrients (e.g., P). Species vary in K absorption; grasses and certain weedy species are luxury consumers of K, and legumes are not. Fertilization to the point of luxury consumption may be economically counterproductive, although a high level of fertilization with K may be desirable if needed to counteract detrimental levels of Na.

The yield response to application of most nutrients follows the law of diminishing returns: each added fertilizer increment produces a progressively smaller yield increase, finally reaching an asymptote. The economic benefit of fertilization is a function of yield response in relation to fertilizer cost (Fig. 5.7). In situation *2* in this example, the greatest return per acre is from dou-

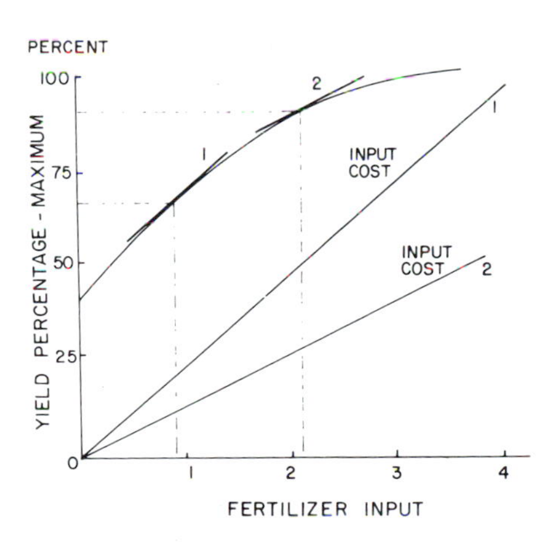

Fig. 5.7. The most profitable rate of application in relation to fertilizer cost. Situation *2*, a rate double that of *1*, gives the greatest return, as indicated by a line tangent to the yield response curve and parallel to the input cost line.

bling the amount of fertilizer added, as measured by gain in yield in relation to lower fertilizer cost.

Nutrient Uptake

A physical or chemical proximity to the root is necessary for nutrient uptake. Engagement between the root and nutrient ions can be made by (1) contact exchange, (2) exchange of soil ions with H in the mucigel, (3) diffusion of ions in response to a chemical gradient, (4) mass flow of ions to the root in response to a moisture gradient, and (5) extension of the root into the ion source (Fig. 5.8). Root extension places newly formed absorptive tissues, particularly the root hair zone, in an unexploited soil medium, enhancing ion uptake opportunity. Aboulroos and Nielsen (1979) found that fertilization with P increased yield and P uptake but also greatly increased root length, fineness, and density. The increase in P uptake may have resulted from more P concentration in the medium, or from increased root extension, or (more likely) both. In any case the root must intercept the nutrient by one or more of the processes given previously. Their relative importance with respect to nutrient uptake by maize (Barber and Olson 1968) varied with the particular nutrient, but mass flow (moving with water) was the primary process in the uptake of most nutrients (Table 5.3). However, chemical diffusion was primary for K in the Mollisol soil of this experiment. Mass flow of K would probably predominate in coarse-textured Spodosols, Entisols, and Ultisols, in contrast to a Mollisol. Interestingly, the nutrient contribution from root extension is relatively low for all except Ca, which is immobile in the plant. Since methods of accurate quantification are difficult with fine roots and especially root hairs, the contribution made by root extension could conceivably have been underestimated.

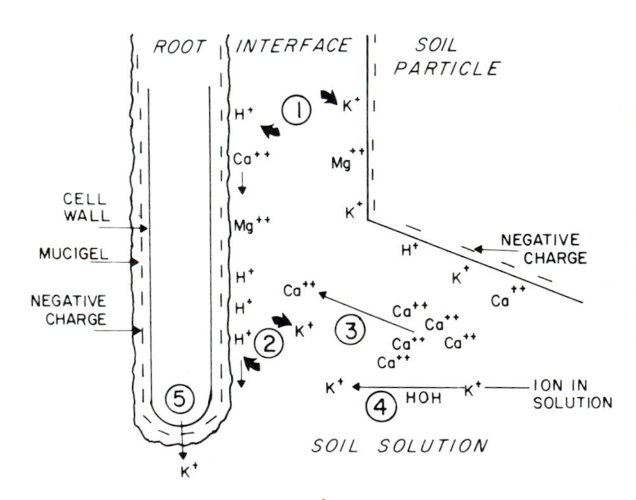

Fig. 5.8. Uptake of plant nutrients. *1.* Contact exchange between H on root and K on soil particle. *2.* Exchange of root H for K in soil solution. *3.* Diffusion of Ca ion from a zone of high concentration to one of low concentration. *4.* Mass flow of K ion in water toward root. *5.* Root extension into zone containing ion.

TABLE 5.3. The relative significance of root interception, mass flow, and diffusion in supplying maize with its nutrient requirements from a typical fertile silt loam soil

Nutrient	Amount Needed for Yield of 9500 kg • ha^{-1}	Approximate Amount Supplied by		
		Root interception (kg • ha^{-1})	Mass flow (kg • ha^{-1})	Diffusion (kg • ha^{-1})
Nitrogen	187	2	185	0
Phosphorus	38	1	2	30
Potassium	192	4	38	150
Calcium	38	66	165	0
Magnesium	44	16	110	0
Sulfur	22	1	21	0
Copper	0.1	0.01	0.4	0
Zinc	0.3	0.1	0.1	0.1
Boron	0.2	0.02	0.7	0
Iron	1.9	0.2	1.0	0.7
Manganese	0.3	0.1	0.4	0
Molybdenum	0.01	0.001	0.02	0

Source: From Barber and Olson 1968, by permission.

The above-mentioned processes only introduce the root to the nutrient, a necessary precondition for absorption. The process of absorption may be active, requiring respiratory energy and aerobiosis, or passive. In *active absorption,* ions cross the cytoplasm membrane, the *plasmalemma,* by the energetics of high-energy phosphate bonds (e.g., ATP) produced in respiration (the ion pump) (Fig. 5.9). Without ion uptake inhibition, Na or K concentration inside the cell may be many times that of the external concentration (Hoagland 1944). Active movement between cells was through living connectors, the *plasmodesmata* (Haynes 1980); hence transport between cells can be active. The vacuole, a storage reservoir inside the cell for water and ions, functions to stabilize supply-demand balance.

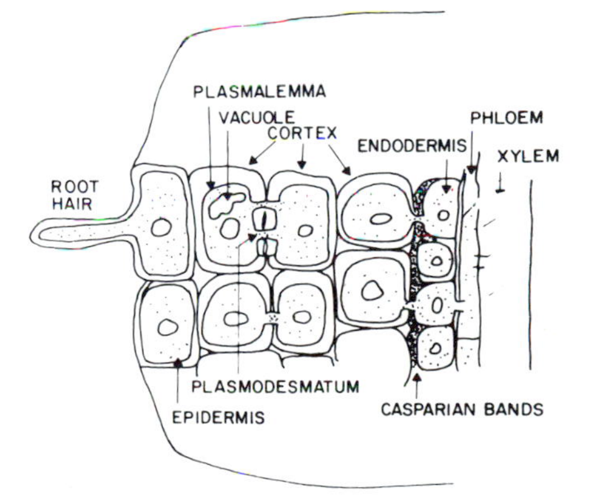

Fig. 5.9. Cross section of a dicot root. The dotted area including the plasmodesmata connections between cells are living (symplasm) and transport actively. The clear areas, cell walls, interspaces, and xylem elements are nonliving and make up the free space (apoplasm) in which passive movement occurs.

TABLE 5.4. Absorption of K from nutrient solution as influenced by temperature and air

Treatment	K Concentration in Cell Sap (milliequivalents per liter)
Temperature of solution (°C)	
0	10
18	50
30	85
Aeration for 24 hr	
Air	90
Nitrogen	25

Source: From Hoagland 1944.

The importance of a high respiration rate for active absorption is illustrated in Table 5.4. After some initial absorption, low temperature inhibited uptake, as did anaerobic conditions.

The importance of radiation and high photosynthetic rate is also suggested. After shading tomato, barley, and wheat plants, root growth was the metabolic process first limited (Crapo and Ketellapper 1981). Potassium uptake was reduced greatly, respiration moderately. Root extension appears to be affected more by anaerobiosis than is nutrient uptake.

Passive absorption is a physical process analogous to absorption of water by a sponge. Ions move with water without metabolic involvement. The total capacity of passive absorption had two components (Epstein 1972): (1) *outer space,* defined as voids in the root tissue such as intercellular spaces but also including any nonliving tissues such as cell walls (Fig. 5.9); and (2) *Donnan free space,* roughly defined as the CEC of cell parts exposed to ions in water in the outer space. The term *apparent free space* or *free space* is used to represent the sum of the two.

Minerals in high concentrations tended to move rapidly in the free space (*apoplasm*) (Pitman 1977), finally pass the endodermis, and enter the transpiration stream of the xylem (Fig. 5.9). The endodermis is a barrier to passive movement in the apoplasm because of the casparian bands, suberin deposits in the endodermis that ostensibly render it impermeable to free water movement. Movement through the endodermis is apparently active in the symplasm, as illustrated in the model presented by Haynes (Fig. 5.10). A severe rate limitation would seem probable, caused by the necessity of rerouting all transport

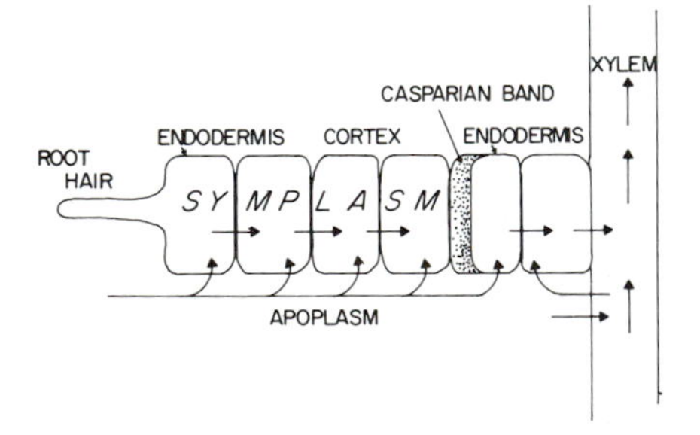

Fig. 5.10. Symplastic and apoplastic system for transport of minerals across a young root. Note that the casparian bands of the root endodermis appear to interfere with symplastic movement directly into the xylem. (From Haynes 1980)

through the endodermal plasmodesmata of the symplasm, but the nature of this process has not been fully clarified.

The CEC of the root, which influenced nutrient uptake (Haynes 1980), varied with species, variety, and age (Ram 1980). Elements held in the mucigel of a young root can exchange with those electrostatically held by the soil particles (Fig. 5.8), a process termed *contact exchange* (Jenny and Overstreet 1939). Once introduced to the root, the ion can move actively in the symplasm or passively in the apoplasm, the passive being faster and resulting in greater uptake.

Two theories are proposed for the movement of hydrophilic ions across lipoprotein membranes (Mengel and Kirkby 1979):

1. The carrier theory holds that molecules in the plasmalemma have binding sites specific to certain ions, causing selectivity. A carrier-ion complex formed at the membrane interface carries the ion across the membrane and later discharges it into the cell. The process is driven by ATP and the enzyme *kinase.*

2. The ion pump theory holds that the energy released by the conversion of ATP to ADP by ATPase brings ions into a cell in response to the change in balance created as other ions leave the cell. The Na-K pump is a common example. Other ions enter the cell by chemical gradient. The ion absorption rate has been observed to be highly correlated to ATPase activity (Fisher et al. 1970).

Interaction of Ions

The availability of an ion was influenced by the presence of other ions in solution (Mengel and Kirkby 1979). Viets et al. (1954) showed that the K-Ca and K-Mg ratios in the plant increased with N fertilization. In alfalfa K uptake interfered with Mg uptake, but not the reverse (Omar and El Kobbia 1965).

Ion uptake may be enhanced by the presence of certain ions, particularly Ca, a phenomenon termed the *Viets effect* (Viets 1944). The uptake of K, rubidium (Rb), bromine (Br), Cl, SO_4, and PO_4 is accelerated by Ca. Calcium is necessary for membrane integrity, but the effect apparently is greater than the membrane effect.

The presence of certain elements is antagonistic to absorption of others. Sodium decreased the uptake of K, although yields of sugar beet were increased by addition of Na (Farley and Draycott 1975). The Ca and Mg percentages decrease concomitantly with increased K uptake. Phosphorus can greatly interfere with Zn and Fe uptake. Normally cations compete with other cations; for example, NH_4^+ decreases other cation uptake. Cation uptake generally increased anion uptake (Leggett and Egli 1980).

Function and Use of Nutrient Elements

For convenience the nutrients essential to crop growth can be grouped into four categories related to their primary role in plant nutrition: (1) basic structure, (2) energy storage and transfer energy bonding, (3) charge balance, and (4) enzyme activation and electron transport. A summary of the information regarding the nutritional aspects of the various elements is given in Table 5.5.

BASIC STRUCTURE: CARBON, HYDROGEN, AND OXYGEN

Carbohydrates $(CH_2O)n$ make up the skeleton or structure of plants and are a source of metabolic energy. Carbohydrates include numerous organic acids, simple and complex sugars, and polymers of sugars, such as starch, cellulose, and hemicellulose. Organic acids are the precursors of amino acids, which polymerize with the peptide linkage to form proteins (see Fig. 5.13). By weight approximately 45, 6, and 43% of a plant is composed of C, H, and O, respectively. Therefore, over 90% of plant dry weight or crop yield is derived from air and water.

ENERGY STORAGE AND TRANSFER ENERGY BONDING

Nitrogen. Nitrogen constitutes 79% of the atmosphere, and even more N is in the soil as organic sediments. Unfortunately neither dinitrogen (N_2) of the atmosphere or combined N in soil sediments is available for plant growth. Only the oxidized (NO_3) or reduced (NH_4) forms (ions) are available. Bonding to hydrogen, which reduces N, can be accomplished by lightning, by nitrogen-fixing organisms, or commercially by the Haber-Bosch process (see Chap. 6). Ammonia is oxidized to nitrate by the nitrifying bacteria: $NH_4 \rightarrow NO_2 \rightarrow NO_3$ (Fig. 5.11). Such N transformations are biological and therefore sensitive to soil pH, temperature, and moisture.

Soil temperatures of 25°C or higher were most favorable for nitrification, while ammonification was less temperature sensitive (Haynes and Goh 1978). Virtually no nitrification occurs in the winter months in temperate climates. During spring, while soils are still cold and wet, nitrification is minimal and usually inadequate for healthy plant growth. Growth of grass plants is stunted and yellow unless an available form of N is applied.

Nitrification was also greatly inhibited in forests and grass climax vegetation due to the presence of natural inhibitors, such as tannins and phenols (Rice and Pancholy 1973). On the other hand, clearing and cultivation greatly enhance nitrification, due to degradation and disappearance of inhibitors. It is of interest to note that N deficiencies in wheat on Oklahoma prairie soils were not widely recognized as a problem until about 1957. Since then these soils have been responsive to N application, necessary because of continued loss of organic matter and N release from cultivation (Tucker 1981). These applications of N fertilizers, however, have resulted in soil acidity and increased lime requirements.

TABLE 5.5. Essential elements absorbed from soil and their representative roles in the plant

Element	Absorption Form	Total Present in Soil, kg • ha^{-1} (general value)[a]	Available, kg • ha^{-1} (general value)[a]	Element in Nutrient Solution, ppm (relative amount needed)[b]	Representative Role in Plant
Nitrogen (N)	NO_3^- NH_4^+	4,000	1–50	100–200	Amino acid; protein synthesis; nucleic acids
Phosphorus (P)	$H_2PO_4^-$ HPO_4^{2-}	1,200	0.01–0.10	63	Utilizing energy from food reserves
Sulfur (S)	SO_4^{2-}	800	1–10	32	Sulfhydryl groups
Potassium (K)	K^+	50,000	5–15	200	Hexokinase
Calcium (Ca)	Ca^{2+}	15,000	10–100	120	Calcium pectate
Magnesium (Mg)	Mg^{2+}	6,000	5–50	24	Chlorophyll; respiration
Iron (Fe)	Fe^{2+}	50,000	Trace	5.6	Cytochromes; ferrodoxin
Manganese (Mn)	Mn^{2+}	1,600	Trace	0.6	Formation of amino acids
Boron (B)	Bo_2^{2-}	100	Trace	. . .	Possibly in sugar translocation
Copper (Cu)	Cu^{2+}	50	Trace	0.02	Nitrate reduction
Zinc (Zn)	Zn^{2+}	50	Trace	0.07	Dehydrogenases
Molybdenum (Mo)	MoO_4^{2-}	Trace	Trace	0.01	Nitrate reductase
Chlorine (Cl)	Cl^-	Trace	Trace	. . .	Photosynthetic phosphorylation

[a]From Iowa State University 1965.
[b]From Schrenk and Frazier 1964.

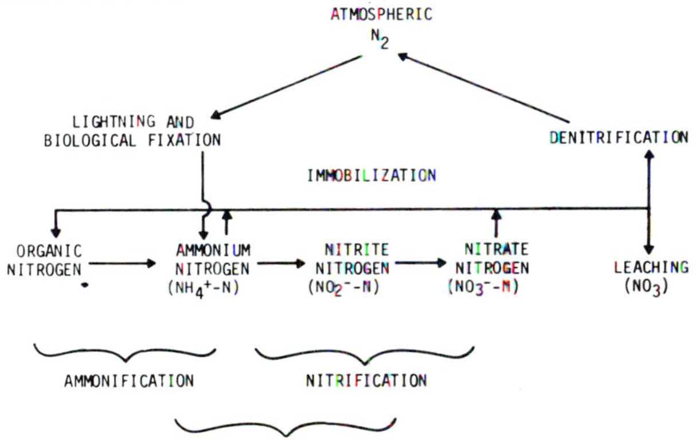

Fig. 5.11. Transformation of N in soils.

Denitrification (Fig. 5.11) is favored by warm temperatures and reducing conditions, such as waterlogging. Warm, aerated soils favor nitrification and nitrate loss by leaching. To guard against nitrate losses, commercial inhibitors such as nitropyrin (2-chloro-6-trichloromethylpyridine) are used to retain the N as ammonium (NH_4), which is adsorbed by soil particles and therefore less subject to leaching. Grain yields of maize were increased significantly and stalk rot incidence was reduced by the use of nitropyrin with fall-applied ammonia (NH_3) (Warren et al. 1975). Nitrification inhibitors can also be used to reduce nitrification in the soil and nitrate uptake and accumulation in leafy vegetables, such as spinach.

The N content of plants averages 2 to 4% and may be as high as 6%. Crop plants can take up NO_3^- or NH_4^+ ions and assimilate them, as summarized in Figure 5.12. The NO_3^- form is absorbed, primarily due to the rapid conversion of NH_4^+ to NO_3^- in the soil. However, maize absorbed NH_4^+ and NO_3^- at the same rate; uptake was linear above the 21 μmol concentration of N and declined below this threshold to a steady state at 4 μmol (Edwards and Barber 1976). In lima bean, dry matter accumulation was consistently higher when NO_3^- was 75% or more of the total available N (McElhannon and Mills 1978), which illustrates genetic differences in ion preference. An interaction with K was demonstrated in maize; yields were depressed with NH_4^+N and the N-K ratios were elevated, compared with NO_3^-N (Dibb and Welch 1976). High tissue levels of NH_4^+ can retard growth and cause elevated levels of Cl in tobacco (Williams and Miner 1982), but NO_3^- is not harmful at high levels. The form used by the plant depends in part on rainfall and pH; acid soils favored NO_3^- uptake and depressed NH_4^+ uptake (Mengel and Kirkby 1982). Because of the above factors, NO_3^- is the predominant ion absorbed by crop plants other than rice.

Assimilation of N into organic molecules (Fig. 5.12) was dependent on the reduction of NO_3^- by the nitrate reductase enzyme in plant tissue (Neyra and Hageman 1975). Nitrate reduction, which must occur before amino acids and other chemical combinations of N can be produced, requires electrons; the primary donors are nicotinamide adenine dinucleotide (NADH) or nicotinamide adenine dinucleotide phosphate (NADPH), which are products of photosynthesis. It follows that high light and photosynthetic rates were found conducive to nitrate reductase activity (Minatti and Jackson 1970). Accumulation of NO_3^- to levels toxic to animals may occur in forages during cloudy conditions. A favorable temperature is also necessary for nitrate reduction,

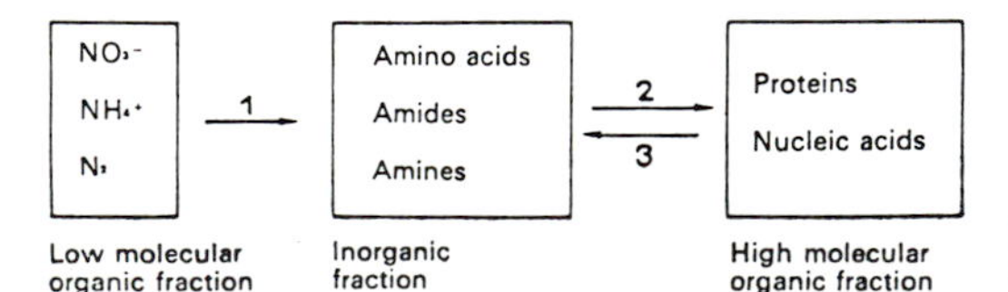

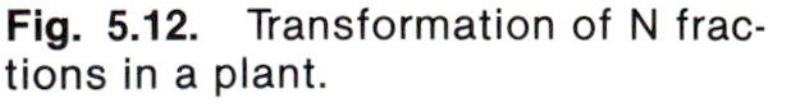
Fig. 5.12. Transformation of N fractions in a plant.

with considerable differences among species. Reduction of NO_3^- is not without an energy cost to the plant.

Woody species essentially limit nitrate reduction to roots, whereas in crop plants reduction occurred in both roots and leaves (Haynes and Goh 1978). Certain vegetable crops, such as spinach and members of the family Chenopodiaceae, are evidently missing nitrate reductase capacity in the roots and may accumulate high amounts of nitrate in leaves. A positive association between nitrate reductase activity and grain yield and protein has been found in corn, wheat, and sorghum by various workers, but in another study it was not highly correlated to high yield and protein in wheat (Deckard and Busch 1978). In two spring wheats nitrate reductase was ample at senescence, but N was not (Hepper 1976).

Nitrogen is a constituent of amino acids, amides, N bases such as purine, and proteins and nucleoproteins (Fig. 5.13). Enzymes contain long-chain, complex protein molecules plus a nonprotein reactive group, which is generally a micronutrient. Proteins are polymers from 20 amino acids joined by the peptide linkage in a myriad of combinations, resulting in a high molecular weight. Amino acids have amino-N attached at the α-carbon position and may also have N in the ring as with tryptophan (Fig. 5.13). Glutamine has N in an

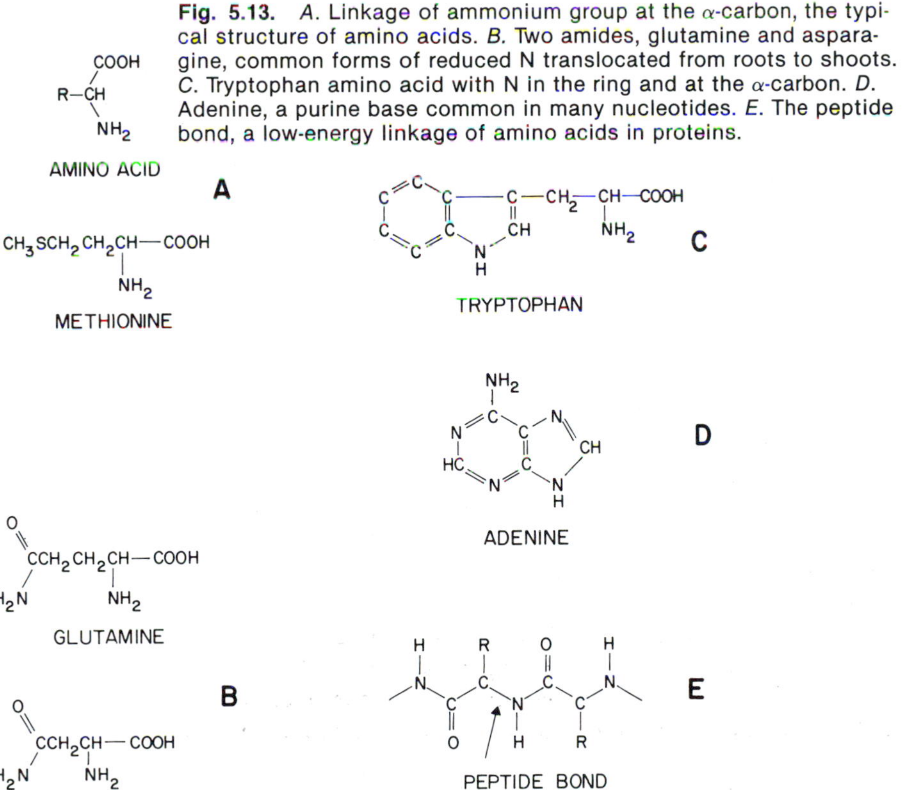

Fig. 5.13. *A.* Linkage of ammonium group at the α-carbon, the typical structure of amino acids. *B.* Two amides, glutamine and asparagine, common forms of reduced N translocated from roots to shoots. *C.* Tryptophan amino acid with N in the ring and at the α-carbon. *D.* Adenine, a purine base common in many nucleotides. *E.* The peptide bond, a low-energy linkage of amino acids in proteins.

amide group, and adenine is a purine base with N in the ring. Adenine is a part of many nucleotides and nucleoproteins, such as DNA and RNA. Nitrogen is also a constituent of a host of compounds termed alkaloids whose function is not well understood and which evidently are not essential metabolites. They are believed to serve as compounds of N storage.

A deficiency of N limits cell expansion and cell division. Deficiency symptoms include general stunting and yellowing, particularly of the older plant parts. The reduction in plant growth can cause accumulation of sugars and in some species, especially maize, cause the basal tissues to turn purple due to anthocyanin formation.

Nitrogen is highly mobile in the plant. Younger leaves and developing organs with strong sink demands, such as fruits and seeds, may draw heavily on N in the older or lower leaves. The result of such redistribution when N uptake is limited is *firing* (yellowing and senescence) of the lower leaves. Nonnodulated, N-starved soybean senesced earlier and had 60% of total N in the seeds, compared with a control group of plants with later senescence and only 20% of total N in the seeds (Egli et al. 1978). Such firing of lower leaves formerly was erroneously attributed to moisture deficiency since it tended to become obvious in midsummer.

In summary, N is an essential constituent of amino acids, amides, nucleotides, and nucleoproteins and is essential to cell division, expansion, and therefore growth. It is mobile in the plant; N moves to young tissues so a deficiency is first visible in older leaves. A deficiency interrupts growth processes, causing stunting, yellowing, and reduced dry matter yields.

Sulfur. Sulfur is derived from soil organic matter and also from inorganic salts, such as calcium sulfate and magnesium sulfate. The atmosphere contains gaseous S (SO_2), as does rainwater (*acid rain*). The atmosphere of geographic areas distant from industrial cities or the sea (i.e., certain areas of Africa, the United States, Australia, and New Zealand), has little S so crop deficiencies are common. Mineralization of S and formation of sulfate ions (transformation) from organic matter are quite similar to N transformations from organic matter. In anaerobic conditions H_2S (reduced) may form and accumulate in toxic concentrations. In poorly aerated conditions this compound is oxidized to elemental S by photosynthetic and chemotrophic bacteria; further oxidation produces sulfuric acid (H_2SO_4) and soil acidity.

Sulfur, absorbed primarily as the SO_4^{2-} ion, is actively and passively translocated. Leaves can absorb appreciable quantities of SO_2 gas. Like N, all oxidized forms must first be reduced enzymatically before assimilation by the plant.

Sulfur, like N, is involved in low-energy bonding and protein synthesis. It forms thiol bonds analogous, energetically, to the N peptide bonds. Sulfhydryl groups (SH) are thought to be important in the hardening of protoplasm to cold and drought. In energy transfer sulfur can function in a manner similar to P.

Sulfur is a constituent of the amino acids cystine, cysteine, and methionine. It also activates certain proteolytic enzymes and is a constituent of coenzyme A, glutathione, and certain vitamins. Members of the Cruciferae family may contain over 1% S, and legumes are also relatively high in S. Maximum alfalfa hay yields were obtained when the S content of leaves was 0.15 to 0.20% (Westermann 1975). For maximum yield, a N-S ratio of 10 to 15 was optimum for sugar cane (Fox 1976), whereas the N-S ratios on soils well supplied with S were 15 to 16 for maize, 20 for soybean, and 8 to 9 for cotton and okra (Gaines and Phatak 1982).

Oils of some plants, particularly those of Cruciferae and onion, are rich in S. Sulfur fertilization has been shown to increase the seed oil content of crops such as flax and soybeans.

Sulfur deficiency, like N, is expressed as stunting and general plant yellowing; stems are thin. Although S is mobile in the plant, redistribution from older to younger leaves is not as pronounced with S as with N and firing of lower leaves does not commonly occur. Bouma (1967) found redistribution occurred from roots and petioles of subterranean clover but not appreciably from the leaves.

In summary S is a part of certain amino acids, glutathione, coenzyme A, and certain vitamins. Sulfur physiology is similar to that of N with regard to mineralization, uptake, reduction, energy bonding, incorporation, and stunted and yellow deficiency symptoms. Redistribution is not as great as that of N and so does not cause firing of lower leaves as does N deficiency.

Phosphorus. Phosphorus is derived from the soil organic and inorganic fractions as follows: (1) soil solution containing extremely minor amounts of soluble P, such as orthophosphate (HPO_4^{2-} or $H_2PO_4^-$); (2) P-containing minerals such as the apatites and the Ca-, Mg-, Fe-, and aluminum (Al)-phosphates; (3) the labile pool consisting of P adsorbed on soil colloids and of Fe- and Al-phosphates in equilibrium with phosphate in solution (Fig. 5.3). The amount of P in solution is extremely low relative to the labile fraction. For this reason, P is generally second to N as the most limiting nutrient for plant growth.

Phosphorus is absorbed primarily as the monovalent ion $H_2PO_4^-$ and less as the divalent ion HPO_4^{2-}, which is more prevalent of the two at neutral pH or above (Fig. 5.14). Roots actively absorb P from very low concentrations in soil solution and hold it in the plant at concentrations of up to 1,000-fold (Russell and Barber 1960). The P absorption capacity of soybean root depended on age (Fig. 5.15); the absorption of 18-day-old roots was four times that at 73 days (Edwards and Barber 1976).

Phosphorus is mobile in the plant, redistributed from older to younger parts. Young leaves or developing fruit can be nourished from the labile P of older plant tissues even though the soil source is interrupted. Depending on geographic location, the critical level in maize was 0.18 to 0.25% in the leaf subtending the ear. Sufficiency levels were 0.25 to 0.41% (Forde 1976).

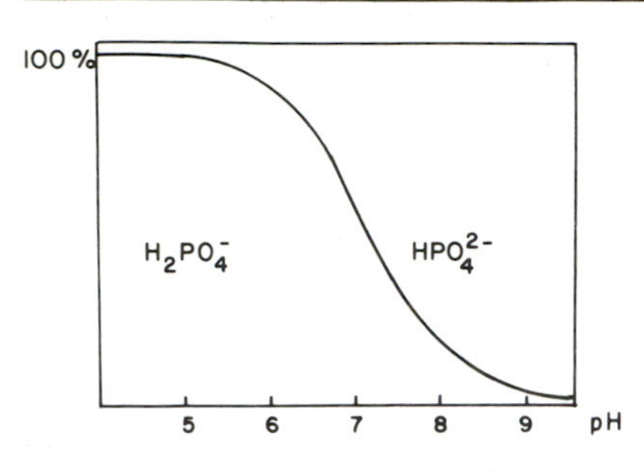

Fig. 5.14. Ratio between $H_2PO_4^-$ and HPO_4^{2-} availability in relation to the pH of the growth medium. (From Mengel and Kirkby 1982)

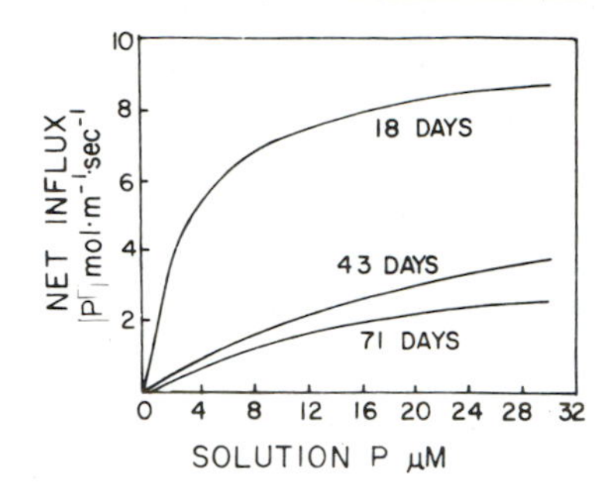

Fig. 5.15. Comparison of the net flux of P per meter of root of soybean plants at three ages. (From Edwards and Barber 1976, by permission)

Phosphorus is a structural component of a number of vital compounds: energy transfer molecules ADP and ATP (adenosine di- and triphosphate), NAD, NADPH, and genetic information system compounds DNA and RNA (desoxyribo- and ribonucleic acid). Phosphate esters are formed with sugars, alcohols, acids, or other phosphates (polyphosphates). The energy-rich bonds of important metabolites that mediate phosphorylation and energy transfer are indicated in Figure 5.16. Phytic acid is an important phosphate storage compound commonly found in seeds. This stored form of P is mobilized to support the high rate of metabolism during seed germination.

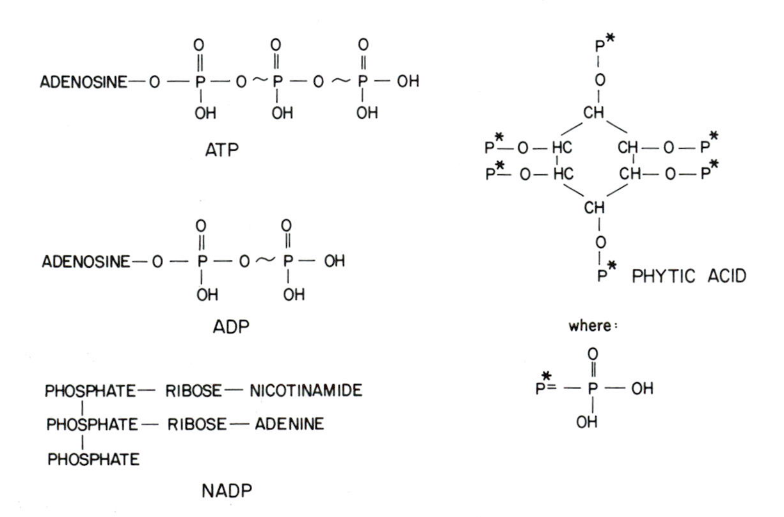

Fig. 5.16. Some important metabolites containing phosphorus. All contain high-energy phosphate bonds, essential energy sources in the synthesis of plant constituents. Phytic acid is a ready reserve of P in seeds.

Phosphorus is also a constituent of phospholipids such as lecithin and choline, which play an important role in membrane integrity. Lecithin, an important by-product in soybean oil extraction, has numerous food and commercial uses.

The visible P-deficiency symptoms, somewhat opposite those of N or S deficiency, are dark green to blue-green leaves rather than yellow. Plants are stunted. Ryegrass root number and length were found to be reduced (Troughton 1977). In P-deficient plants sugar accumulates, reflected by anthocyanin pigmentation in the base of stems and veins, particularly in maize. As with N deficiencies, older leaves show P deficiencies first, due to redistribution of P to young tissues.

In summary, P is usually present in extremely low concentations in the soil solution. It is an essential component of the energy transfer compounds (ATP and other nucleoproteins), the genetic information system (DNA and RNA), cell membranes (phospholipids), and phosphoproteins. Phosphorus is mobile and is redistributed from old to young tissues, so older leaves first show deficiency symptoms.

CHARGE BALANCE

Potassium. Potassium is derived from primary minerals and from secondary minerals such as clays. Generally, soils high in clay content tend to be relatively high in K, while organic and sandy soils are generally low. The main source of K for plants comes from weathering of K-containing minerals. Soil K exists in three fractions: (1) chemically bound in primary and secondary soil minerals; (2) exchangeable, adsorbed to soil particles; and (3) in the soil solution (Fig. 5.3). In mineral soils (e.g., soils high in montmorillonite), most of the K is in the mineral lattices. Only about 1 to 3% of the total is adsorbed or exchangeable and even a smaller fraction is in the soil solution (Wiklander 1954). Exchangeable and soil solution K are in equilibrium. Uptake was primarily from the soil solution (Mengel and Kirkby 1979), but K can come, to some extent, from nonexchangeable forms. Most soils are highly buffered for K; fluctuations from year to year are minor.

Uptake of K is in the form of the monovalent cation K^+. Uptake was active and translocation can be against strong electrical and chemical gradients (Hoagland 1944). Soil temperature affects absorption; the optimum for most species was about 25°C, but species vary (Worley et al. 1963). For example, sudangrass absorbed K at 30 to 35°C, whereas pea lost K at 35°C. Soybean lost K from the roots at low temperatures (i.e., 5°, 13°, and 15°C). Hall and Baker (1972) showed that K constituted 80% of the cations found in the phloem. Transport is primarily acropetal (upward), and K enhanced transport of nitrate (Blevins et al. 1978). Redistribution of K from older organs to younger is the rule; K is the most mobile of the plant nutrients.

While K is essential to all higher and lower plants, it is not a part of any known plant constituent. It is stored in large quantities in the vacuoles. It did not form ligands (complex organic molecules), serving primarily as an enzyme activator or cofactor for some 46 enzymes (Evans and Sorger 1966). Cofactor use only partially explains the high requirement for K as, in addition to K, micronutrients and Mg act as enzyme activators for certain enzymes. This aspect is of particular interest, considering that neither enzyme nor cofactor is used up in the chemical reaction; a little should go a long way.

Potassium also aided in the maintenance of osmotic potential and water uptake (Epstein 1972). Plants well supplied with K lost less water since K increased osmotic potential and had a positive influence on stomatal closure as well (Humble and Hsiao 1969). Potassium also serves to balance the charges of anions and influence their uptake and transport. It was reported to have reduced the incidence of certain diseases and associated lodging in maize (Liebhardt and Munson 1976) for physiological reasons yet to be explained. For example, K significantly decreased the presence of *Verticillium* wilt on cotton (Hafez et al. 1975).

Potassium was found to serve a vital role in photosynthesis by directly increasing growth and leaf area index, and hence CO_2 assimilation and increasing the outward translocation of photosynthate (Wolf et al. 1976). The latter appears to result from formation of more ATP, essential for loading assimilate into the phloem. Sodium can partially substitute for K in a number of crops, especially in sugar beet and cotton; substitution is minimally effective in others, such as maize and sorghum (Marschner 1971).

The critical level of K in plant tissue is relatively high, usually about 1.0% or 4-fold that of P. Nearly all the K is absorbed during vegetative growth; little is transferred to the fruits or grains. An application of K to wheat during the reproductive stage had little effect on grain yield (Chapman and Keay 1971). A deficiency of K resulted in increased root and stalk lodging in maize (Liebhardt and Murdock 1965), indicating a probable association with diseases. The number of brace roots decreased and stalk parenchyma disintegrated when K fertilizer was omitted, as in N-O-O or N-P-O treatments. Severe K deficiencies caused small necrotic spots between veins and the firing of leaf tips and margins of the older leaves of many species (Fig. 5.17).

Calcium. Because of the number of Ca-bearing minerals, the earth's crust is relatively high in Ca. Apatite (Ca-phosphates), calcite ($CaCO_3$), and dolomite ($CaCO_3$, $MgCO_3$) are especially common, but soils derived from these minerals under humid conditions may be leached and actually low in Ca. Young soils from marl, chalk, or limestone may contain over 10% Ca. Mineralization of N to nitrates and the formation of carbonic acid in humid regions leads in time to acid soils low in Ca and Mg and to a degraded soil structure due to the replacement of these adsorbed cations with Al^{3+} and H^+ on the soil colloids. In modern agricultural practice, dolomitic limestone is used extensively as a soil amendment to raise the pH and to supply Ca and Mg as nutrients.

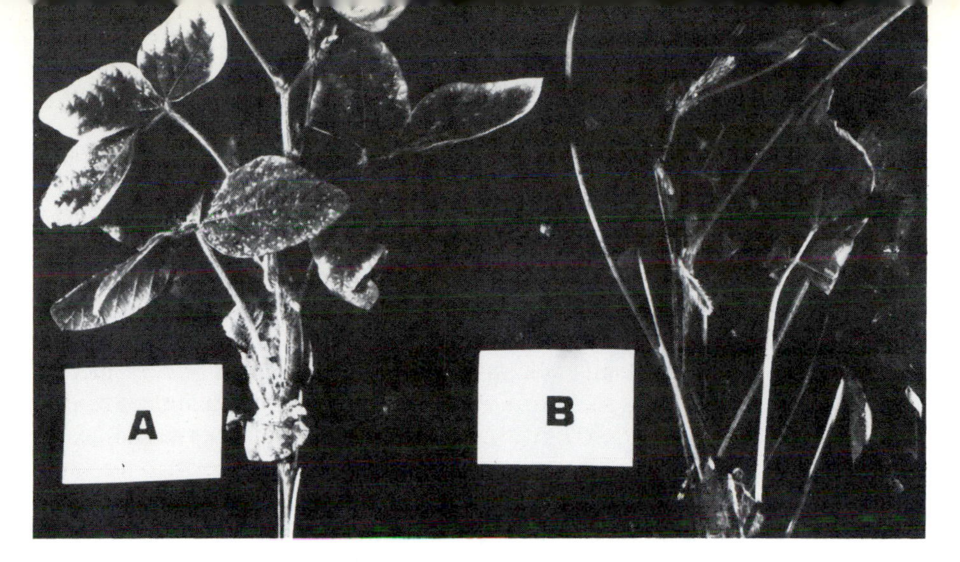

Fig. 5.17. *A*. Soybean plant with a severe K deficiency; *B*. a healthy plant. Note symptoms of firing of leaf margins and stunted growth.

Calcium is absorbed as the bivalent cation Ca^{2+}. It is the most immobile of the essential elements. Uptake and transport are passive; entry into the stele was via the free space, and upward movement was with the transpiration stream (Epstein 1972). Compared with other ions, there is little or no movement in the phloem. Calcium is highly adsorbed on exchange sites of the free space, which is probably a limiting factor in Ca delivery to other plant organs. Peanut required a high Ca content in the pegging zone for normal pod development, which is absorbed directly by the pegs and fruits (Harris 1948). It is believed that the transpiration stream to these underground fruits is negligible and therefore insufficient for the required Ca delivery.

Calcium is a component of the cell wall, particularly of the cementing substance, Ca-pectate. It is also found as Ca-oxalate and Ca-carbonate in the vacuoles; these salts supposedly immobilize the constituent organic acids to a nontoxic level. It is essential for cell division and elongation. A Ca deficiency caused plant meristem (root, shoot, fruit, and nodule) malformation and dieback (McKently 1981), presumably due to lack of phloem transport and immobility in the plant; growth of bunch bean (phaseolus vulgaris) stopped almost immediately when Ca was withdrawn from the nutrient culture. Calcium is also essential for the selective regulatory function of cell membranes.

The Ca status of a plant is highly related to pH, which affects more than Ca availability. As indicated previously, Ca affects availability of other nutrients and growth of soil microflora, especially bacteria. The optimum pH range for a number of crops is listed in Table 5.2. Many legumes of a temperate climate origin have a high pH and presumably a high Ca requirement. If grown at a low pH, a legume such as alfalfa soon becomes stunted and

Fig. 5.18. Bean plant (*Phaseolus vulgaris*) showing a Ca deficiency. Note deformed and reduced growth of young leaves; pods did not form or were aborted.

chlorotic. The author (Gardner) clearly demonstrated that such plants were also nonnodulated and N deficient, since they turned green with N topdressing. The major sensitivity was apparently in the *Rhizobium meliloti* rather than in the alfalfa per se. It seems conclusive that a Ca deficiency in many legumes, indicated by stunting and yellowing, is primarily N deficiency resulting from pH sensitivity of the bacterium symbiont. Sensitivity varies widely among rhizobia, temperate species being the most sensitive.

Deficiencies of Ca are first seen in the younger plant parts (Fig. 5.18) as deformed and chlorotic leaves, while deficiencies are seldom observed in older organs. Calcium is not redistributed to younger tissues; hence, young leaves and developing fruits are totally dependent on Ca delivery in the transpiration stream of the xylem. In fruit growth, a Ca deficiency caused blossom-end rot in tomato, and brown heart in peanut (Harris 1948). There is evidence that Florida citrus fruit size may be limited by Ca deficiency even though roots are well supplied (Koch 1982).

Magnesium. Soil Mg is derived primarily from the weathering of primary minerals (e.g., biotite, serpentine, hornblende, dolomite, and olivine). It is also present in secondary minerals (e.g., montmorillonite, illite, and vermiculite). Arid soils are generally high in dolomite and $MgSO_4$. As with other cations, Mg^{2+} is in the soil solution, adsorbed to soil particles, and in primary and secondary minerals. Normally Mg constitutes about 4 to 20% of the CEC, compared with as much as 80% for Ca and 5% for K. As expected, in humid soils Al may readily replace Mg^{2+}.

Uptake of Mg^{2+} is both active and passive. Transport is primarily in the transpiration stream. However, Mg is more mobile in the plant than Ca; more Mg than Ca was demonstrated to be present in the phloem (active transport) by autoradiogram studies (Steucek and Koontz 1970). Developing fruits and storage organs are dependent on Mg redistribution from older leaves via phloem transport, but deficiencies developed slowly compared with Ca (McKently 1981).

Magnesium is the center of the chlorophyll molecule, a Mg-chelate in the chloroplast. It also chelates with ADP, ATP, and organic acids and hence is essential for hundreds of enzymatic reactions. Magnesium forms a bridge between ATP and the enzyme molecule and is required for photophosphorylation in synthesis and breakdown reactions of photosynthesis and in oxidative phosphorylation in respiration. It was a cofactor for many enzymes activating phosphorylation in the glycolysis and in the tricarboxylic acid cycle (Hewett and Smith 1975). Since it is required to activate RuBP carboxylase, it is rate limiting in the photosynthetic process. Nitrogen metabolism and protein synthesis are also dependent on the presence of Mg, and it is believed to enhance the integrity of the ribosomes.

Deficiencies of Mg are generally observed first as interveinal chlorosis in the older leaves but may progress to affect younger leaves (Chapman 1966). Like K, and unlike Ca, Mg is somewhat mobile in the plant. Older leaves are affected first. Deficiencies were shown to affect chloroplast substructure of bunch bean (Thomson and Weier 1962), causing a reduction in grana number and size. Chlorosis started at leaf margins and tips and progressed inward in the leaf parenchyma cells. The veins remained green. In severe cases leaf necrosis occurred (McKently 1981) and reproductive phases were delayed.

In summary, Mg is a part of the chlorophyll molecule and an activator of photosynthesis and respiration enzymes and necessary for protein synthesis. It is redistributed in the plant, so a deficiency shows first in older leaves as interveinal chlorosis.

ENZYME ACTIVATION AND ELECTRON TRANSPORT

Iron. Iron constitutes about 5% of the earth's crust and is universally present in soils. It is derived from primary minerals of ferromagnesium silicates, which include olivine, augite, hornblende, and biotite. Iron oxides common in many soils include hematite (Fe_2O_3), magnetite (FeO_4), and siderite ($FeCO_3$). Iron may also be present in the lattices of secondary minerals (e.g., montmorillonite). Highly weathered ferromagnesium minerals produce hydrous Fe-oxides in the soil; these oxides along with the clay and aluminum oxides are

Fig. 5.19. Oat plant showing iron chlorosis (*left*) and one with Fe-chelate (sequestered Fe) treatment (*right*). (Photo courtesy of McDaniel and Dunphy 1978)

concentrated in lateritic soils (e.g., Oxosols), which generally predispose severe management problems. Probably all soils have an ample content of Fe, but the solubility, which is regulated primarily by pH, may be so low as to cause Fe deficiency, especially in Fe-inefficient species and cultivars (Fig. 5.19). Solubility may decrease 1000-fold per unit change in pH (Lindsay 1972a), as indicated by the following equation:

$$\underset{\text{acid soil}}{Fe^{3+}\ (\text{ion})} \rightarrow \underset{\text{alkaline soil}}{Fe_2(OH)_3\ (\text{solid})}$$

In poorly drained soils, reduced or ferrous forms (Fe^{2+}) of Fe predominate and increase Fe availability, even to the extent of toxicity (known as *bronzing* in paddy rice).

The uptake of Fe is primarily as Fe^{2+}, although Fe^{3+} and Fe-chelates are also present in the rhizosphere. Reduction is essential for absorption, the electron source probably being the cytochromes or flavins at the plasma membrane (Chaney et al. 1972). Uptake is in competition with, and is influenced by, other cations. Good aeration, high pH, and Ca, phosphate, and nitrate ions depressed uptake, whereas ammonium ions did not (Mengel and Kirkby 1979).

Iron is a constituent of the electron transport enzymes, for example, the cytochromes and ferredoxin, which are active in photosynthesis and in mitochondrial respiration. It is a constituent of the enzymes catalase and peroxidase, which catalyze the breakdown of H_2O_2 into H_2O and O_2, preventing

H_2O_2 toxicity. Iron, along with Mo, is an element of the nitrite and nitrate reductase enzymes and of the N_2-fixation enzyme nitrogenase. Although Fe is not a part of the chlorophyll molecule, it affects chlorophyll levels because it must be present for chloroplast ultrastructure formation. An Fe deficiency reduces the number and size of chloroplasts. Grana and lamella of the chloroplast were reduced in Fe-deficient maize plants (Stocking 1975).

Iron is highly immobile in the plant and not redistributed. Brown (1961) recognized that plant genotypes varied widely in Fe uptake efficiency. Deficient plants develop chlorosis (Fig. 5.19), which seemed to be a problem more of metabolism than of uptake (Foy et al. 1977). Species and cultivars that were high secretors of OH^- ions were Fe-inefficient. Cereals and grasses like wheat were high secretors of OH^- (Van Egmond and Aktas 1977). Iron-efficient tomato plants under Fe stress acidified the rooting medium and secreted reductants; one, caffeic acid, enhanced the solubility of Fe (Olsen et al. 1981). Young leaves depend on current uptake. The addition of inorganic Fe compounds to the soil may have little or no value in correcting deficiencies except at high rates, since such compounds are rapidly converted to unavailable forms. Ferrous-Fe ($FeSO_4$) applied to leaves has been used with some success. Fertilization with Fe-chelates as a soil amendment or foliar spray is more effective; Fe-EDDHA and Fe-montmorillonite clay were more effective than Fe-EDTA as a soil treatment in calcareous soils (Navrot and Banin 1976). Success of leaf application of $FeCl_3$ on a number of crop species was related to the number of stomata, application in light rather than at night, and use of surfactant (Eddings and Brown 1967).

Manganese. Ferromanganese minerals in soils supply Mn as well as Fe, the two being closely associated. Oxides of both are common. The Mn content of most soils ranges from 200 to 3,000 ppm, more than adequate in total but not necessarily adequate in available Mn. Manganese exists in soils as the divalent Mn^{2+} ion in the soil solution, as exchangeable Mn^{2+}, and as Mn^{3+} and Mn^{4+} oxides in equilibrium with other Mn forms. The Mn^{2+} ion predominates at low pHs, in natural chelates, and in reducing conditions such as waterlogging. Waterlogging, as in rice paddies and some soils low in pH, can produce soluble Mn to toxic levels. Uptake of Mn^{2+} is active; it may compete with other cations, particularly with NH_4^+ and Fe^{2+}. It is believed to be passively transported. Manganese is an activator of several enzymes, especially those involved in fatty acid and nucleotide synthesis and is essential in respiration and photosynthesis. In photosynthesis Mn^{2+} is oxidized to Mn^{3+} with the transfer of one electron from water to the chlorophyll molecule. Manganese can substitute for Mg in certain reactions; both ions are capable of bridging with certain enzymes (e.g., phosphokinase and phosphotransferase). It also activates indoleacetic acid (IAA) oxidase, which results in less IAA concentration in tissues. Like Fe, Mn is relatively immobile and preferentially translocated to young or meristematic tissues. These parts cannot depend on transfer from older leaves and hence are the first to show Mn deficiency as lesions on younger leaves. About 10 ppm was the critical level in young sorghum leaf tissue (Ohki 1975). Oats are prone to Mn deficiency, which appears as spotting, or grey specks, on younger leaves. Soybean, pea, and sugar beet are also susceptible to Mn deficiencies. Soybean cultivars differ significantly in tolerance to Mn deficiency; for example, 'Bragg' was resistant and 'Forrest' susceptible (Brown and Jones 1975). Deficiencies usually occur on calcareous peat soils, due to reduced availability at high pHs (Fig. 5.4) and immobilization by microorganisms. They can be corrected by Mn foliar sprays of $MnSO_4$ or chelated Mn (Mn-EDTA). On leached Podzol soils, which are naturally low in total Mn, a soil treatment with $MnSO_4$ is effective in correcting deficiencies.

Zinc. Zinc in soils is derived from the ferromagnesium minerals augite, hornblende, and biotite, which are found in basic igneous rocks. It is also present in the secondary minerals sphalerite (ZnFe)S, zincite (ZnO), and smithsonite ($ZnCO_3$). Zinc sulfide may be present under reducing conditions.

As with other cations, Zn^{2+} and $ZnOH^+$ may occupy exchange sites on soil colloids. Generally zinc levels are positively correlated with increasing organic matter and negatively correlated with increasing pH (Fig. 5.20) (Thorne 1957). Zinc interacts with organic matter to form Zn-organic complexes. About 60% of these chelates are soluble and constitute a major source of Zn in soils (Hodgson et al. 1966). Zinc availability was negatively related to P solubility (Terman et al. 1975). Leggett (1952, cited by Thorne 1957) experienced 30 to 50% reduction of Zn in maize plants fertilized with approximately 300 kg P

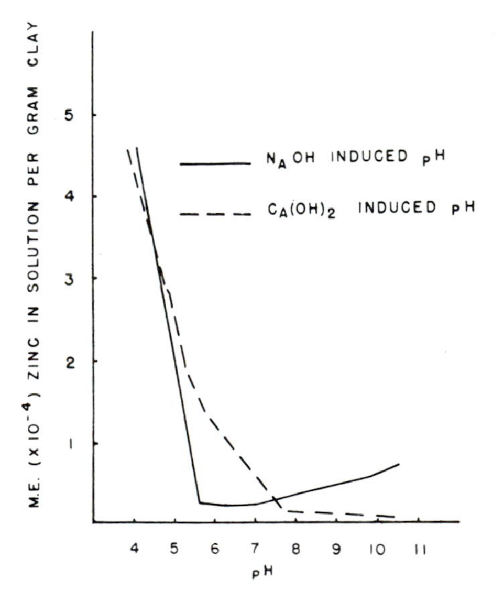

Fig. 5.20. Zinc in solution as a function of pH in a bentonite clay suspension. (Adapted from data of Jurinak and Thorne [1955]. From Thorne, *Advances in Agronomy* 9[1957]:31-65, by permission of Academic Press, Inc.)

• ha^{-1}. Similar results were obtained with citrus in California and maize in Nebraska.

Uptake of Zn is primarily from the divalent Zn^{2+} ion, but probably some is from $ZnCl^+$ and $ZnOH^+$. Iron and Mn were found to be antagonistic to Zn uptake (Reddy et al. 1978).

Zinc was found to be essential for the enzymes in the synthesis of tryptophan, the precursor to IAA (Nason 1958; Lindsay 1972b). Plants deficient in Zn are low in tryptophan and IAA and exhibit small leaves and early abscission. Zinc is also a constituent of carbonic anhydrase, which catalyzes the reaction $H_2CO_3 \rightarrow H_2O + CO_2$. Zinc and Cu are both constituents of superoxide dismutase, which can split molecular O_2. In soybean a Zn level below the critical $12 \mu g \cdot g^{-1}$ in leaf 3 (from the top) reduced photosynthesis and reduced the carbonic anhydrase activity (Ohki 1978). A deficiency of Zn caused a reduction of RNA synthesis and ribosome stability (Prask and Ploke 1971). In soybean small leaf size was the first visible symptom, followed by chlorosis in the youngest leaves (Kapur and Gangwar 1975). Acute deficiencies showed initial chlorosis in the interveinous parenchyma, then retarded leaf growth, and finally dieback. A Zn deficiency can be corrected by foliar sprays or soil treatment with Zn-chelates, preferably Zn-EDDHA on calcareous soils, since Ca replaces the Zn in the complex. A $ZnSO_4$ application of 4 to 5 kg • ha^{-1} every 5 to 8 years has been used with success. Soil treatment around individual fruit trees such as pecan and orange is the usual practice. Liming corrected a Zn deficiency in alfalfa but induced a B deficiency (Brown and Graham 1978). Cultivars of maize varied in Zn response (Safaya and Gupta 1979), and soybean was more Zn efficient than maize. Zinc deficiency did not relate to Zn uptake but rather to the P-Zn ratio, evidently because of competition with Zn in metabolism (Mengel and Kirkby 1982).

Boron. Boron is derived from primary minerals, such as the borosilicates. It is in the soil solution at very low levels as boric acid or borate (HBO_3) and adsorbed to soil particles as borate. Soils derived from sedimentary rock such as shales may have 100 ppm B, compared with about 15 ppm in soils from igneous rocks (Taylor 1964). Boron deficiency was found to be the most widespread of micronutrient deficiencies (Gupta 1979).

Adsorption of borate decreases with increasing soil pH, and availability is low in alkaline soils (pH 7–9). Heavy liming, as for alfalfa, frequently induces a B deficiency.

Uptake is believed to be as undisassociated boric acid; it seems to be primarily passive, based on B-polysaccharide complexes observed in the free space, although a small amount of active transport of B has been demonstrated (Bowen and Nissen 1976). Passive transport is via the transpiration stream. It follows that B is relatively immobile in the plant, and young organs are dependent on current uptake.

Boron is believed to influence cell development by control of sugar transport and polysaccharide formation. Another function attributed to this element is combining with the active site of phosphorylation to inhibit starch formation, which prevents excessive sugar polymerization at the sites of sugar synthesis. Further, it appears that B may determine whether sugars are decomposed for energy release via the glycolytic pathway or the pentose phosphate shunt, the two alternate pathways of sugar decomposition to pyruvic acid. Requirements for B and Ca often go hand in hand; this has suggested that B, like Ca, may be needed for cell wall formation and for the metabolism of pectic compounds. It is of interest that numerous physiological (nonpathogenic) diseases, such as brown heart of turnip, heart rot of sugar beet, and leaf roll of potato (all indicating cell wall integrity problems), have been traced to B deficiency. A wide range of B fertility levels had no effect on the vegetative growth of maize, but tassels from deficient plants had no viable pollen; also, silks were not receptive to the pollen taken from high-B-fertility plants (Vaughan 1977). In fenugreek, a B deficiency caused flowering and reproductive failure, rosetting of terminal buds, small leaves, and chlorosis (Molgarrd and Hardman 1980). Deficiencies were usually corrected by application of a borated fertilizer mixture at 0.5 to 3 kg • ha^{-1} broadcast, 0.1 to 0.5 kg • ha^{-1} leaf spray, or 2 kg • ha^{-1} row application (Gupta 1979).

Copper. Copper is found in primary and secondary minerals but occurs primarily in organic complexes. It is present as an exchange ion on soil particles and minutely in the soil solution. Adsorbed Cu is held tightly, replaceable to some extent by other cations. For this reason, correction of a Cu deficiency is best accomplished by Cu-chelates. Total Cu in soils is usually less than 50 ppm, and leached, sandy soils are inherently low in Cu. In Florida, Cu deficiencies have normally been associated with organic soils; however, cucumber yields from a fine sandy loam were nearly doubled by the addition of 2.24 kg • ha^{-1} and increased further by 8.96 kg • ha^{-1} of $CuSO_4$ (Navarro and Locascio 1980). Liming induced a Cu deficiency in alfalfa on an acid soil (pH 5.3), which was corrected by adding Cu (Brown and Graham 1978). The Cu content of most plants varies from 2 to 20 ppm (Mengel and Kirkby 1979).

Copper plays a role in photosynthesis, as it is a part of the chloroplast enzyme plastocyanin in the electron transport system between photosystems I and II. Most of the Cu in plants is found in the organelles. Copper is a part of several oxidases, such as ascorbic acid oxidase and polyphenol oxidase. As previously mentioned, Cu and Zn are found in superoxide dismutase, which can split O_2 in aerobic organisms. It is a cofactor for the synthesis of certain enzymes.

Cereal crops, such as oat, are more prone to Cu deficiency. At the tillering stage, leaf tips become white and twisted and give an overall bushy appearance. Panicles may fail to develop and form grain. In fruit trees, terminal shoots are stunted and may suffer summer dieback. Species and cultivars differ in tolerance; soybean is highly tolerant to Cu deficiencies.

Copper may become toxic in soils with a history of Cu sprays, such as Bordeaux mixture. However, most soils are strongly buffered against free Cu in quantities sufficient to be toxic by their strong Cu adsorption. Deficiencies can be corrected by chelates such as Cu-DTPA (diethyltriaminepentacetate).

Molybdenum. Molybdenum may be derived from weathering of a number of minerals that include MoS_2 (reduced), oxycomplexes such as Ca-MoO_4, and hydrated forms. Absorbed primarily as a divalent anion (MoO_4^{2-}), Mo occurred in the soil solution at low concentrations, 2×10^{-8} M to 8×10^{-8} M (Lavy and Barber 1964). The average content for agricultural soils was 2 ppm (Swaine 1955). Deficiencies occur at concentrations below 1 ppm in the Ivory Coast (Eschbach 1980), and the critical level in leaf 17 of oil palm is 0.1 to 1 ppm.

Molybdenum availability increases with increasing pH (Fig. 5.4), so liming increases availability. The saying attributed to Australians, "An ounce of molly is worth a ton of lime," refers to the light or aerial application of Mo fertilizer to rangelands that is as effective in promoting clover growth as tons of lime per hectare. Microbes in high-organic soils immobilize Mo, and highly leached, sandy Podzols may be deficient.

The only known use of Mo is in the enzymes nitrite reductase and nitrate reductase, where it acts as an electron carrier between oxidized and reduced states. Mo-deficiency symptoms include whiptail disease and dieback on cauliflower and broccoli. Interveinal chlorosis often occurs. Visible deficiency symptoms could not be demonstrated on oil palm in nutrient culture (Eschbach 1980). Deficiencies of Mo can generally be corrected by adding lime to the soil or by adding Na_2MoO_4.

Chlorine. Chlorine is the most common anion in nature and may be super optimal in areas near the sea, old lake beds, and slightly leached soils in arid areas and in arid irrigated lands. The ocean sodium chloride content is high enough to preclude life for higher plants. Plants may derive sufficient chlorine from the chlorine gas of the atmosphere (Johnson et al. 1957). This source increases in importance near the sea. In soil, chlorine is adsorbed by colloids as the Cl^- anion. Hoagland (1944) demonstrated that plants can take up Cl at a concentration many times that of the external solution, which indicated active uptake. Normal accumulation is in the vacuole, and the tonoplast membrane becomes rate limiting (Cram 1973). Uptake is in competition with other anions, especially NO_3^-. Chlorine is not mobile in the plant and accumulates in the older parts.

Chlorine is not a constituent of any known plant metabolite but was found essential for the evolution of oxygen in photosystem II (Salisbury and

Ross 1978). Deficiency symptoms appear first as wilted leaves that subsequently can become chlorotic and bronze colored. The contribution of Cl to reduction of lodging has been a frequent question, since observations of reduced lodging were made when the fertilizer KCl was used. Chlorine had no benefit on lodging of maize when KCl and NH_4Cl were compared, and it was concluded that the reduction in lodging was due to the benefits from K (Liebhardt and Munson 1976).

Correction of Cl deficiency is seldom, if ever, needed, since adequate amounts can come from the air, rain, and from animal urine and perspiration.

Summary

Sixteen elements—C, H, O, N, S, P, K, Ca, Mg, Fe, Zn, Cu, B, Mn, Mo, and Cl—are required by all crop plants. Silicon is also required by rice, sugarcane, and probably by several other grass species. The three elements C, H, and O, obtained from air or water, compose about 95% of plants. The mineral elements are obtained from weathering of primary and secondary soil minerals, from biodegradation of organic matter, and from gases in the atmosphere (SO_2 and Cl). Often the native supply is deficient in one or more essential elements, and supplementation with commercial fertilizers is required for economic production. Plant nutrients recycle to the extent that plant materials are returned to the soil.

Usually only minute amounts of an element are available in the soil solution, large amounts adsorbed on fine soil particles are exchangeable and partially available, and vast quantities are essentially unavailable as mineral and organic soil fractions. Availability of most elements is more highly related to soil pH than to the total quantity. The quantitative requirement for micronutrients is small, a millionth of that of certain macroelements such as C. Usually fertilization with microelements is not required except on organic or sandy soils.

Uptake of minerals by plant roots, whether from the soil solution or by contact exchange, is as ions, and uptake and transport can be active or passive. Transport can be apoplastic, symplastic, or both, depending on the mineral element and concentration. Calcium transport is restricted to the apoplast.

Nitrogen is generally the most limiting nutrient in crop production, except for well-nodulated legumes, and most often establishes yield levels. Along with water, N supply is a major factor in global production levels. Phosphorus may be extremely deficient for crop production, particularly on acid and native soils with little or no fertilization history. It is not lost from the soil like N, that is, by leaching and denitrification. Both N and P are mobile in the plant and are distributed from older to younger parts, which results in deficiency symptoms that first show in the older parts. Potassium is often limiting in crop production, particularly on sandy soils. Clay or fine-textured soils may hold large amounts as exchangeable K. Potassium is absorbed primarily during vegetative growth and, if sufficiently available, in luxury amounts. It is highly mobile in the plant, and deficiencies first show in older parts (necrosis of lower leaf margins). Potassium is not a constituent of any known plant metabolite, despite the large requirement for normal growth. Magnesium is a part of the chlorophyll molecule and a cofactor of the enzymes involved in phosphorylation. It is mobile in the plant and deficiencies show first as interveinal chlorosis in the older leaves. Calcium, on the other hand, is highly immobile and not redistributed in the plant. Deficiency symptoms are aberrant growth of both fruits and terminal buds and dieback. Peanut requires Ca in the pegging (fruiting) zone, and pods absorb Ca independent of the roots.

The available micronutrients, which are required in minute amounts, are generally adequate for crop production. However, high- and low-pH, organic, and sandy soils are frequently deficient in certain micronutrients, depending on the crop. Some genotypes are more tolerant of deficiencies or toxicities than others. Micronutrients are constituents of enzymes or are activators. Most are mobile in the plant, but B is highly immobile, which causes an abnormal growth of active tissues similar to that caused by Ca.

References

Aboulroos, S. A., and N. E. Nielsen. 1979. Acta Agric. Scand. 29:326–36.
Allaway, W. H. 1968. In Advances in Agronomy, vol. 20, ed. A. G. Norman. New York: Academic Press.
Arnon, D. I., and Stout, P. R. 1939. Plant Physiol. 14:371–75.
Arnon, I. 1974. Mineral Nutrition of Maize. Bern-Warblaufen: International Potash Institute.
Barber, S. A., and R. A. Olson. 1968. In Changing Patterns in Fertilizer Use, ed. L. B. Nelson et al. Madison, Wis.: Soil Science Society.
Blevins, D. G., A. J. Hiatt, R. H. Lowe, and J. E. Leggett. 1978. Agron. J. 70:393–96.
Bonner, J., and J. E. Varner. 1965. Plant Biochemistry. New York: Academic Press.
Bouma, D. 1967. Aust. J. Biol. Sci. 20:613–21.
Bowen, J. E., and P. Nissen. 1976. Plant Physiol. 57:353–57.
Boyer, T. C., A. B. Carlton, C. M. Johnson, and P. R. Stout. 1954. Plant Physiol. 29:526–32.
Brown, J. C. 1961. Adv. Agron. 13:329–69.
______. 1977. Agron. J. 69:399–404.
Brown, J. C., and W. E. Jones. 1975. Agron. J. 67:468–72.
Brown, J. C., and J. H. Graham. 1978. Agron. J. 70:367–73.
Brownell, P. F. 1965. Plant Physiol. 40:460–68.
Brownell, P. F., and C. J. Crossland. 1975. Plant Physiol. 49:794–97.
Chaney, R. L., J. C. Brown, and L. O. Tiffin. 1972. Plant Physiol. 50:208–13.
Chapman, H. D. 1966. Diagnostic Criteria for Plants and Soils. Berkeley: University of California, Division of Agricultural Science.
Chapman, M. A., and J. Keay. 1971. Aust. J. Exp. Agric. Anim. Husb. 11:223–28.
Cram, W. J. 1973. Aust. J. Biol. Sci. 26:757–79.
Crapo, N. L., and H. J. Ketellapper. 1981. Am. J. Bot. 68:10–16.
Deckard, E. L., and R. H. Busch. 1978. Crop Sci. 18:289–93.
Dibb, D. W., and L. F. Welch. 1976. Agron. J. 68:89–94.
Eddings, J. L., and A. L. Brown. 1967. Plant Physiol. 42:15–19.
Edwards, J. H., and S. A. Barber. 1976. Agron. J. 68:17–19.
Egli, D. B., J. E. Leggett, and W. G. Duncan. 1978. Agron. J. 70:43–47.
Elawad, S. H., G. J. Gascho, and J. J. Street. 1982. Agron. J. 74:481–84.

Epstein, E. 1972. Mineral Nutrition of Plants: Principles and Perspectives. New York: Wiley.
Eschbach, J. M. 1980. Oleagineux 35:291–94.
Evans, H. J., and G. J. Sorger. 1966. Annu. Rev. Plant Physiol. 17:47–76.
Farley, R. F., and A. P. Draycott. 1975. J. Sci. Food Agric. 26:385–92.
Fisher, J. D., D. Hanson, and T. K. Hodges. 1970. Plant Physiol. 46:812–14.
Forde, S. C. 1976. Trop. Agric. 54:273–79.
Fox, R. L. 1976. Agron. J. 68:891–96.
Foy, C. D., P. W. Voigt, and J. W. Schwartz. 1977. Agron. J. 69:491–96.
Gaines, T. P., and S. C. Phatak. 1982. Agron. J. 74:415–18.
Gauch, H. G. 1972. Inorganic Plant Nutrition. Stroudsburg, Pa.: Dowden, Hutchinson and Ross.
Gupta, U. C. 1979. Adv. Agron. 31:273–307.
Hafez, A. A. R., P. R. Stout, and J. E. DeVay. 1975. Agron. J. 67:359–61.
Hall, S. M., and D. A. Baker. 1972. Planta 106:131–40.
Harris, H. C. 1948. Plant Physiol. 23:150–60.
Haynes, R. J. 1980. Bot. Rev. 46:75–99.
Haynes, R. J., and K. M. Goh. 1978. Biol. Rev. 53:465–510.
Hepper, C. M. 1976. Crop Sci. 18:584–87.
Hewett, E. J., and T. A. Smith. 1975. Plant Mineral Nutrition. London: English University Press.
Hoagland, D. R. 1944. Lectures on the Inorganic Nutrition of Plants. Waltham, Mass.: Chronica Botanica.
Hodgson, J. F., W. L. Lindsay, and J. F. Trierweiler. 1966. Soil Sci. Soc. Am. Proc. 30:723–26.
Humble, G. D., and T. C. Hsiao. 1969. Plant Physiol. 44 [Suppl.]:21.
Iowa State University. 1965. Cooperative Extension AG-26.
Jenny, H., and R. Overstreet. 1939. Soil Sci. 47:257–72.
Johnson, C. M., P. R. Stout, T. C. Broyer, and A. B. Carlton. 1957. Plant Soil 8:337–53.
Jurinok, J. J., and O. W. Thorne. 1955. Soil Sci. Soc. Am. Proc. 19:446–48.
Kapur, O. C., and M. S. Gangwar. 1975. Indian J. Agric. Sci. 45:559–60.
Koch, K. 1982. Private communication.
Lavy, T. L., and S. A. Barber. 1964. Soil Sci. Soc. Am. Proc. 28:93–97.
Leggett, J. E., and D. B. Egli. 1980. In World Soybean Conference II, ed. F. T. Corbin. Boulder, Colo.: Westview.
Liebhardt, W. C., and R. D. Munson. 1976. Agron. J. 68:425–26.
Liebhardt, W. C., and T. J. Murdock. 1965. Agron. J. 57:325–28.
Lindsay, W. L. 1972a. In Micronutrients in Agriculture. Madison, Wis.: Soil Science Society of America.
______. 1972b. Adv. Agron. 24:147–86.
McDaniel, M. E., and D. J. Dunphy. 1978. Crop Sci. 18:136–38.
McElhannon, W. S., and H. A. Mills. 1978. Agron. J. 70:1027–32.
McKently, A. H. 1981. M.S. thesis, University of Florida, Gainesville.
Marschner, H. 1971. In Potassium in Biochemistry and Physiology. Bern: International Potash Institute.
Mengel, K., and E. A. Kirkby. 1982. Principles of Plant Nutrition. 3d ed. Bern: International Potash Institute.
Molgaard, P., and R. Hardman. 1980. J. Agric. Sci. [Camb.] 94:455–60.
Nason, A. 1958. Soil Sci. 85:63–77.
Navarro, A. A., and S. J. Locascio. 1980. Soil Crop Sci. Soc. Fla. 39:16–19.
Navrot, J., and A. Banin. 1976. Agron. J. 68:358–61.
Neyra, C. A., and R. H. Hageman. 1975. Plant Physiol. 56:692–95.
Ohki, K. 1975. Agron. J. 67:30–32.
______. 1978. Crop Sci. 18:79–82.
Okuda, O., and E. Takahashi. 1964. In The Mineral Nutrition of the Rice Plant. Baltimore: International Rice Research Institute and the Johns Hopkins University Press.

Olsen, R. A., R. B. Clark, and J. H. Bennett. 1981. Am. Sci. 69:378–84.
Omar, M. A., and T. El Kobbia. 1966. Soil Sci. 101:437–40.
Pierre, W. H., J. Meisinger, and J. R. Birchett. 1970. Agron. J. 62:108–12.
Pitman, M. G. 1977. Annu. Rev. Plant Physiol. 28:71–88.
Prask, J. A., and D. J. Plocke. 1971. Plant Physiol. 48:150–55.
Ram, L. C. 1980. Plant Soil 55:215–24.
Reddy, K. R., M. C. Saxena, and U. R. Pal. 1978. Plant Soil 49:409–15.
Rice, E. L., and S. K. Pancholy. 1973. Am. J. Bot. 60:691–702.
Russell, R. S., and D. A. Barber. 1960. Annu. Rev. Plant Physiol. 11:127–40.
Safaya, N. M., and A. P. Gupta. 1979. Agron. J. 71:132–36.
Salisbury, F. B., and C. W. Ross. 1978. Plant Physiology. 2d ed. Belmont, Calif.: Wadsworth.
Schrenk, W. G., and J. C. Frazier. 1964. Plant Food Rev., Fall 1964.
Shrift, A. 1969. Annu. Rev. Plant Physiol. 20:475–94.
Steucek, C. G., and H. V. Koontz. 1970. Plant Physiol. 46:50–52.
Stocking, C. R. 1975. Plant Physiol. 55:626–31.
Swaine, D. J. 1955. Soil Sci. Tech. Comm., no. 48. York, Eng.: Herald.
Taylor, S. R. 1964. Geochim. Cosmochim. Acta 28:1273–86.
Terman, G. L., P. M. Giordano, and N. W. Christensen. 1975. Agron. J. 67:782–84.
Thomson, W., and T. E. Weier. 1962. Plant Physiol. 37:xi.
Thorne, W. 1957. In Advances in Agronomy, vol. 9, ed. A. G. Norman. New York: Academic Press.
Troughton, A. 1977. Ann. Bot. n.s. 41:85–92.
Truog, E. 1961. In Mineral Nutrition of Plants, ed. E. Truog. Madison, Wis.: University of Wisconsin Press.
Tucker, B. B. 1981. Personal communication.
Van Egmond, F., and M. Aktas. 1977. Plant Soil 48:685–703.
Vaughan, A. K. F. 1977. Rhod. J. Agric. Res. 15:163–70.
Viets, F. G. 1944. Plant Physiol. 19:466–80.
Viets, F. G., C. E. Nelson, and C. L. Crawford. 1954. Soil Sci. Soc. Am. Proc. 18:297–301.
Warren, H. L., D. M. Huber, D. W. Nelson, and O. W. Mann. 1975. Agron. J. 67:655–60.
Westermann, D. T. 1975. Agron. J. 67:265–68.
Wiklander, L. 1954. Forms of Potassium in the Soil. Bern: International Potash Institute.
Williams, L. M., and G. S. Miner. 1982. Agron. J. 74:457–62.
Williams, M. C. 1960. Plant Physiol. 35:500–505.
Wolf, D. D., E. L. Kimbrough, and R. E. Blaser. 1976. Crop Sci. 16:292–94.
Worley, R. E., R. E. Blaser, and G. W. Thomas. 1963. Crop Sci. 3:13–16.
Yoshida, S., Y. Onishi, and K. K. Tagishi. 1959. Soil Plant Food [Tokyo] 5:127–33.

6 Biological Nitrogen Fixation

NITROGEN (N) is generally the major limiting factor in crop production. Plant biomass averages 1 to 2% N and may contain as much as 4 to 6%. In terms of total quantity required for production of a crop, N ranks fourth among the 16 essential elements.

Ironically there is no scarcity of N anywhere. Nearly 79% of the atmosphere is N_2, which amounts to tons over each hectare. Unfortunately gaseous N_2, relatively inert, is unavailable to plants. Soil sediments and rocks contain even more N than the atmosphere, but this too is unavailable until decomposed by weathering. Only combined N in the ionic forms (NH_4^+ and NO^{-3}) are available to higher plants. Some bacteria, actinomycetes, and blue-green algae (Cyanobacteria) are the only known plant life that can utilize gaseous N_2. The N_2-fixing activity of these organisms is essential to the global N balance because fixed forms of N are continually subject to losses by denitrification and leaching (Table 6.1).

Agriculture has always relied heavily on N produced by N_2-fixing orga-

TABLE 6.1. Nitrogen fixation and losses in the earth's N balance

	Area (ha × 10^6)	Kg N_2 fixed per ha × yr	Metric tons per yr × 10^6
Biological fixation			
Legumes	250	55–140	14–35
Nonlegumes	1,015	5	5
Rice fields	135	30	4
Other soils and vegetations	12,000	25–30	30–95
Marine	36,100	0.3–1	10–36
Industrial fixation			30
Atmospheric fixation			7.6
Juvenile addition			0.2
Denitrification			
Terrestrial	13,400	3	43
Marine	36,100	1	40
Loss to sediments			0.2

Source: Quispel 1974, by permission.

TABLE 6.2. Representative estimates of N_2 fixation by various legumes

Legume	Estimated N_2-fixed ($kg \cdot ha^{-1}$)	Source[a]
Lucerne	90	Waksman 1952
	260	Alexander 1961
	40–350	Nutman 1965; Bell and Nutman 1971
Clovers	50	Fred et al. 1932
	250–350	Russell 1950
	65	Waksman 1952
	50–200	Nutman 1965
Legumes in pasture	10–550	Nutman 1965
	30–170	Williams 1970
Peas	400–500	Russell 1950
	30–140	Nutman 1965
Soybean	20–200	Russell 1950
	65	Waksman 1952
	40–120	Sundara Rao 1971
Tropical	40–80	Wetselaar 1967
	40–360	Whitney 1967
	20–260	Henzell 1968

Source: Vincent 1974, by permission.
[a]Including data quoted from other workers.

nisms for crop production. Rhizobial bacteria associated with host legumes are generally the most important. These associations can fix 100 kg $\cdot$ ha^{-1} of N per season and frequently triple this amount, which is much more than other biological N_2-fixing systems (Table 6.2). Legumes have two important advantages over other crops: (1) they are autotrophic for N as well as for carbon (C) (i.e., the association requires no N in the growth medium) and (2) legumes add N for subsequent crops. In natural ecosystems (forests and grasslands), free-living N_2-fixing bacteria and certain nonlegume N_2-fixing associations are more important than legumes in the N balance.

Industrial Production of Ammonia

The world requirement of N for crop production is about 125 $\times$ 10^6 Mg per year and is expected to increase to 200–245 $\times$ 10^6 Mg by 2000 A.D. (Gibson 1977). Industrial production, principally from the Haber-Bosch process, accounts for approximately 40% (30 $\times$ 10^6 Mg) of this requirement (Table 6.1). This process requires large inputs of fossil energy as hydrogen feeder stock and to produce the high temperatures (5,000°C) and pressures (200 bars) of the process. There is also an enormous initial energy investment in manufacturing the materials and constructing an industrial plant, which costs about $150 million.

The chemical reaction in the Haber-Bosch process is as follows:

$$3N_2 \text{ (air)} + H_2 \text{ (natural gas)} \xrightarrow[K_2O;Al_2O_3]{Fe} 2HN_3 + 860 \text{ cal}$$

Atmospheric N_2 Fixation

An estimated 10 × 10^6 Mg ammonia is fixed each year by lightning charges (Table 6.1). Lightning has sufficient energy to ionize water vapor into H^+ and OH^- ions, and these plus oxygen (O_2) can react with N_2 molecules to produce nitric acid, which is brought to earth in precipitation. Compared with biological N_2 fixation, the amount fixed by lightning is small but still important in natural ecosystems, in which the fixed N accumulates in the biomass and is mineralized and recycled during the course of time. Shifting-cultivation agroecosystems take advantage of such accumulated N to produce crops. A long fallow of tropical forest vegetation (over a period of 12 or more years) can accumulate as much as 500 kg • ha^{-1} N from lightning and from biological N_2-fixing sources.

Biological N_2 Fixation

Numerous free-living bacteria and associations between bacteria and higher plants have the capacity to reduce atmospheric N_2 to ammonia (NH_3). The following reaction, mediated by the enzyme *nitrogenase,* is common to all organisms.

$$N_2 + 6H^+ + 6e^- + n\text{MgATP} \xrightarrow{\text{nitrogenase}} 2NH_3 + n\text{MgATP} + nP_i$$

The N_2-fixing organisms are primarily bacteria but also blue-green algae.

Classification of N_2-fixing organisms is somewhat in a state of flux, but the outline below (Quispel 1974) will provide a guide in the discussion of biological N_2-fixing systems that follows.

I. Asymbiotic (free living)
   A. Bacteria
      1. Aerobic. Three genera of Azotobacteraceae, *Azotobacter, Azospirillum,* and *Beijerinckia,* are of primary importance.
      2. Anaerobic. The agriculturally important and widely distributed *Clostridium pasteurianum* and two photosynthetic genera, *Rhodospirillum* and *Chromatium,* are anaerobic.
   B. Blue-green algae (Cyanobacteria). Two genera, *Anabaena* and *Nostoc,* are the commonest. Cyanobacteria is a relatively new classification (Buchanan and Gibbons 1974) and is not in common use, so the term *blue-green algae* will be used in this discussion.
II. Symbiotic (nodulating)
   A. Root nodulating
      1. *Rhizobium.* Associated with legumes.
      2. Actinomycetes (e.g., *Frankia*). Associated with woody angiosperms. *Alnus* (alder) is a well-known host plant.

3. Blue-green algae. Associated with gymnosperms. Nodules are formed on the surface roots (light required) of gymnosperm species.

B. Leaf nodulating (phyllosphere). A number of bacteria, including some free-living types, produce nodules on leaves of woody species in the humid tropics.

III. Symbiotic (nonnodulating; associative)

A. Blue-green algae. Associated with fern (*Azolla*) and with fungi (lichens).

B. Bacteria (Azotobacteraceae). Associated with grasses. Bacteria include *Azospirillum brasilense, Spirillum lipoferum,* and *Azotobacter paspali,* which are commonly found in pastures of tropical and semitropical C_4 grasses.

FREE-LIVING ORGANISMS

On the evolutionary scale, the first N_2-fixing organisms were free-living organisms, which probably included certain heterotrophic and photosynthetic bacteria, and the blue-green algae. These three plant forms can fix N_2 without aid or cooperation from another organism.

Heterotrophic N_2-fixing bacteria can be either aerobic, anaerobic, or facultative (Mulder and Brotonegoro 1974). All are widely distributed in nature, and the first two may contribute significant amounts of fixed N to the N balance of agricultural and natural ecosystems.

Bacteria. Azotobacteraceae is an important family of free-living aerobic bacteria, especially the genera *Azotobacter, Azospirillum,* and *Beijerinckia; Azotobacter* is more important in temperate agricultural soils. In well-drained, cultivated soils, N_2 fixation by free-living organisms is predominantly by Azotobacteraceae. However, fixation of appreciable amounts of N_2 requires large amounts of C, such as those in crop residues with a high C-N ratio. Approximately 100 kg organic material is required to supply the reductants to fix 1 kg N. Because large inventories of C cannot be maintained in warm, well-aerated soils, it is usually the limiting factor in the system.

Azotobacter cells contain PBH (poly-β-hydroxybutyrate), used for encystment of the bacteria, and a cytochrome system for electron transport, indicative of a high oxidative respiration rate. Efficiency of *Azotobacter* (an obligate aerobe) is estimated at 10 to 15 mg N_2 fixed per g glucose, essentially the same as that of the anaerobe *Clostridium pasteurianum,* but the latter only partially oxidizes sugar (Mulder and Brotonegoro 1974). Much of the energy utilized by *Azotobacter* goes into producing new cell mass and much apparently goes into the protection of the enzyme nitrogenase against O_2 inactivation. *Clostridium,* being anaerobic, does not expend energy on O_2 protection, and thus achieves parity in efficiency despite only partial substrate oxidation. With large amounts of C in the form of plant residues, such as straw, and the additional advantage of waterlogged soils, the anaerobic *C. pasteurianum* multiplies rapidly, increasing soil N content. Facultative bacteria such as *Klebsiella* spp. also may fix N_2 in soils.

Three requirements for effective N_2 fixation apply to aerobic, anaerobic, and facultative organisms: (1) a plentiful supply of C, (2) a low level of combined N (e.g., ammonia or nitrate) in the medium, and (3) protection of the nitrogenase complex against excessive O_2.

The photosynthetic bacteria capable of N_2 fixation are found primarily in saltwater, fresh water, and marine mud. They may be green or purple, the latter responsible for the so-called "red tides." The purple bacteria are further subdivided into purple sulfur (S) and purple non-S bacteria. The purple S bacteria substitute H_2S for H_2O as the electron donor in photosynthesis. Sulfur is ionized as follows:

$$CO_2 + H_2S \xrightarrow{\text{light}} \text{carbohydrate} + 2S + H_2O$$

The free S or sulfates so liberated can form deposits of elemental S, such as those currently mined in coastal areas.

Blue-Green Algae (Cyanobacteria). Blue-green algae date to the precambrian period, when they presumably were the dominant vegetation. Their N_2-fixing capability has been known since the turn of the century. They contribute significantly to soil formation on rock surfaces, in part as the associative component of lichens.

Normally blue-green algae consist of chains of cells, with some cells having elongated and thickened cell walls (Fig. 6.1). These specialized cells, the *heterocysts,* were found to be the sites of nitrogenase activity (Stewart 1974). Other cells are vegetative and may contain polyphosphate bodies, while a third type is reproductive and contains *akinetes* (spores). Some 40 species of blue-green algae are known to fix N_2, particularly *Nostoc* and *Anabaena.* As free-living organisms, they contribute to the N balance in wet soils and in flooded environments, such as rice paddies. There are associations between blue-green algae and other organisms, such as the small fern *Azolla,* which are symbiotic and normally more N productive than free-living algae alone. Nitrogen fix-

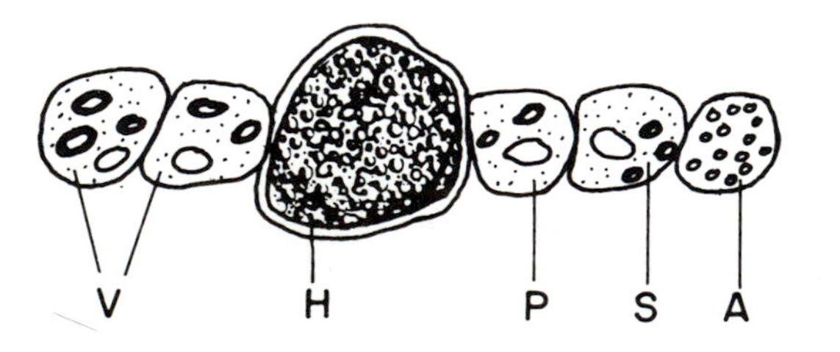

Fig. 6.1. Filament (transection) of a blue-green algae (*Anabaena cylindrica*) showing vegetative cells (*V*), heterocyst (*H*), polyphosphate body (*P*), structured granules of vegetative cells (*S*), and akinetes (spores in a reproductive cell) of the filament (*A*). (From Stewart 1974)

ation by blue-green algae has been effectively quantified by the use of the acetylene (C_2H_2) reduction and the ^{15}N-dilution methods. *Azolla* can fix up to 50–100 kg • ha^{-1}.

Blue-green algae are most active in tropical and subtropical environments and are probably the main N_2 fixers in flooded systems, growing freely in the floodwater and attached to the surfaces of submerged soils and plants. However, Yoshida (1981) observed that their contribution to the N balance in Japanese rice paddies was small compared with rhizosphere bacteria. Blue-green algae were active only in the early growth phases, before shading by the rice canopy.

Most estimates of the amount of N_2 fixed per growing season in flooded soil systems have varied from 3 to 30 kg • ha^{-1}, depending on sampling technique and the amount of combined N present (Buresh et al. 1980). Yoshida (1981) estimated an annual rate as high as 30 kg in flooded rice in Japan but did not attribute this rate exclusively to blue-green algae. Jones (1974) estimated 462 kg • ha^{-1} N was fixed in an English salt flat, which was neutral to slightly alkaline and probably favored fixation. Light is essential for algae growth and N_2 fixation. Fixation of N_2 is repressed if ammonia or other combined forms of available N are present. In a N-poor soil, Huang (1978) found that inoculation with blue-green algae increased N_2 fixation and rice grain production in a pot experiment by 34 to 41%, depending on the rice cultivar. However, in paddy fields there was no advantage of blue-green algae inoculation over the control; both produced 20% less than paddies treated with N fertilizers. Evidently, there is little value from N_2 fixation by blue-green algae in dense canopies of productive rice fields. Low fixation rates probably result from shading but could also be due to phenolic compounds liberated from decomposing rice straw. These compounds were observed to inhibit growth of the blue-green algae *Anabaena cylindrica* (Rice 1980).

Cultures of *Anabaena* and *Nostoc* cells have recently been marketed as commercial inoculants for cultivated field crops such as maize. Considering that these organisms are adapted to tropical flooded environments, claims of increased soil N and crop yields in temperate environments from inoculation of cultivated fields would appear to be greatly exaggerated.

NODULATING ORGANISMS

Actinomycete-Angiosperm Associations. Like legumes, certain other angiosperms produce nodules and fix N_2 symbiotically in association with a small bacterium, genus *Frankia,* known as an *actinomycete.* The host plants are woody, nonleguminous species. Alder (*Alnus*) is the best known example; more than a dozen genera and some 33 species of *Alnus* have been observed to nodulate.

Nodules containing actinomycetes are initiated by lateral swellings of the root after infection via the root hairs (Newcomb et al. 1978). As a result of the formation of new meristems at the base, nodules branch extensively or produce clusters. Bacteroids in the nodules have a pinkish color, thought to be due to anthocyanin rather than leghemoglobin (Bond 1974). Bond demonstrated a 10- to 15-fold difference in growth between inoculated and control plants of *Alnus* and *Myrica*. Another nonlegume, *Trema cannabina,* was nodulated by *Rhizobium* isolates from cowpea and soybean (Trinick 1976), the first observation of N_2 fixation by *Rhizobium* with a nonleguminous species.

The role of actinomycete-angiosperm associations appears to be minor in crop production. It probably is important to the N balance in some natural ecosystems.

Leaf Nodule Organisms. Ruinen (1956) reported that the aerobic bacterium *Beijerinckia* spp. were commonly found on the surfaces or the phyllosphere of vegetation in the wet Indonesian tropics. The genus was isolated from 192 of 198 samples. This epiphyte, which grows typically on the luxuriant vegetation surfaces of the wet tropics, appears to contribute significantly to N balance, especially in low-N soils. Bartholomew et al. (1953, cited in Ruinen 1974), observed that a Congo forest accumulated 95 kg • ha^{-1} • yr^{-1} N for the first 2 years in fallow, 129 kg • ha^{-1} • yr^{-1} for the next 3 years, and then leveled off at about 13 kg • ha^{-1} • yr^{-1}. Several epiphytes, such as *Beijerinckia, Azotobacter,* and *Mycoplana,* are found in phyllosphere N_2-fixing associations. In addition, yeasts and fungi usually are involved, providing a moist habitat. A tropical milieu is excellent for phyllosphere growth of organisms: (1) a green leaf surface 10- to 20-fold that of temperate plants, (2) tripled primary production, and (3) N uptake 3- to 10-fold that of temperate plants (Ruinen 1974). Leaves provide support, water, organic nutrients, and low levels of combined N. Dew and up to 80% of the light showers are intercepted by the canopy, so organic nutrients wash down to lower leaf strata, enhancing the phyllosphere habitat. Phyllosphere organisms also wash into the soil but apparently do not survive there. Phyllosphere N_2 fixation probably contributes significantly to the N balance in tropical forest ecosystems and indirectly to the agroecosystems that may follow them.

***Rhizobium*-Legume Associations.** Legumes, ranking second or third in number of species among flowering plants, are distributed worldwide and contribute significantly to filling human needs for food, forage, oil, and timber. A great many legume species fix N_2 symbiotically and, hence, are autotrophic for N as well as for C and contribute greatly to the global N balance. A great many herbaceous legumes evolved in temperate climates during a period of abundant calcium; these legumes and especially their *Rhizobium* coworkers are well adapted to such conditions. The largest number of legume species, both woody and herbaceous, have evolved in tropical climates combined with leached, acid soils, and they thrive under these conditions. Many of these legumes associate with the cowpea type of *Rhizobium.*

The amount of N_2 fixed by legume associations is highly variable, depending on the legume, cultivar, bacterial species and strain, and growing conditions, especially the pH and soil N. Common values cannot be stated (Vincent 1974), but estimates for a number of crops are given in Table 6.2. It is interesting to note that estimates as high as 500 kg • ha^{-1} • yr^{-1} are reported for pea, an herbaceous annual. The record reported yields of soybean and alfalfa might indicate rates higher than 500 kg • ha^{-1}. The problem with most estimates of this type is assigning an appropriate value to the N component obtained from the soil. The total amount of N_2 fixed per season is a function of rate and time. Assaying the rate by C_2H_2 reduction can measure the rate at a specific time but does not estimate the duration or season fixation. Total biomass of N accumulated for the season, less the soil contribution, is an estimate of the seasonal N yield.

NODULE FORMATION. Following colonization of the root by compatible *Rhizobium* strains, the infection and nodulation process is more or less as follows:

1. Root hair deformation (i.e., curling or branching), possibly as a response to indoleacetic acid (IAA), the production of which is stimulated by the bacterium, or perhaps as a response to ethylene, stimulated by IAA
2. Formation of the infection thread for transference of the bacterial cells into the root cortex
3. Release of bacteria into the cortex cells
4. Formation of the nodule meristem and expansion of the nodule by division of the cortex cells
5. Enlargement of the infected cortex cells inside the nodules (Fig. 6.2)
6. In older nodules, loss of the *bacteroid* (nodule bacteria) envelope and nitrogenase activity as senescence commences.

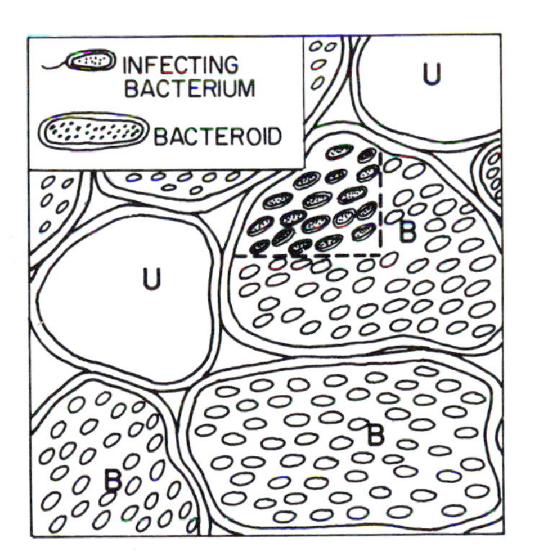

Fig. 6.2. Transection of a legume nodule showing cells infected with bacteroids (*B*) and uninfected cells (*U*). Infected cells are enlarged. Inset shows a bacterium cell (*Rhizobium*) with flagellum and a bacteroid, bacterium enlarged in an envelope. Nitrogenase activity is in the bacteroids.

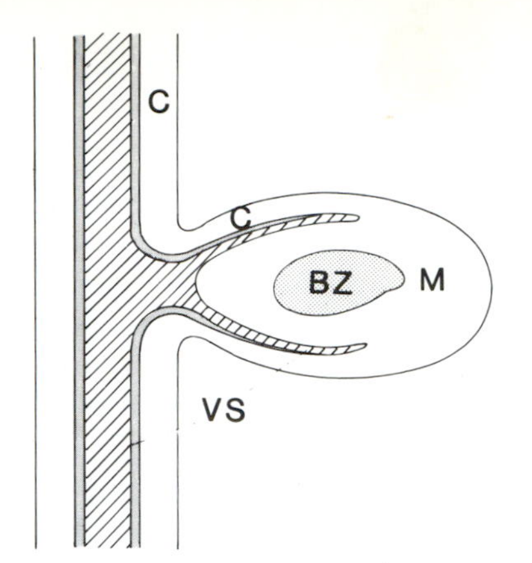

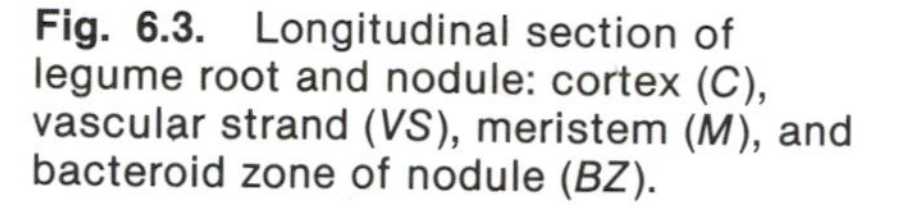
Fig. 6.3. Longitudinal section of legume root and nodule: cortex (*C*), vascular strand (*VS*), meristem (*M*), and bacteroid zone of nodule (*BZ*).

The vascular system (Fig. 6.3) transports sugars, water, and minerals to bacteroids and removes fixed N_2 as amino acids, amides, and ureides (allantoin). In addition to these nutritional aspects, the nodule functions to provide a favorable environment for the bacteria, including oxygen protection of nitrogenase.

The four major components of nodule structure (Fig. 6.3) are cortex, meristem, vascular system, and bacteroidal zone. Nodule sizes and shapes vary greatly, depending on the meristem characteristic of the legume species (Fig. 6.4). Some (e.g., alfalfa and pea) have terminal and indeterminate nodule meristems that produce oblong nodules. Terminal but determinate meristems produce spherical nodules with limited longevity, such as found on soybean, peanut, and birdsfoot trefoil. The nodules of some species, such as velvet bean, develop lateral meristems, which result in branches and irregular shapes (Fig. 6.4). Nodules smaller than normal are generally an indication of infection by an ineffective strain of *Rhizobium*. Nodules from ineffective strains may also lack leghemoglobin.

Infected cortex cells are much larger and frequently mixed with smaller, uninfected cells (Fig. 6.2). The bacteroids in infected cells are enclosed in plant-derived membranes (Tu 1974). Nearly all nodule cells of cowpea and peanut were infected, compared with 50% or less in garden bean (Vincent 1974). The cowpea type of *Rhizobium* appears to be ubiquitous in tropical and semitropical areas and exhibits a high nitrogenase activity.

CROSS-INOCULATION GROUPS. Legumes can be divided into groups based on specificity for *Rhizobium* spp. (Fred et al. 1932). For example, pea (*Pisum*) and vetch (*Vicia*) are inoculated by *R. leguminosarum* (Table 6.3). In addition to alfalfa, sweet clover and several legumes can be inoculated with *R. meliloti*. On the other hand, *R. japonicum* is specific to soybean, and the bacterium that infects birdsfoot trefoil is specific to that host.

The cross-inoculation classification is problematic because of the wide variability in specificity among strains of bacterial species. Strains can be effective on one legume but ineffective on other legumes in the same group.

Mungbean

Pigeonpea

Showy Crotalaria

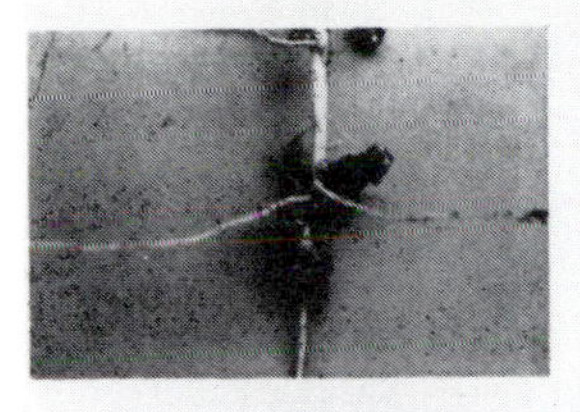

Crotalaria sp.
PI 436527-1

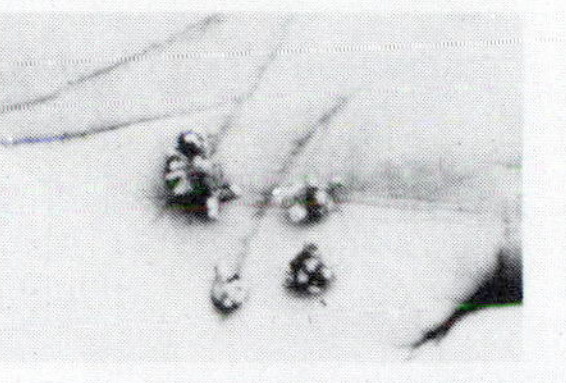

Crotalaria sp.
PI 436527-2

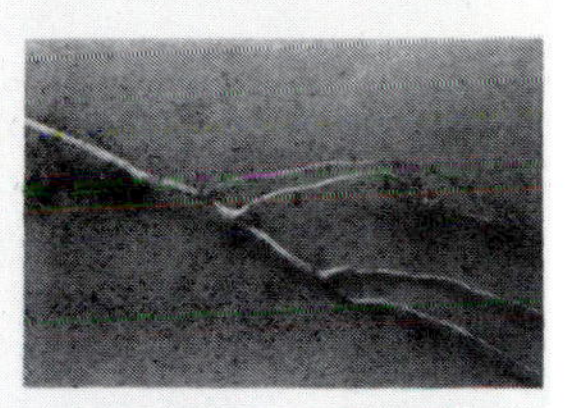

Hairy Indigo

Joint Vetch

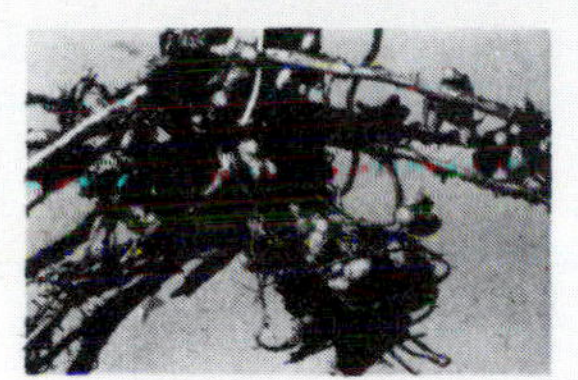

Velvetbean
1

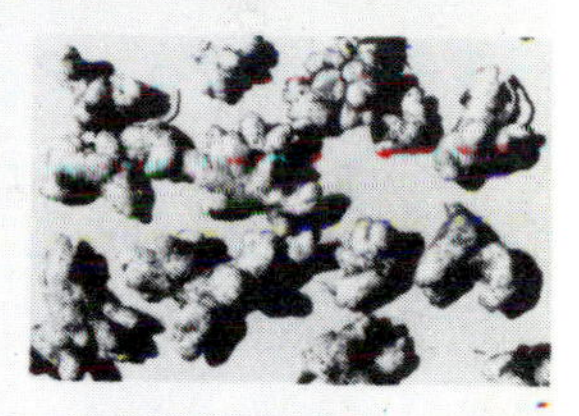

Velvetbean
2

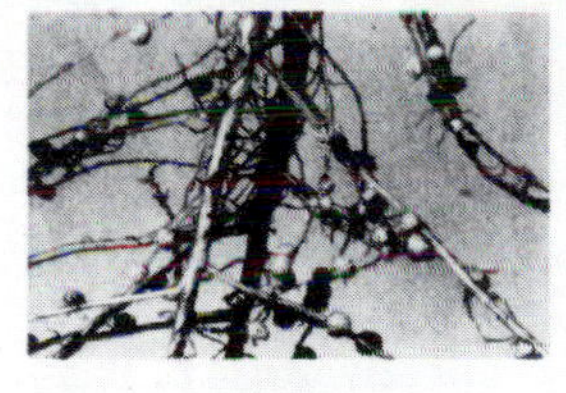

Soybean

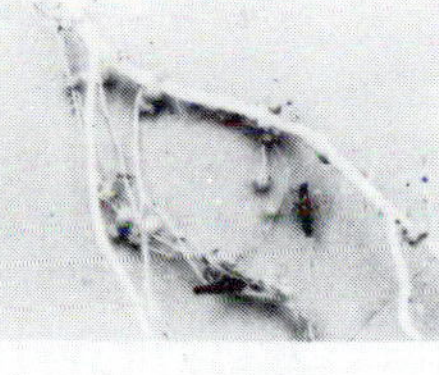

Alyceclover

Fig. 6.4. Nodulation in several legume species.

TABLE 6.3. Cross-inoculation groups of legume species

Host Type	*Rhizobium* Species	Some Members of Cross-inoculation Group: Legume	Genus and species	Comments
Pea	*leguminosarum*	Pea Vetch Sweet pea	*Pisum* sp. *Vicia* sp. *Lathyrus* sp.	
Clover[a]	*trifolii*	White Red Crimson Subterranean	*Trifolium repens* *T. pratense* *T. incornatum* *T. subterraneum*	May be ineffective on tropical *Trifolium* genera
Bean	*phaseoli*	Common garden Scarlet runner	*Phaseolus vulgaris* *P. coccineus*	
Alfalfa	*meliloti*	Alfalfa Black medic Sweet clover	*Medicago sativa* *M. lupulina* *Melilotus* sp.	
Soybean	*japonicum*	Soybean	*Glycine max*	Specific to soybean
Lupine	*lupini*	Lupine Trefoil	*Lupine* sp. *Lotus* sp.	Ineffective on *Lotus corniculatus*
Cowpea	(not defined)	Cowpea Peanut Lespedeza	*Vigna sinensis* *Arachis hypogaea* *Lespedeza* sp.	Cowpea type extremely heterogeneous, inoculates numerous genera and species

[a]Only temperate species.

***Rhizobium* STRAINS.** Strains of *Rhizobium* spp. vary widely in specificity relative to legume species, cultivar nodulation habit, and nitrogenase activity. Some strains may vary in N_2 fixation even with varieties of the same species (Fig. 6.5). Out of 24 pea-group strains (*R. leguminosarum*), only one was compatible with all seven host legumes of the pea group (Vincent 1974). Six strains were ineffective on all the legumes in the pea group. Strain effectiveness can be attributed to lack of infection of the root and/or lack of nitrogenase activity or low efficiency.

Some strains of *R. japonicum* are theoretically more efficient than others, due to a capacity of recycling H_2 generated by nitrogenase to produce ATP. Electrons not recycled produce H_2 rather than reducing N_2 to NH_4^+, the desired end product. Strains capable of using the released H_2 as a source of energy have a lower cost of N_2 fixation in terms of ATP (Fig. 6.6). Loss of H_2 was a loss of energy to the system (Schubert and Evans 1976).

INOCULATION. Adequate *Rhizobium* numbers can be lost due to soil acidity, as with *R. meliloti* in acid soils, or may never have been present. The purpose of seed or soil inoculation is to establish an adequate population of effective *Rhizobium* strains for colonization and infection of the legume roots. Com-

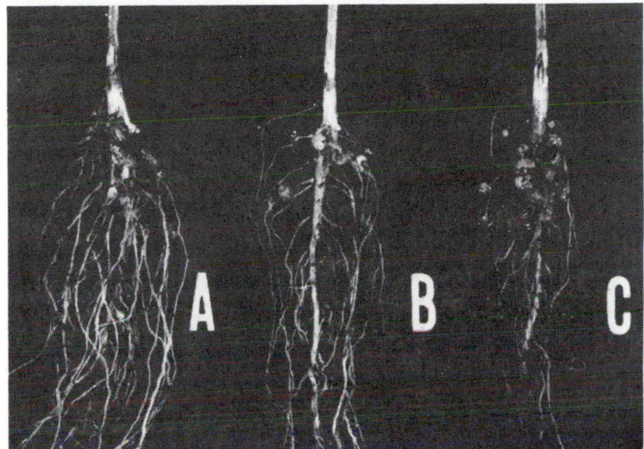

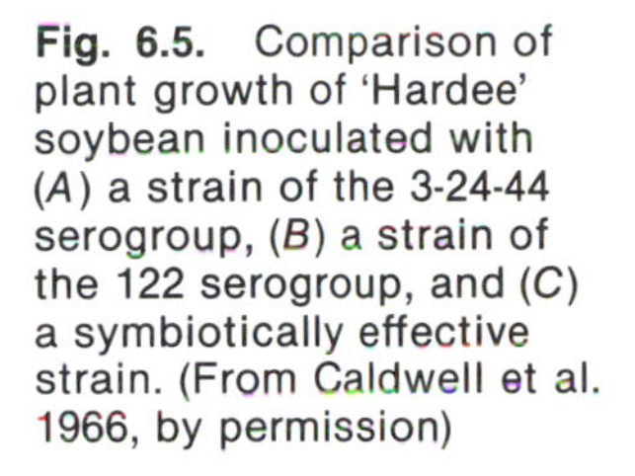

Fig. 6.5. Comparison of plant growth of 'Hardee' soybean inoculated with (*A*) a strain of the 3-24-44 serogroup, (*B*) a strain of the 122 serogroup, and (*C*) a symbiotically effective strain. (From Caldwell et al. 1966, by permission)

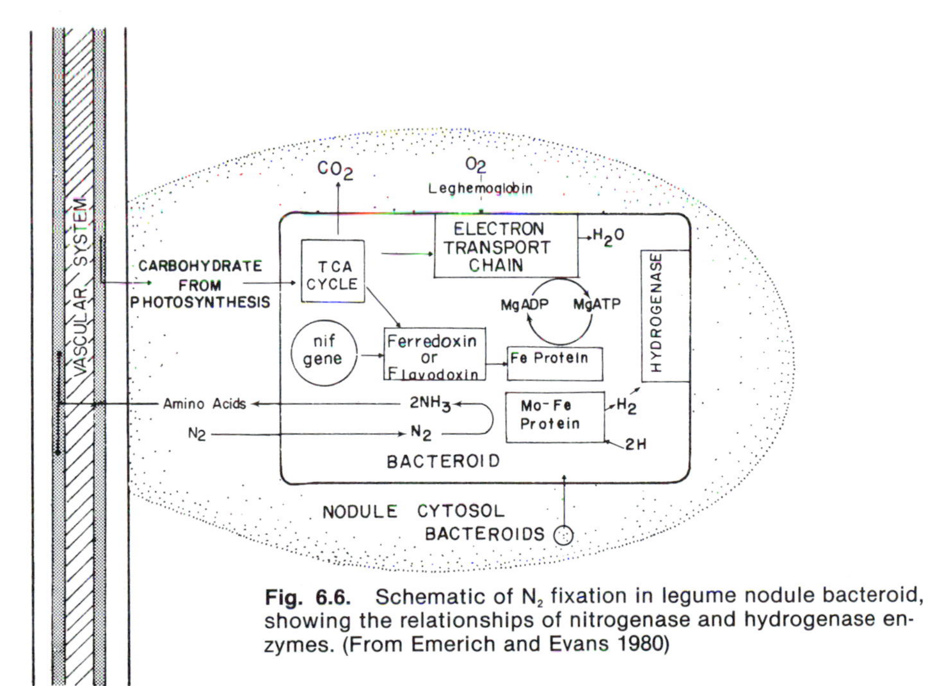

Fig. 6.6. Schematic of N_2 fixation in legume nodule bacteroid, showing the relationships of nitrogenase and hydrogenase enzymes. (From Emerich and Evans 1980)

mercial inoculants presumably contain selected strains of viable organisms. If, however, a large indigenous population of the specific *Rhizobium* is already present, it may swamp introduced (alien) or improved strains. Hence success in establishing a population of improved strains is uncertain, at best. Results sometimes are improved by applying massive dosages of the inoculant, such as using a row application of granules containing the inoculant. However, a 25-fold increase of inoculant on soybean seeds resulted only in a 5% increase in the new strain in soybean nodules (Johnson et al. 1965).

The basis for differential nodulation by strains is complex, related to genotypes of the host and bacteria and to the conditions of the environment. Failure of infection may be due to lack of (1) colonization of roots, (2) invasion of the root hairs, or (3) nodule formation. A strain that effectively nodulates a legume increases its population and becomes the dominant strain in subsequent associations.

The need for inoculation is indicated if the *Rhizobium* species is absent, sparse, or ineffective. In such situations, inoculation can establish populations of effective strains that produce well-nodulated legume plants. For example, in acid soils liming and inoculation of alfalfa or sweet clover may be highly beneficial, as was demonstrated at Rothamsted, England, where the liming and inoculation of an acid soil (pH 4.8–5.7) increased forage yield of *Medicago lupulina* 7- to 8-fold (Nutman 1962). The effects of inoculation were particularly pronounced the first year of the legume crop, since there had been an insufficient indigenous population of the bacteria to colonize roots. Effect of inoculation generally is less evident with time, as the bacterial population increases.

Improved methods of inoculation have emerged over the years, from the practice of transferring soil from old stands to new seedings to the current commercial preparations of improved *Rhizobium* strains, which are broth cultured and preserved for use in a finely ground peat. This peat culture can be stored for several months under cool conditions. It can be applied effectively as a slurry to the seeds or as granules at seeding but should not be applied directly to seeds that have been treated with fungicides. A sticking agent such as sugar may be added to the slurry. The addition of lime ($CaCO_3$, $MgCO_3$) to the pellets may improve early survival, especially of high-lime *Rhizobium* species (e.g., *R. meliloti*). *Preinoculation* (the impregnation of the seeds with inocula during processing) is usually inferior to the slurry method. Survival of organisms on preinoculated seeds is less than with the peat slurry method because of more exposure to heat and dessication; possible toxic factors in the seed coat (e.g., phenols); pesticides, especially heavy metals; and fertilizers (salts).

SURVIVAL OF RHIZOBIA IN SOIL. Rhizobial survival in nature depends largely on soil conditions, especially pH, moisture, organic matter, and length of the interval between host crops. Coarse sands dry easily and may lose indigenous rhizobia, and acid soil may lose rhizobia requiring a high pH.

Elkins et al. (1976) observed that southern Illinois soils had sufficient *R.*

japonicum populations for soybean nodulation after at least 10 yr of cropping corn without an intervening soybean crop. Bacterial suspensions were prepared from soils with different cropping histories of 0 to 11 yr between soybean crops, and these suspensions were used to inoculate seedlings in pots of sterilized sand. Generally, no differences in soybean growth, nodule mass, or nitrogenase activity could be related to cropping history. The same serogroups (strains), 125/126 and 123, predominated regardless of inoculation treatment, including a suspension from fields that had never been in soybeans. Rhizobia live as heterotrophs in the absence of host crops. Survival in Illinois soils might be higher than in some areas because of generally more favorable soil and climatic conditions.

Caldwell and Vest (1970) tested 28 strains and two commercial *R. japonicum* preparations on five soybean cultivars. Significant yield differences were found with three soils free of rhizobia. In soils containing *R. japonicum,* there were no differences in seed yield due to inoculation, and only 5 to 10% of the nodules were derived from the seed-applied inoculum. None of the 28 strains was able to compete with the established soil rhizobial strains. No advantage was gained from using each inoculum at 25 times the recommended rate or from increasing its adherence to the seed with gum arabic.

RHIZOSPHERE FIXATION IN GRASSES

The world food deficit and increasing N requirements for food and forage production has created a keen interest in the possibility of N_2 fixation in grass plants. The production of cereals, which supply 75% or more of the calories and protein consumed by humans, is dependent on high N levels, principally from commercial fertilizers in developed nations. However, N fertilizers are becoming increasingly costly and are dependent on finite sources of fossil energy. Also, commercial fertilizers are often not available in developing nations. The long-range goal of the scientific community is to transfer the capability of N_2 fixation (*nif* gene) to cereal and forage grasses. But many complex obstacles will have to be overcome before this can be accomplished.

The discovery that certain species of Azotobacteraceae colonize and form loose associations with roots of grass plants (Fig. 6.7) and fix N_2 in the *rhizosphere* (root environment) under favorable conditions (Dobereiner and Day 1976; Neyra and Dobereiner 1977) is a step in the progress toward this goal. *Azospirillum brasilense, Azotobacter paspali, Beijerinckia* spp., and *Spirillum lipoferum* have been isolated from rhizosphere associations with tropical (C_4) grasses. In a good light environment the C_4 grasses supposedly (as a result of a higher photosynthetic rate than that of C_3 grasses) excrete more C compounds into the rhizosphere to support the energy requirement of N_2 fixation.

The first grass–rhizosphere bacterium association studied in detail was that of *Paspalum notatum* (bahiagrass)–*A. paspali* (Neyra and Dobereiner 1977). Of 33 bahiagrass cultivars or ecotypes only five, all tetraploids, formed rhizosphere associations with *A. paspali.* Several months were required to establish large populations of the bacteria; this time was not affected by inocu-

lation. Plants transferred from the field to vermiculite pots supplied with a N-free nutrient solution fixed 80 mg N per pot in 2 mo, enough to support normal growth. Mycorrhiza were found to enhance establishment of the association. The bacteria were found concentrated in the mucagel layer of the root. Nitrogen fixation rates of about 0.1 to 0.3 kg • ha^{-1} day^{-1} were estimated by the C_2H_2-reduction and the ^{15}N-dilution methods. These rates suggested N_2 fixation of as much as 50 to 75 kg • ha^{-1}, enough to support moderate production of bahiagrass. These rates reported by Neyra and Dobereiner use preincubated excised grass roots. However, wheat (C_3) cores assayed in Oregon showed low activity, 2g • ha^{-1} • day^{-1} (Neyra and Dobereiner 1977), or no more than 1 kg • ha^{-1} for the season. It should be noted that these researchers employed preincubation and C_2H_2 reduction of excised roots, a procedure known to give much higher estimates of N_2 fixation from the total N in the biomass.

Hybrids and inbreds of pearl millet (C_4) were assayed for N_2 fixation after inoculation with the *Azospirillum brasilense* Sp 13T strain (Bouton et al. 1979). Dry matter yield and total plant N for the pearl millet hybrid 'Gahi 3' were improved by more than 30% by inoculation, but the increased N_2 fixation was not observed by C_2H_2 reduction. A N-balance study in the greenhouse showed no significant gain in N from inoculation with the Sp 13T strain.

Albrecht et al. (1981b) observed *A. brasilense* nitrogenase activity on isolated maize roots as assayed by acetylene reduction. In 50% of their field experiments inoculation increased plant weight and N content. The N gain from inoculation with *A. brasilense* was estimated at 15 kg N • ha^{-1}, but the benefits were not consistent in terms of either plant yield or N content.

Digitgrass (*Digitaria* spp.), a C_4 on a low-N soil in Australia, produced 23% more dry matter and more total N per plant when inoculated with *A. brasilense.* On a high-N soil, production was increased by 8.5% (Schank et al. 1981).

In Israel two C_4s, sweet corn (maize) and millet (*Setaria italica*), showed a marked response to inoculation with improved *A. brasilense* strains Sp 7, Sp 80, and CO where the soil N content was low (Cohen et al. 1980) (Table 6.4).

TABLE 6.4. Effect of inoculation of *Seteria italica* with *Azospirillum brasilense* SP 7 on dry matter yield and N content

	Dry Weight (%)		N (%)	
Soil	Inoc	Control	Inoc	Control
Sand	180	100	250	100
Sandy loam	175	100	195	100
Loess	125	100	140	100

Source: From Cohen et al. 1980.

Fig. 6.7. Scanning electron micrograph of *Azospirillum brasilense* colonizing a young sorghum root. The bacteria are intermingled with sloughed epidermis cells (*foreground*) and with intact epidermis (*background*). (Courtesy of H. Berg and S. Schank)

Percentage of marketable ears was increased by inoculation. Root branching was increased by inoculation, which could have been due to hormone production by *Azospirillum* rather than to N_2 fixation. Root branching during the flowering stage may have been particularly advantageous for nutrient uptake to support fruiting. The effect appears analogous to that produced by mycorrhiza. The magnitude of response in the Israeli experiments is impressive and has not been matched elsewhere, perhaps due to exceptional conditions in Israel (e.g., low-N, high-lime soils; high temperatures; high radiation).

The theory of hormone production by *A. brasilense* suggested above has been confirmed by Tien et al. (1979). In 2 weeks the IAA content of the culture solution quadrupled, cytokinins increased, and some gibberellin (a third growth regulator, see Chap. 7) was found. Roots of pearl millet showed a striking growth response to root solution without the *A. brasilense* bacteria. Lateral roots became densely covered with root hairs and had more branching.

The observations that *A. brasilense* and probably other rhizosphere associations produce growth hormones and resultant root proliferation leaves unanswered the question of whether the benefits from the association are from N_2 fixation as such or from increased root proliferation and resultant nutrient uptake. Some recent work seems to indicate that hormone effects may be the primary stimulus.

Longtime observations and experience with tropical-grass pastures lead many to believe that rhizosphere associations are contributing to the N balance in these pastures. However, more research will be required before good quantification is available and especially before inoculation of cereal crops produces sufficiently consistent results for it to become a standard crop production practice.

GENETICALLY CONTROLLED FACTORS

With any species or association, the reduction of N_2 to ammonia (N_2 fixation) is regulated by the enzyme complex nitrogenase. Nitrogenase production is controlled by the *nif* gene, positioned on the chromosome adjacent to the *his* gene, which controls histidine synthesis (Dixon and Postgate 1972). Pure nitrogenase extract does not fix N_2, so it can be assumed that many more genes of the organism and of the host plant are involved in the total process.

A gene, *gln,* controlling glutamine synthase, which allows direct utilization of NH_4 rather than glutamate, was identified by Shanmugan et al. (1978). Ammonia normally inhibits (represses) *nif* expression. Some mutant strains of *Rhizobium* with derepression capabilities have been identified, that is, strains that can express *nif* despite the presence of NH_3.

The fact that the *nif* gene has been successfully transferred from a N_2-fixing bacterium, *Klebsiella pneumoniae,* to *Escherichia coli,* enabling *E. coli* to fix N_2 also, is confirmation of its existence. Strains have been identified that are ineffective in N_2 fixation (Brill 1974). Ineffective strains of *R. trifolii* on red clover were found to be the result of two genes (Nutman 1968). Effectiveness of red clover varieties as host plants resulted from four genes. A nonnodulating line of soybean was differentiated by a single gene (Caldwell et al. 1966) (Fig. 6.5).

A large number of essential factors under genetic control were found to be common in N_2-fixing organisms (Ljones 1974):

1. Nitrogenase enzyme complex. The enzyme complex consists of two proteins: an Fe protein, the smallest, with a molecular weight of 50,000–70,000; and an Mo-Fe protein, the largest, with a molecular weight of 200,000–220,000 (Winter and Burris 1976). Both proteins contain sulfur amino acids and the Mo-Fe protein (Fig. 6.6).

It is believed that N_2 first is bound to the Mo-Fe protein in the reduction to ammonia and that MgATP binds with the Fe protein. Nitrogenase has been isolated in pure form (Carnaham et al. 1960). The enzyme complex does not fix N_2 in vitro unless several specialized requirements are satisfied. Both components of nitrogenase are inactivated by O_2; therefore the level of O_2 near nitrogenase must be strictly regulated (Albrecht and Gaskins 1982).

2. Reductants. The conversion of N_2 to N_4^+ is a reduction process requiring an input of electrons. It is believed that the electron donors, pyridine nucleotides (ATP and NADPH), are reduced via ferredoxin or flavodoxin (Benemann and Valentine 1972). Pyruvate was utilized in large quantities by *C. pasteurianum* to generate electrons for the reduction of ferredoxin, donating $2e^-$ per molecule of pyruvate (Mortenson 1964). In cell-free systems, $Na_2S_2O_4$ has been used to donate electrons directly to the system.

3. Leghemoglobin (in legume nodules). Hemoglobin is present in nodular tissue of legumes but not in other N_2-fixing systems. Studied extensively by Virtanen et al. (1947), this pigment has been found in some fungi and yeasts but not in higher plants. The leghemoglobin in nodules was once believed to be in the nodule tissue rather than in the bacteroids, but recent evidence indicates it is in the bacteroid envelope derived from the host cells (Bergersen 1971). Host cells containing bacteroids are larger than uninfected cells (Fig. 6.2). Nitrogenase activity and N fixation in legumes is highly correlated with the leghemoglobin content (pink to red pigmentation). Yellow to brown coloration denotes senescence or dysfunction of the bacteroids, usually because of adverse conditions. White or green nodule mass generally indicates ineffective nitrogenase activity. The physiological significance of the leghemoglobin pigment is not fully understood but is probably required in O_2 transport to support nodule respiration and ATP production. It follows that visual sampling of the number and size of nodules (nodule mass) and color can indicate the status of N_2 fixation in a legume crop.

4. ATP. This pyridine nucleotide is essential since other compounds cannot substitute for it, but the actual cofactor is MgATP. Usually 20 to 30 mol

ATP are considered necessary to convert 1 mol N_2 to NH_4^+ and then to glutamic acid (Fig. 6.6). Other amino acids are subsequently produced from glutamic acid by transamination. Overall, six electrons are required for conversion of one molecule of N_2 to $2NH_3$. A high photosynthetic rate or other C source is mandatory for substrate oxidation and ATP from respiration.

5. Protection from O_2. While nodule formation on legumes and most N_2-fixing organisms require O_2, it was inhibitory to nitrogenase activity (Bond 1951), apparently blocking the binding sites for N_2 and MgATP on the Mo-Fe protein and the Fe protein, respectively (Albrecht and Gaskins 1982). In some organisms, N_2 fixation is greatest at low O_2 potentials (0.2–0.8 atm), but the presence of O_2 completely inhibits *nif* expression in *C. pasteurianum*. Other compounds besides O_2, such as H_2, CO, and NO, may also inhibit nitrogenase activity (Ljones 1974). Oxygen levels even near ambient inhibit nitrogenase activity. It is believed that the envelope of the bacteroids serves to exclude O_2 (Fig. 6.2).

ENVIRONMENTAL FACTORS

The following environmental factors influence N_2 fixation:

1. Carbon-N ratio. High amounts of N in relation to C in the soil or other media represses *nif* expression, thus reducing nodulation and/or nitrogenase activity (Fig. 6.8). Nodulation and N_2 fixation in legumes are inhibited by available N (Fred et al. 1932). While NH_3 greatly decreases N_2 fixation in legumes, it is completely inhibitory to fixation in the free-living organisms (Brill 1980).

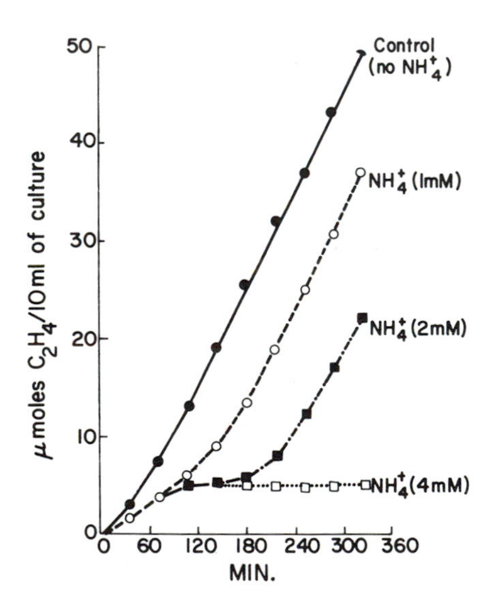

Fig. 6.8. Nitrogenase activity indicated by C_2H_2 reduction in time with different concentrations of NH_4^+ as ammonium acetate in the medium. (From Quispel 1974, by permission)

The repressive effect of available N depends to an appreciable extent on the N_2-fixing species and the environment. Moderate amounts of available N have been reported to benefit the establishment of some legumes. However, the practical benefits from N fertilization of legumes are usually small to nil. For example, Hinson (1975) obtained significantly more tops and roots from N-fertilized soybean in a pot experiment, but the nodule number and mass were reduced. In field experiments, vegetative growth was greater on N-fertilized plots, but there was no effect on seed yield. However, improved seed yields from added N have been reported on cowpea and garden pea (Mahon and Child 1979; Minchin et al. 1981). In general, garden bean (*Phaseolus vulgaris*), unlike most legumes, responds to N fertilization if the soil N is low.

2. Mineral nutrients. Mineral requirements of N_2-fixing organisms are essentially the same as those of other plants. Particular attention should be given to soils deficient in molybdenum, iron, or sulfur, since these are mineral components of nitrogenase. Nitrogenase activity is also responsive to other macroelements. Lynd et al. (1981) found that nodule growth, nitrogenase activity, and the support enzymes of nitrogenase (e.g., glutamate synthetase, or GOGAT) in hairy vetch were significantly increased by potassium fertilization (Table 6.5). Legume growth and N_2 fixation is also responsive to phosphorus. Copper was observed to be necessary for nodulation, probably because of its role in the cytochrome system and oxidative respiration (Cartwright and Hallsworth 1970). Generally *Rhizobium*-legume associations are sensitive to low pHs, particularly legumes adapted to temperate climates.

3. Pesticides. Certain pesticides, especially mercury fungicides for seed treatment, can reduce the number of N_2-fixing organisms and decrease nodulation (Vincent 1974).

4. Weather factors. Heat and drought can reduce the population of bacteria and reduce N_2 fixation, as suggested by the wide spread of reported N yields from legumes (Table 6.2). Cold (5°C) reduces N_2 fixation virtually to zero. This effect was primarily due to reduced nodulation rather than to reduced nitrogenase activity (Roughley 1970; Lie 1974). Pea nodulated well at 26°C but not at 20°C. Temperature effects on N_2 fixation varied greatly, de-

TABLE 6.5. Effects of potassium (K) and calcium (Ca) on growth and N_2-fixing of hairy vetch

	Fertilizer[a]			
Parameter Measured	0	K	Ca	K + Ca
Top dry weight (g • plant^{-1})	1.78	1.85	1.91	2.38
Fresh nodule weight (mg • plant^{-1})	552	745	678	926
Nitrogenase activity (μmol C_2H_4 • g^{-1} • hr^{-1})	22.5	32.5	20.9	36.1
Nodule enzyme activity[b] (μg • g^{-1} • nodule^{-1})	3,361	5,154	2,563	5,208

Source: From Lynd et al. 1981.

[a]Soil phosphorus was at a high level.

[b]Enzymes associated with nitrogenase in production of glutamate from the primary substrate α-keto-gluterate.

pending on *Rhizobium*-legume association (Lie 1974). *Rhizobium*–temperate legume associations were effective at temperatures as low as 7°C (Roughley 1970), whereas tropical associations ceased fixation at temperatures lower than 20°C. Temperate legume symbiosis has an optimum temperature of 20 to 25°C; that of tropical legumes is about 35 to 40°C. Most temperate legumes have a Mediterranean origin, evolving in a climate characterized by mild, moist winters and dry summers conducive to cool-season growth.

A soil moisture equivalent to 25 to 75% of field capacity was found to be optimum for soybean and alfalfa symbiosis (Fred et al. 1932). Nodule moisture content must remain above 80% for nodule survival. With soybean, soil moisture near field capacity was optimum at a high greenhouse temperature, but neither soil moisture nor depth of inoculum placement had an effect at moderate temperatures (Wilson 1975). Excess moisture or waterlogging generally reduced N_2 fixation, probably by reducing root respiration and ATP production. Flooding did not reduce nitrogenase activity in *Aeschynomene,* whereas a moisture deficit did (Albrecht et al. 1981a).

5. Calcium (Ca) and pH. The effect of pH on N_2 fixation can be either direct or indirect. Acid soils became devoid of *Rhizobium* (Mulder and Van Veen 1960). Also, the nodules formed in acid soils were often from ineffective strains of bacteria (Holding and Lowe 1971). The pH of the medium also directly affects nodule formation; nodules began to appear on roots in a nutrient solution in 3 to 5 days at a relatively high pH (Lie 1974). The pH-sensitive phase in nodulation may be that of root curling prior to infection (Munns 1969).

Calcium is essential for plant and nodule meristem growth. The Ca requirement for the legume itself is much lower than the requirement for symbiosis, especially in bacteria–temperate legume associations. Without adequate Ca, aberrant growth and aborted nodule meristems occurred (Sorokin and Sommer 1940). *R. meliloti,* microsymbiont for sweet clover and alfalfa, has a high pH requirement.

6. Carbon dioxide. Normally, the atmosphere of N_2-fixing bacteria is much richer in CO_2 (10× –100×) and lower in O_2 than ambient air. Rhizobia in pure culture required the presence of CO_2 for optimal growth of the culture (Lowe and Evans 1962); a CO_2 content of 4% enhanced N_2 fixation (Mulder and Van Veen 1960). Conditions favoring good root growth and respiration probably remove the need for additional CO_2, but rhizobia are known to be capable of fixing some CO_2, probably via pep carboxylase.

Summary

Plant biomass contains about 1% N, and N is a major limiting factor in crop production. Except for the relatively small contribution by lightning fixation, about half the global N requirement comes from biological N_2 fixation and half from commercial fixation by the Haber-Bosch process.

Numerous species of lower plants, including bacteria, actinomycetes, and blue-green algae (Cyanobacteria), have the capacity of biologically fixing N_2 either as free-living organisms or as a component of plant associations. Most important agriculturally are the *Rhizobium*-legume associations, which can fix 100 kg • ha^{-1} N per season. Free-living bacteria (e.g., *Azotobacter* and *Clostridium*) and blue-green algae generally fix approximately one-fourth this amount; however, they are highly important in maintaining the N balance in natural ecosystems (forests and grasslands). Some species of bacteria of the family Azotobacteraceae (e.g., *Azospirillum*) colonize and form loose rhizosphere associations capable of N_2 fixation, particularly with C_4 grasses, which apparently excrete more photosynthate to the rhizosphere than do C_3 grasses and thus support the process. To date, results of experiments on the N_2-fixing capacity of these bacteria and the value of inoculating grass plants have been highly variable and often disappointing.

Regardless of the system, biological N_2 fixation is mediated by the enzyme complex nitrogenase, which consists of an Fe protein and an Mo-Fe protein. Production of nitrogenase is governed by the *nif* gene. Requirements for a high nitrogenase activity include (1) an O_2-free environment; (2) low levels of available substrate N, such as ammonia; and (3) high levels of available C to energize the system and protect nitrogenase against O_2 inactivation. Legumes maintain a continuous supply of C by photosynthesis, but free-living heterotrophic organisms have no photosynthetic source. Carbon supply is probably the most limiting factor to N_2 fixation with free-living bacteria. *Rhizobium*–temperate legume associations require lower cardinal temperatures and a higher pH for N_2-fixing activity than do tropical legume (e.g., peanut, cowpea, and pigeonpea) associations.

Blue-green algae (especially *Nostoc* and *Anabaena*) contribute to the N balance in wet environments. The blue-green algae–fern (*Azolla*) association is of current interest for rice culture. Recent evidence suggests that the contribution of blue-green algae to the N balance in productive rice paddies may be relatively small, primarily due to shading by the rice canopy.

Legumes produce nodules that vary in size and shape, depending on the type and duration of nodule meristematic activity characteristic of the species. A *Rhizobium* species generally can colonize and infect a number of legume species (cross-inoculation), but certain *Rhizobium* species are specific (e.g., *R. japonicum* to soybean). Strains of *Rhizobium* vary widely in effectiveness, ranging from no fixation to effective fixation.

Rhizobia can persist heterotrophically in soil for many years without an intervening host-legume crop. Populations of indigenous strains nodulate new seedings of legumes after causing inoculation with additional new strains to be in vain. Soil levels of Ca, phosphorus, and potassium aid in *Rhizobium* survival, nitrogenase activity, and enzyme activity associated with nitrogenase.

Inoculation can be beneficial to establishment and N_2 fixation of new seedings of legumes in areas where a legume of the cross-inoculation group has not been grown previously or where *Rhizobium* populations may have been severely reduced by adverse soil conditions.

References

Albrecht, S. L., and M. H. Gaskins. 1982. Univ. Florida–USDA. Unpublished report.
Albrecht, S. L., J. M. Bennett, and K. H. Quesenberry. 1981a. Plant Soil 60:309–15.
Albrecht, S. L., Y. Okon, J. Lonnquist, and R. H. Burris. 1981b. Crop Sci. 21:301–6.
Alexander, M. 1961. Introduction to Soil Microbiology. New York: Wiley.
Bell, F., and P. S. Nutman. 1971. Plant Soil Spec. Vol., pp. 231–34.
Benemann, J. R., and R. C. Valentine. 1972. Adv. Microbiol. Physiol. 8:59–104.
Bergersen, F. J. 1971. Annu. Rev. Plant Physiol. 22:121–40.
Bond, G. 1951. Ann. Bot. n.s. 15:95–108.
______. 1974. In The Biology of Nitrogen Fixation, ed. A. Quispel. Amsterdam, Oxford: North-Holland.
Bouton, J. H., R. L. Smith, S. C. Schank, G. W. Burton, M. E. Tyler, R. C. Littell, R. N. Gallaher, and K. H. Quesenberry. 1979. Crop Sci. 19:12–16.
Brill, W. J. 1974. In The Biology of Nitrogen Fixation, ed. A. Quispel. Amsterdam, Oxford: North-Holland.
______. 1980. In The Biology of Crop Production, ed. P. S. Carlson. New York: Academic Press.
Buchanan, R. E., and N. E. Gibbons. 1974. Bergey's Manual of Determinate Bacteriology. 8th ed. Baltimore: Williams and Wilkins.
Buresh, R. J., M. E. Casselman, and W. H. Patrick, Jr. 1980. Adv. Agron. 33:149–92.
Caldwell, B. E., and G. Vest. 1968. Crop Sci. 8:680.
Caldwell, B. E., K. Hinson, and H. W. Johnson. 1966. Crop Sci. 6:495–96.
Carnaham, J. H., L. E. Mortenson, N. F. Mower, and J. E. Castle. 1960. Biochim. Biophys. Acta 39:188–89.
Cartwright, B., and E. G. Hallsworth. 1970. Plant Soil 33:685–98.
Cohen, E., Y. Okon, J. Kigel, I. Nur, and Y. Henis. 1980. Plant Physiol. 66:746–49.
Dixon, R. A., and J. R. Postgate. 1972. Nature 237:102–3.
Dobereiner, J., and J. M. Day. 1976. In Proc. Int. Symp. Nitrogen Fixation I, ed. W. E. Newton and C. J. Nyman. Pullman: Washington State University Press.
Elkins, D. M., G. Hamilton, C. K. Y. Chan, M. A. Briskovich, and J. W. Vandeventer. 1976. Agron. J. 68:513–17.
Emerich, D. W., and H. J. Evans. 1980. In Biochemical and Photosynthetic Aspects of Energy Production, ed. A. San Pietro. New York: Academic Press.
Fred, E. B., I. L. Baldwyn, and E. MacCoy. 1932. Root Nodule Bacteria and Leguminous Plants. Madison: University of Wisconsin Press.
Gibson, A. H. 1977. CSIRO Div. Plant Ind. Annu. Rep., pp. 33–39.
Henzell, E. F. 1968. Trop. Grassl. 2:1–17.
Hinson, K. 1975. Agron. J. 67:799–804.
Holding, A. J., and J. F. Lowe. 1971. Plant Soil Spec. Vol., pp. 153–66.
Huang, C. 1978. Bot. Bull. Acad. Sin. 19:41–52.
Johnson, H. W., U. M. Means, and C. R. Weber. 1965. Agron. J. 57:179–85.
Jones, K. 1974. J. Ecol. 62:553–65.
Lie, T. A. 1971. Plant Soil 34:663–73.
______. 1974. In The Biology of Nitrogen Fixation, ed. A. Quispel. Amsterdam, Oxford: North-Holland.
Lowe, R. H., and H. J. Evans. 1962. Soil Sci. 94:351.
Ljones, T. 1974. In The Biology of Nitrogen Fixation, ed. A. Quispel. Amsterdam, Oxford: North-Holland.
Lynd, J. Q., E. A. Hanlon, Jr., and G. V. Odell, Jr. 1981. Soil Sci. Soc. Am. J. 45:302–6.

Mahon, J. D., and J. J. Child. 1979. Can. J. Bot. 57:1687–93.
Minchin, F. R., R. J. Summerfield, and M. C. P. Neves. 1981. Trop. Agric. [Trinidad] 58:1.
Mortenson, L. E. 1964. Proc. Natl. Acad. Sci. [U.S.] 52:272–79.
Mulder, E. G., and S. Brotonegoro. 1974. In The Biology of Nitrogen Fixation, ed. A. Quispel. Amsterdam, Oxford: North-Holland.
Mulder, E. G., and W. L. Van Veen. 1960. Plant Soil 13:91–113.
Munns, D. N. 1969. Plant Soil 30:117–19.
Newcomb, D., R. L. Peterson, D. Cullaham, and J. G. Torrey. 1978. Can. J. Bot. 56:502–31.
Neyra, C. A., and J. Dobereiner. 1977. Adv. Agron. 29:1–38.
Nutman, P. S. 1954. Heredity 8:35–46.
______. 1962. Soil Microbiol. Dep., Rothamsted Exp. Stn., Annu. Rep., pp. 79–80.
______. 1965. In Ecology of Soil-borne Plant Pathogens, ed. K. F. Baker and W. C. Snyder. Berkeley and Los Angeles: University of California Press.
______. 1968. Heredity 23:537–51.
Quispel, A., ed. 1974. The Biology of Nitrogen Fixation. Amsterdam, Oxford: North-Holland.
Rice, E. L. 1980. Bot. Bull. Acad. Sin. 21:111–17.
Roughley, R. J. 1970. Ann. Bot. n.s. 34:631–46.
Ruinen, J. 1956. Nature 177:220.
______. 1974. In The Biology of Nitrogen Fixation, ed. A. Quispel. Amsterdam, Oxford: North-Holland.
Russell, E. W. 1950. Soil Conditions and Plant Growth. London: Longmans, Green.
Schank, S. C., K. L. Wier, and I. C. McRae. 1981. Appl. Environ. Microbiol. 41:342–45.
Schubert, K. R., and J. H. Evans. 1976. Proc. Natl. Acad. Sci. [U.S.] 73:1207–11.
Shanmugan, K. T., R. O'Gara, K. Andersen, and R. C. Valentine. 1978. Annu. Rev. Plant Physiol. 29:263–76.
Sorokin, H., and A. L. Sommer. 1940. Am. J. Bot. 27:308–18.
Stewart, W. D. P. 1974. In The Biology of Nitrogen Fixation, ed. A. Quispel. Amsterdam, Oxford: North-Holland.
Sundara Rao, W. V. B. 1971. Plant Soil Spec. Vol., pp.287–91.
Tien, T. M., H. M. Gaskins, and D. H. Hubbell. 1979. Appl. Environ. Microbiol. 37:1016–24.
Trinick, M. J. 1976. In Proc. Int. Symp. Nitrogen Fixation, ed. W. E. Newton and C. J. Nyman. Pullman: Washington State University Press.
Tu, J. C. 1974. J. Bacteriol. 119:986–91.
Vincent, J. M. 1974. In The Biology of Nitrogen Fixation, ed. A. Quispel. Amsterdam, Oxford: North-Holland.
Virtanen, A. I., J. Jorma, H. Linkola, and A. Linnasalmi. 1947. Acta Chem. Scand. 1:90–111.
Waksman, S. A. 1952. Soil Microbiology. London: Chapman and Hall.
Wetselaar, R. 1967. Aust. J. Exp. Agric. Anim. Husb. 7:518–22.
Whitney, A. S. 1967. Agron. J. 59:585.
Williams, C. H. 1970. J. Aust. Inst. Agric. Sci. 36:199–205.
Wilson, D. O. 1975. Agron. J. 67:76–78.
Winter, H. C., and R. H. Burris. 1976. Annu. Rev. Microbiol. 110:207–13.
Yoshida, T. 1981. Unpublished seminar paper, University of Florida, Gainesville.

7 Plant Growth Regulation

PLANT GROWTH and development are controlled by extremely low concentrations of chemical substances called *plant growth substances, growth hormones, phytohormones,* or *plant growth regulators* (PGRs). The concept that plant growth and development are regulated by a substance produced in minute quantities in one organ that elicits a response in another was first suggested by Julius von Sachs, the father of plant physiology, in the latter half of the nineteenth century. His observations were confirmed by Charles Darwin in 1880 in his experiments on the effect of light and gravity on plant growth. Darwin observed that canarygrass seedlings bent toward the light source (phototropism) unless the seedling tips were covered with an opaque foil. He concluded that the light stimulus was perceived in the shoot tip (coleoptile), but the response was in lower or more basal tissues.

The full impact of PGR manipulation on modern agriculture began with the use of auxin-type herbicides at the close of World War II. Presently PGRs are used to control a host of physiological processes in crop production, including flowering and fruiting (fruit set and parthenocarpy), partitioning of assimilate, germination, propagation, growth suppression, defoliation, and postharvest ripening. Cloning and tissue culture would not be possible without PGRs. Most of the commercial acreage of tobacco in the United States is treated with a PGR to control suckering (growth of new shoots from buds in leaf axils).

PGRs are used as herbicides on nearly all cultivated crop acreage in industrialized countries, and their production is a multibillion-dollar industry. They are used extensively in horticultural crops to control growth and development, particularly in fruit production. Field crop plants have a relatively short breeding cycle, and it has been possible to obtain genetic control by breeding and selection for endogenous levels of hormones that produce the desired physiological responses. Tobacco is an exception (as are barley and wheat, on which a PGR is used in Europe to control tiller growth). As more effective PGRs are produced, their activities better understood, and method of delivery of needed concentrations to the response organs over time developed, the use of PGRs in field crop production may increase.

Terminology and Classification

The term *plant growth regulator* covers the broad category of organic substances (other than vitamins and microelements) that in minute amounts promote, inhibit, or otherwise modify physiological processes (Wareing and Phillips 1978). *Endogenous* (produced internally) PGRs are referred to as *plant hormones* or *phytohormones.* The term *hormone* originated in animal physiology, where it means a substance synthesized in one organ that in turn stimulates a response in another. Plant hormones are not as specific as to organ of synthesis or organ of response as are animal hormones, but they tend to follow this general pattern of behavior.

PGRs, whether endogenous or *exogenous* (originating externally), elicit essentially the same plant responses. For example, two synthetic PGRs—2,4-dichlorophenoxyacetic acid (2,4-D) and picloram, a substituted picolinic acid (Tordon)—are equally effective with in vitro tissue cultures. A synthetic auxin is necessary because the tissue is detached from the natural auxin source (Collins et al. 1978). Synthetic growth promoters, in an appropriate mix, stimulate *callus* (formation of undifferentiated cell mass), organ differentiation, and whole plant morphogenesis from a single parenchyma cell of, for example, tobacco pith, carrot root, or potato leaf.

PGRs are currently divided into five classes: auxins, gibberellins, cytokinins or kinins, growth inhibitors, and ethylene. Two hormones (brassinalide, a steroid, and triacontanol, an alcohol, the latter reportedly producing striking growth stimulation), chemically do not fit in the above five categories. Both substances have been recently isolated from rape seed (*Brassica napus*) and certain higher plants, respectively (Thomas 1976). These substances and others that will likely be discovered may require revision of the current classification. Numerous analogs of most hormones in the five classes have been produced synthetically, and many have important agricultural applications.

Certain properties are necessary for a compound to be considered a phytohormone: (1) the site of synthesis differs from the site of activity (e.g., synthesis is in buds and young leaves but response is in stems, roots, or other organs); (2) responses are produced by microquantities (i.e., concentrations as low as 10^{-9} M; (3) unlike vitamins and enzymes, responses may be formative and *plastic* (irreversible) (e.g., tropic responses).

Often the natural supply of phytohormone is suboptimal, and an exogenous source is required to produce a desired response. Supraoptimal amounts of auxins typically behave as herbicides. Generally a phytohormone acts synergistically with other hormones in eliciting responses.

Auxins

Auxin is the generic term for growth substances that typically stimulate cell elongation, but auxins also cause a wide range of growth responses (Table

TABLE 7.1. Activities of phytohormones in plant growth and development

Plant Process	Auxin	Gibberellin	Cytokinin	Growth Inhibitor (Abscisic Acid)	Ethylene
Cell division		X	X		
Cell wall loosening	X				
Cell enlargement	X	X	X		
Root initiation	X		X		
Callus formation	X		X		
Xylem formation	X		X		
Increased respiration and K^+ uptake					
RNA and protein synthesis	X	X	X		
Stem elongation	X	X			
Lateral bud growth	X		X	X	X
Release of α-amylase		X			X
Dormancy		X	X	X	X
Juvenility	X	X			
Growth rate	X	X	X	X	
Flower initiation	X	X	X	X	X
Sex determination	X	X	X		X
Fruit set	X	X	X		X
Fruit growth	X	X	X		X
Fruit ripening	X	X	X		X
Tuberization	X	X	X	X	X
Abscission	X	X	X	X	X
Rooting	X	X	X		X
Senescence	X	X	X	X	X
Seed germination		X	X		X

Source: Leopold and Kriedemann 1975.

7.1). A number of natural substances exhibit auxin activity, but the dominant one, the first isolated and identified, is indoleacetic acid (IAA).

About 50 years ago, Paal and Boysen-Jensen demonstrated that the growth stimulus was in fact produced in the coleoptile tip and was translocated down to the area of bending, as theorized by Darwin (Wareing and Phillips 1978). Paal and Boysen-Jensen observed that if the tip was removed and placed on one side of the decapitated coleoptile, growth and curvature were induced directly below that side. Further, the stimulus could be translocated through a layer of agar gel inserted between the tip and the stump (the zone of bending) but not through a layer of mica.

A major advancement in knowledge of plant growth regulation came with the research findings of F. W. Went in the 1920s, working in Utrecht, Holland (and more recently in the United States) (Went and Thimann 1937). He extracted into agar gel the active substance from coleoptile tips and, by placing a small block of the agar gel containing the extract on one side of a decapitated coleoptile stump, obtained curvature proportional to the extract concentration in the agar. This led to the development of the first quantitative assay for auxins, the *Avena* curvature test. An effective assay quite naturally gave impetus to auxin research. IAA was isolated in pure state and identified by Kogl, Haagen-Smith, and Erxleben in 1931. The agricultural potential of IAA was soon explored but, due to the relative instability of IAA, practical

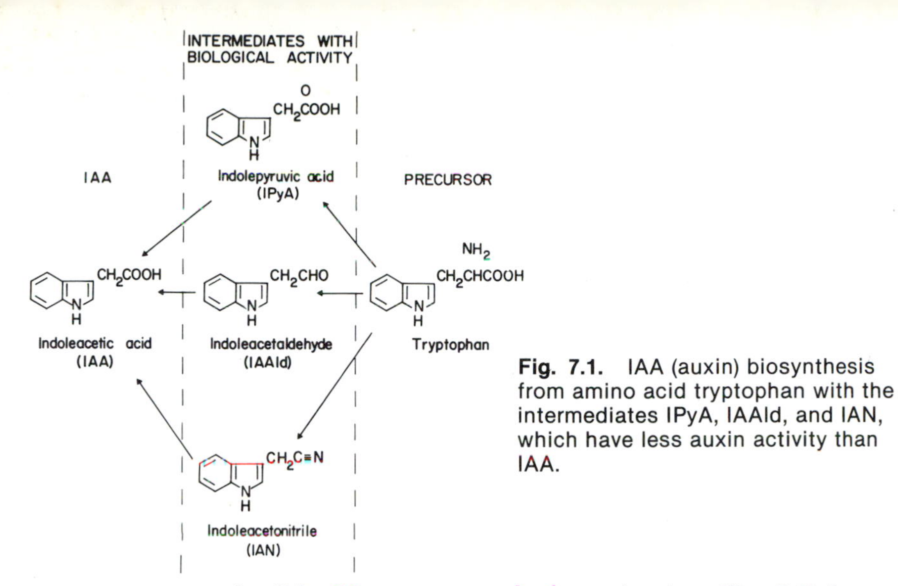

Fig. 7.1. IAA (auxin) biosynthesis from amino acid tryptophan with the intermediates IPyA, IAAld, and IAN, which have less auxin activity than IAA.

uses were not feasible. Numerous synthetic auxins (see Fig. 7.2) have since been developed and are used extensively in agriculture.

NATURAL AND SYNTHETIC AUXINS

While IAA is recognized as the principal auxin in plants, a number of natural auxinlike substances (analogs) are converted to IAA (Fig. 7.1). Indoleacetonitrile (IAN), indolepyruvic acid (IPyA), and indoleacetaldehyde (IAAld) are intermediates in the synthesis of IAA from the amino acid precursor tryptophan (Fig. 7.1). IAN was the first hormone extracted from leaves and stems of higher plants (Cruciferae) (Jones et al. 1952).

IAA usually does not exist naturally in the free state; rather it is conjugated with ascorbic acid, sugars, amino acids, and other organic compounds (bound forms). Bound forms are readily converted to free IAA by enzymatic hydrolysis.

The phenoxyacetic, naphthaleneacetic, picolinic, and benzoic acids and dinitrophenols are synthetic auxins with important agricultural uses, especially as herbicides (Fig. 7.2). The herbicide 2,4-D was developed in the United States

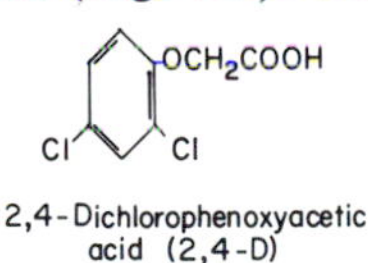

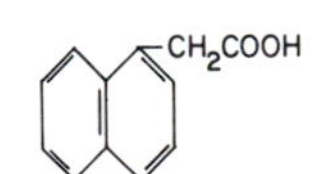

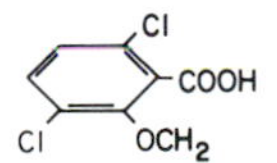

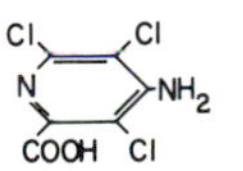

4-amino-3,5,6-
trichloropicolinic
acid (picloram)
(Tordon)

Fig. 7.2. Structural formulas of some common synthetic auxins used commercially in agriculture; 2,4-D and picloram (Tordon) are used as herbicides and in tissue culture.

during the 1940s and the analog 2-methyl,4-chlorophenoxyacetic acid (MCPA) was developed concurrently in England. The 2,4-D analog 2,4,5-trichlorophenoxyacetic acid (2,4,5-T, or *agent orange*) has been a popular herbicide for brush control, but its use is presently prohibited because it was found to contain an impurity, dioxin, that can be carcinogenic. Dicamba, a benzoic acid derivative, and picloram, a picolinic acid derivative, are auxins and potent herbicides.

Hundreds of auxin analogs have been synthesized by chemists, but not all exhibit auxin activity. Certain molecular chemical, structural, and spatial characteristics were found necessary (Leopold and Kriedemann 1975): an unsaturated ring, an acid side chain, and the proper spatial relationship between the ring and side chain.

AUXIN METABOLISM

The endogenous auxin level and activity in tissues is related to the balance between synthesis and loss in transport and metabolism. Auxins are produced in active meristematic tissues (i.e., buds, young leaves, and fruits). Immobilization by photooxidation and enzyme oxidation (IAA-oxidase) occurred throughout the plant, especially in older tissues (Wareing and Phillips 1978). Peroxidation (H_2O_2) occurs throughout the plant in the presence of oxygen (O_2) to reduce auxin activity. Auxin is also conjugated with organic compounds (e.g., ascorbic acid, amino acids, and sugars), which reduces activity.

Transport of auxins is *basipetal,* that is, tip to base (Fig. 7.3). Reversal of the ends of the stem piece does not change this polarity of movement. However, modern studies using radioactive isotopes of IAA have shown some *acropetal* (base to tip) movement (Wareing and Phillips 1978).

Rate of transport of IAA is linear, occurring at about 6 mm • h^{-1}. The rate of 2,4-D is about 1 mm • h^{-1}. Generally, auxin transport is *symplastic* (in phloem) and active, that is, the rate declines without O_2 or in the presence of carbon dioxide (CO_2). Supraoptimal levels may also cause *apoplastic* (in xylem) as well as symplastic transport. Since transport did not cease in a nitrogen (N) atmosphere, a passive as well as an active transport was indicated (Leopold and Kriedemann 1975). Cytokinins and especially gibberellins accelerate auxin transport, whereas growth inhibitors impede it. Sodium fluoride and triiodobenzoic acid are transport inhibitors.

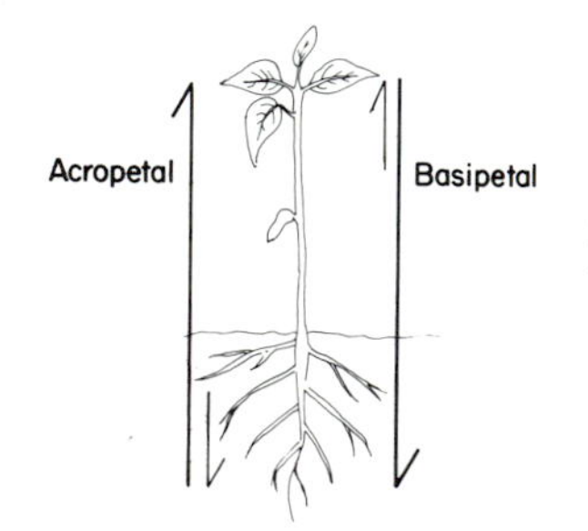

Fig. 7.3. Relative daily polar transport of IAA in young shoot and root. Most of the IAA produced in the buds and young shoot tissues is transported downward; transport in roots is upward or away from root tips.

AUXIN ASSAY

As mentioned previously, the challenge of quantitative determination of a chemical present in such minute concentrations (10^{-7} or 10^{-8} M) was first met by Went, using the *Avena* curvature test. A source of unknown concentration placed asymmetrically on an oat, maize, or wheat coleoptile stump (Weaver 1972) resulted in differential growth and curvature proportional to the concentration. The angle of the new growth in relation to normal is an index of the concentration.

The *Avena* coleoptile straight growth test is another bioassay, also based on cell expansion. This involves determining growth response in terms of the increase in length of young, etiolated, decapitated shoot segments in a solution of the test growth substance. Chromatography has added a new dimension by providing a method of effective separation of hormones and analogs. Light and mass spectroscopy are effective tools for identification and quantification by chemical methods.

RESPONSES TO AUXINS

Responses to auxins range from influences on cellular metabolism to coordination of plant morphogenesis, including abscission and senescence (Table 7.1). Cellular effects included (1) increases in the nucleotides DNA and RNA, and protein and enzyme synthesis; (2) increases in proton exchange, membrane charge, and potassium uptake (Marre 1977); and (3) influences on the phytochrome reaction with red and far-red light (Ali and Fletcher 1971; Leshem 1973).

Auxin response is related to concentration. A high concentration is inhibitory, which has been explained as competition for attachment on receptor sites (Fig. 7.4), that is, increasing the concentration increases the probability of partially attached molecules occupying receptor sites, rendering the complex less effective. Also, responses vary greatly, depending on sensitivity of plant organ. Stems respond to a wide range of auxin concentrations. Roots are essentially inhibited over most of the hormone range (Fig. 7.5).

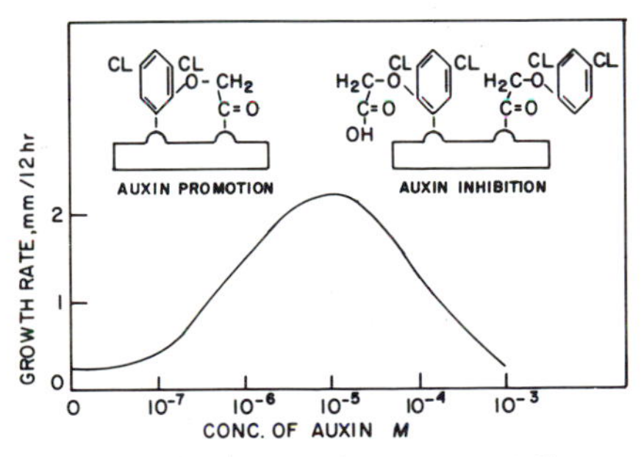

Fig. 7.4. Diagramatic representation of inhibition from high auxin concentrations. (From Leopold 1964, after Foster et al. 1952)

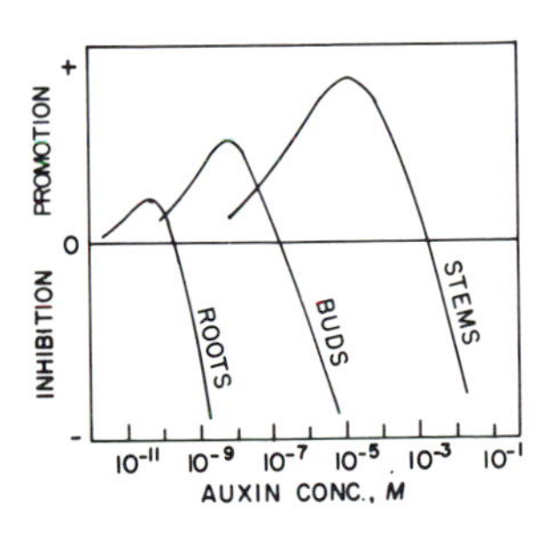

Fig. 7.5. Growth response of roots, buds, and stems to auxin concentrations. Roots were inhibited over most of the concentration range, whereas stems were stimulated. (From Thimann 1937)

Until recently geo- and phototropic responses have been explained, respectively, by asymmetric, gravity-induced shoot levels due to redistribution of auxin and asymmetric levels due to light destruction of auxin on the lighted side (Audus 1972). In geotropic or gravitropic responses, auxin moves to the cells on the lower side of a horizontal organ, stimulating cell elongation and curvature asymmetrically; this is known as the classic Cholodny-Went theory. It is theorized that movement of auxin to the lower side of a root inhibits growth on that side, with resultant downward curvature. A number of physiologists have raised doubts as to the validity of this theory (Wilkins 1977; Wheeler and Salisbury 1980). It has been suggested that the root cap rather than the growing point is the gravity-sensing tissue, and movement of abscisic acid (inhibitor) acropetally and to the lower side may explain the tropic root response. Likewise the Cholodny-Went theory has been questioned because of observations that suggest that ethylene diffusing upward and inhibiting the upper part of a horizontally placed stem is the cause of upward bending (Wheeler and Salisbury 1980) (see Fig. 7.22). IAA seems to move too slowly to initiate geotropism and may be only incidentally associated with it, rather than the causative factor.

Whatever the causative factor, one-sided expansion of the stem or root was associated with cell wall extensibility, which appears to result from loosening of the polysaccharide matrix (Masuda 1977). Auxins bind to the plasmalemma, particularly to lecithin, inducing increased respiration and potassium uptake. These effects may explain the plastic expansion of the cell wall by the deposit of additional polysaccharides in the loosened matrix.

Auxins were necessary for callus growth (Wain and Faucett 1969), whether in tissue culture or in gall and nodule tissues. Auxin was believed to induce curl of root hairs, a prerequisite for *Rhizobium* infection (Allen 1973).

Auxins coordinate plant processes in morphogenesis (Table 7.1, Fig. 7.6).

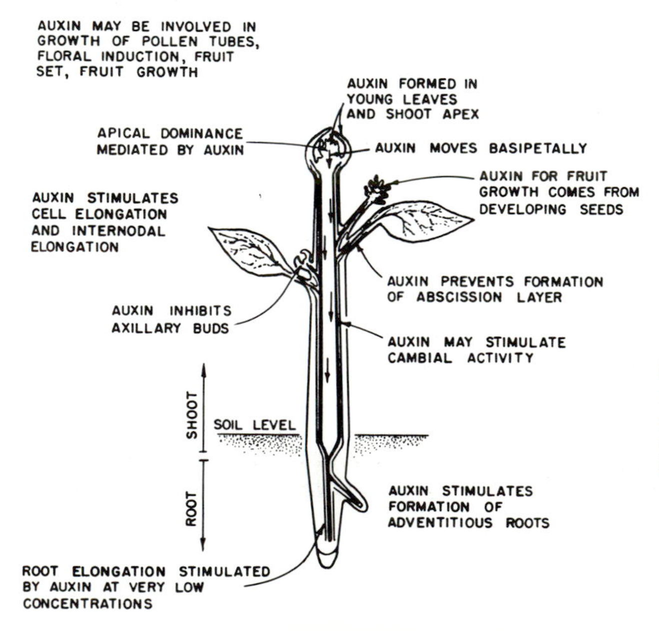

Fig. 7.6. Auxin activity in growth, development, and coordination in morphogenesis. (From Steward 1964)

For example, lateral bud and root growth are inhibited by auxins, but new root initials are promoted on callus tissue formed on cuttings. On hard-to-root species or cultivars (Fig. 7.7), an exogenous source of auxin was nearly always essential (Hart and Carlson 1967). Callus tissue forms first on the cutting, and roots are differentiated from the callus. Cuttings of numerous species rooted readily only if active bud or young leaf tissue was left on the cutting (often referred to as the *leaf factor*) (Weaver 1972).

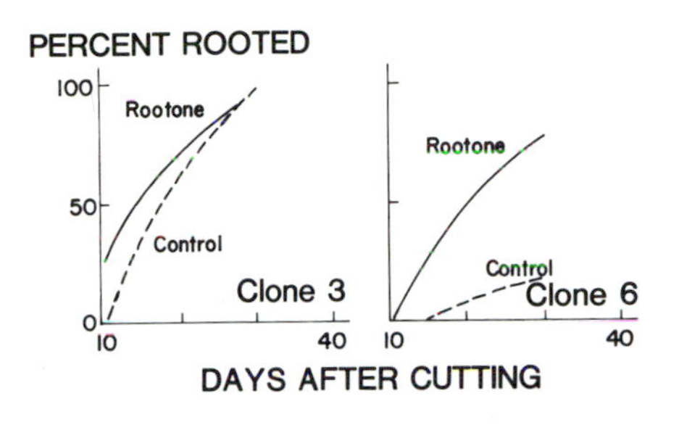

Fig. 7.7. Effect of auxin Rootone on easy-to-root (clone 3) and hard-to-root (clone 6) crownvetch cuttings. (From Hart and Carlson 1967)

Auxins delay leaf and fruit abscission. They induced parthenocarpy (seedlessness) in fruits; for example, strawberry fruits grow without seeds if treated with naphthaleneacetic acid (NAA) (Nitch 1950) or with picloram (Tordon) (Wareing 1976). Normally the presence of seeds or an exogenous source of auxin is necessary for fruit growth (see Chap. 12).

Excessive concentrations of auxins cause abnormalities, such as *epinasty* (malformation of leaf due to differential growth of upper and lower leaf midvein), onion leaf, fused brace roots, and brittle grass stalks. Even vapors from a distant source can cause epinasty in sensitive species like tomato or grape. Supraoptimal concentration of auxins may kill certain species and not affect others; thus auxins are used as selective herbicides. The reasons for such a high degree of selectivity have not been completely resolved.

AGRICULTURAL USES OF AUXINS

Some of the most valuable and widely used selective herbicides in the weed control arsenal are auxins, particularly the phenoxyacetic acid analogs (e.g., 2,4-D, 2,4,5-T, and MCPA) (Fig. 7.2). One of the earliest selective herbicides, 2,4-D probably still ranks as the single most important one. It is highly selective, noncorrosive, effective at low concentrations, safe to handle, relatively easy to formulate, and economical to use. Several benzoic acid analogs (e.g., dicamba, chloramben, and substituted picolinic acid, picloram [Tordon]) are also important herbicides (Fig. 7.2).

Auxins have other important commercial uses, as reviewed extensively by Weaver (1972). Using the principle of inhibition of abscission layer formation, auxins (e.g., NAA or 2,4-D) are effective in prevention of fruit drop in apple

TABLE 7.2. Effect of auxin on fruit drop of apple

Auxin	Total Fruits Borne	Average Drop (%)
NAA	344	10.7
IBA	240	33.9
IAA	228	33.9
IPA	375	44.6
Control	347	52.4

Source: From Mitchell and Marth 1947.

and pear (Table 7.2). Auxins, including 2,4-D, induced ethylene formation and fruit set in pineapple (Burg and Burg 1966).

Biennial bearing (light and heavy crops in alternate years) is common to many tree crops. This problem can be corrected by thinning in the heavy years by a timely application of NAA or other auxins (Luckwill 1976).

Commercial preparations of rooting compounds are available that promote callus and root formation, which can improve establishment from cuttings. Species and cultivars difficult to root are enhanced by dipping the cut surface into a rooting compound (Fig. 7.7). Commercial nurserymen also recognize the importance of selecting cuttings with some active bud development to supply endogenous auxin. Auxins were also effective in the prevention of sprouting of stored potatoes (Mitchell and Marth 1947). The potato may be dipped in the auxin solution (such as NAA), sprinkled with talc or fuller's earth containing an auxin, or stored with strips of paper impregnated with an auxin solution. Newer and more effective PGRs are now available for this purpose.

Gibberellins

Before the discovery of gibberellins (GAs), Japanese farmers had long been aware of the presence in their fields of uncommonly tall rice seedlings that rarely flowered or lived to maturity. They concluded that these plants were diseased and named the disease *bakanae* (foolish seedling disease), since the plants that looked so promising early in the season did not fruit. In 1926, the fungus *Gibberella fujikuroi* (*Fusarium moniliforme*) was isolated and identified as the causative pathogen. From this fungus Japanese workers extracted a cell-free extract of GA-like substances that produced the same height increase as in diseased plants.

Following the discoveries of the Japanese, a flurry of research activity in growth regulator research in general, and with GAs in particular, emerged in the 1950s in Great Britain (Brian 1958) and the United States (Thimann 1963), especially after the observation that GAs were also present in higher plants.

GAs are diterpenoids, which places them in the same chemical family as chlorophyll and carotene. The basic chemical moieties of the GAs are the *gibbane skeleton* and the free carboxyl group (Fig. 7.8). The various forms of

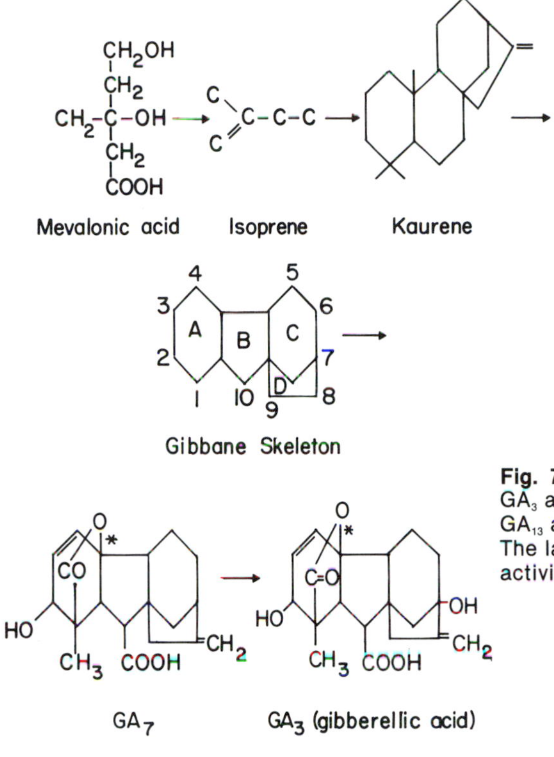

Fig. 7.8. Pathway of biosynthesis of GA_3 and GA_7 with the lactone ring and GA_{13} and GA_{14} without the lactone ring. The latter two have low biological activity, compared with GA_3 and GA_7.

GA differ primarily by substitution of hydroxy, methyl, or ethyl groups onto the gibbane skeleton and by the presence of the lactone ring, which is produced by the condensation of carbon 20 to carbon 19 in the gibbane structure (Fig. 7.8). The lactone ring present, for example, in GA_3, GA_4, and GA_9 is responsible for the greater biological activity of these analogs, compared with GA_{12} and GA_{13} and similar analogs that do not have the lactone ring.

Different GAs are designated by a letter-number code (GA_1, GA_2, GA_3, . . ., GA_{52}). The number of distinctly different GAs was reported to be 52 (Hedden et al. 1978). *Gibberellic acid* (GA_3), the first identified, is the best known and most widely researched. It was first crystallized from the fungus *Gibberella fujikuroi*. Interestingly, GA_3 has the broadest range of biological activity. Commercial sources of GA_3 are obtained from fungal cultures, although GA_3 and most other GAs are widely distributed among higher plants.

NATURAL OCCURRENCE OF GIBBERELLINS

A large number of GAs with requisite chemical structure and biological activity occur naturally, and many have been isolated from bacteria, fungi, mosses, ferns, and/or vascular plants and identified as GA-like substances (Krishnamoorthy 1975). The GAs have a common chemical structure but exhibit a broad spectrum of biological activity. The GA-like substances were less specific chemically than GAs and had a narrower range of activity (Phinney and West 1960).

All plant organs contain GAs at varying levels, but the richest sources and probable sites of synthesis were found to be the fruits, seeds, buds, young leaves, and root tips (Carr 1972). Seeds are especially rich in GAs; immature seeds are rich in GAs, but these occur in bound forms as seeds mature (Paleg 1965).

Plant species and type and age of tissue vary in GA kind and concentration. In general, intercalary meristems have lower than normal levels and respond to GAs from exogenous sources. For example, young stems of genetic dwarfs, certain other intercalary meristems, and seeds of some species are responsive to exogenous GA, probably due to suboptimal endogenous levels.

GIBBERELLIN METABOLISM

Biosynthesis of GAs occurred principally in immature fruits and seeds, buds, leaves, and roots (Wareing and Phillips 1978). Although GAs are known to inhibit root growth, roots are a source of GAs to other organs. Generally seeds are the single richest source, as evidenced by the rapid growth of the fruit that surrounds them.

Three chemical metabolites were commonly involved in GA biosynthesis (Leopold and Kriedemann 1975) (Fig. 7.8):

1. Mevalonic acid acts as a precursor to the formation of isoprene, the basic moiety in the 19- and 20-carbon gibbane skeletons.
2. Kaurene is formed from isoprene.
3. GA is formed from kaurene, the major GA precursor.

The breakdown of GAs in plant tissues from either endogenous or exogenous sources is less well understood. It appears evident that bound and free forms are readily reversible. Seeds are high in bound form, but soaking and chilling seeds released GAs in free form (Aung et al. 1969). Cold exposure (*vernalization*) of soaked seeds and stratification of dormant buds increases GAs in free form, the result being induction of flowering and breaking of dormancy, respectively (see Chap. 12). In breaking dormancy, GAs can substitute for red light. Internode and leaf growth evidently has a GA-light interaction (see Chap. 11). These findings appear to illustrate the rate of interconversion between free and bound forms and the interaction with light via the phytochrome pigment receptor (Loveys and Wareing 1971).

GA activity can be inhibited chemically, presumably if the receptor sites are blocked by molecules structurally similar to GAs. Abscisic acid (ABA), an inhibitor of GAs, prevented GA_3 reversal of dwarfism (Thomas et al. 1965). It is chemically similar to GA. Although not structurally similar, ethylene may also inhibit GA activity (Scott and Leopold 1967).

A number of synthetic chemicals from exogenous sources, termed *growth retardants,* effectively blocked GA_3 activity (Cathy 1964; Lang 1970). Synthetic growth retardants such as AMO-1618, CCC, SADH or daminozide, Phosfon-D, and morphactins act as anti-GAs (see Fig. 7.16).

GAs are assumed to be translocated symplastically, but their presence in both phloem and xylem under certain conditions suggests both symplastic and apoplastic transport (Krishnamoorthy 1975). Phloem transport rate was observed to be similar to that of carbohydrate, about 5 cm • h^{-1}. While auxin movement is polar and basipetal, GA moved freely basipetally and acropetally (Chlor 1969).

GIBBERELLIN ASSAY

The minute concentration of GAs in plant tissues makes identification and quantification difficult, and until recently, quantification was restricted to bioassay. Recent advances in chromatography (gas and liquid column and thin layer chromatography) are effective in separation. Nuclear magnetic resonance and mass- and fluorospectrometry are used to assay GAs and other growth substances physiochemically.

Weaver (1972) has listed the more successful bioassay tests.

1. Barley aleurone. Sterile, embryoless seeds are treated with GAs, releasing α-amylase and converting starch to sugar, which can be quantified. This test is simple and rapid (see Fig. 7.10).
2. Dwarf pea. Genetic dwarf pea plants are treated with GAs and grown under red light for observation of height change, compared with controls.
3. Other expansion growth tests based on the principle of increased internode elongation from GAs. Other assays include the lettuce and cucumber hypocotyl test and the dwarf rice test.

RESPONSES TO GIBBERELLINS

A wide range of responses of numerous woody and herbaceous plants to GAs have been reported (Paleg 1965) (Table 7.1). GA acts synergistically with auxins, cytokinins, and probably with the other hormones, in what might be called a systems approach, or *synergism.* For example, apical dominance, cambium growth, geotropism, abscission, and parthenocarpy are attributed to auxin activity, but GAs also influence or are essential for these responses. GA_3 was highly effective in increasing fruit-set, even in apple and pear, which responded poorly to auxins (Thimann 1972). Parthenocarpy can be induced in stone fruits that fail to respond to IAA.

The best known GA response is the stimulation of internode growth. Dwarf plants of maize, pea, and bush bean became normal after treatment with GAs (Fig. 7.9) (Phinney 1956). The requirement of a period of cold to

Fig. 7.9. Response of bush bean (*left*) and cabbage (*right*) to a leaf spray (GA), compared with controls (O). (Courtesy of S. H. Wittwer, Michigan Agricultural Experiment Station)

induce flowering in certain biennials (e.g., beet and cabbage) was replaced by treatment with GA_3 (Wittwer 1958) (Fig. 7.9).

Release of α-amylase and resultant starch hydrolysis and germination require GAs (Fig. 7.10).

Flowering has not been linked to a specific hormone, but GAs were shown to be active in flowering and maintaining the indeterminate growth habit (nonflowering) in a photoperiod-sensitive pea cultivar under long days (Prob-

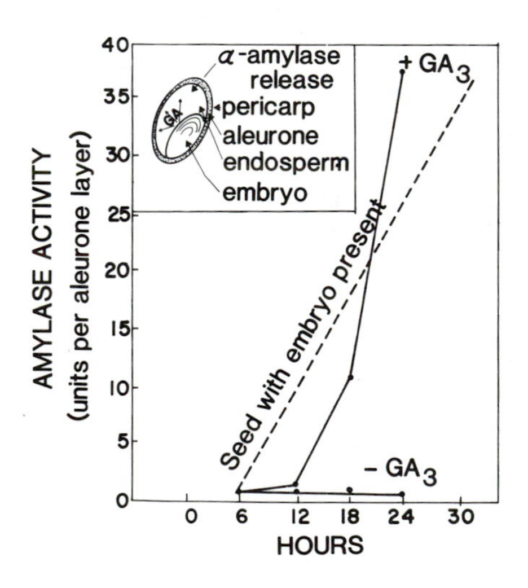

Fig. 7.10. The α-amylase activity of barley aleurone in embryoless seed, with GA ($+GA_3$) and without GA ($-GA_3$), compared with normal seed. (From Bailey et al. 1976) Inset is a schematic of GA and of α-amylase release. (From Paleg 1965)

sting et al. 1978). A graft-diffusible GA_9 metabolite was isolated that delayed flowering. Long days, which promoted flowering in all cultivars, resulted in a 10-fold drop in the GA_9 metabolite, evidence of direct action of a growth hormone in flowering. These findings are significant in providing a plausible explanation for the cause of indeterminate and determinate growth habits. Indeterminate flowering plants, such as northern latitude soybean cultivars, produce flowers and fruits from axillary buds in response to photoperiod but maintain a vegetative terminal bud. All buds flower nearly simultaneously on determinate types. The control mechanism may be the GA level in buds, which could repress flowering of the terminal bud of indeterminate types or all buds under long days.

GA responses can be generally summarized as follows.

1. Whole-plant, genetic dwarfs elongate stem internodes to normal plant height if treated with GAs, whereas excised parts generally do not respond.
2. Most plant species and cultivars have sufficient endogenous levels of GAs and do not respond to exogenous sources. Genetic dwarfs, especially single-gene dwarfs, responded to GA_3 as an exogenous source (Phinney 1956).
3. Positive responses to GAs occur over wide concentration ranges, in contrast to auxin responses over only a narrow concentration range. Thus even high GA levels are not toxic and elicit no positive or negative responses, except on sensitive dwarf plants, whereas high concentrations of auxins are effective herbicides.

The GAs vary greatly in biological activity. GA_7 and GA_3 had the widest range of activity, although GA_4, GA_7, and GA_9 were more active than GA_3 in cucumber hypocotyl elongation (Paleg 1965). Dwarf maize responded to GA_4, while dwarf beans did not, but both responded to GA_7. GA_4 is many times more active on test plants than GA_8. A common characteristic of GA_4, GA_7, and GA_9 is the absence of the hydroxy radical at carbon 7 (Fig. 7.8). It is likely all GAs are synergistic with auxin.

AGRICULTURAL USES OF GIBBERELLINS

In the 1950s expectations for GAs to improve crop production were high. Control of flowering and enhancement of growth and productivity were visualized. A great many researchers around the world initiated research on GAs, investigating their effect on growth habit and yield parameters of a host of economic species. Germination and emergence were enhanced in certain genotypes, but the effects on wheat were negative. Yield of dry matter generally was not affected despite an increase in height (Krishnamoorthy 1975).

The discovery that GAs can produce sterile male plants created interest in GAs in the hybrid seed industry. GA_3 caused a high degree of male sterility in maize, but results were inconsistent, being highly dependent on dosage and time of application (Nelson and Rossman 1958). Use of GAs to produce male sterility has not become a common practice, due to inconsistent response.

The use of GA_3 on 'Thompson Seedless' grape is a success story. GA_3 treatment at 200 ppm at the *calypta* (floral bracts) fall produced larger grapes with improved table quality (Weaver 1972). GAs are also used in the malting industry to promote α-amylase activity and the resultant starch hydrolysis in embryoless barley seeds (Fig. 7.10). Despite these uses, in general the high expectations for GAs remain largely unfulfilled, primarily due to the fact that modern crop cultivars have been selected for desired growth and reproductive habits that indirectly assure adequate endogenous GA levels and therefore no need for exogenous sources.

Cytokinins

Cytokinin (*kinin*) is the generic name for growth substances that typically stimulate cell division (*cytokinesis*). Cytokinin was discovered in the 1950s from observations in the Skoog laboratory on cell division in callus grown from tobacco pith or carrot root phloem. This work clearly demonstrated there was little growth of parenchyma cells of excised tissues unless a growth factor found in coconut milk or yeast extract was added to the culture medium (Miller 1961). Using only IAA, which was formerly considered to be the growth hormone, only cell elongation occurred. However, rapid callus growth from cell division and elongation occurred in the presence of IAA plus coconut milk, yeast extract, or the purine base adenine,6-aminopurine. In 1961 a compound that stimulated cell division was isolated from a degraded autoclaved sample of DNA. The active substance, an artifact present only in autoclaved samples, was identified as 6-furfurylaminopurine and named *kinetin* (Miller 1961).

Cytokinin was first isolated from higher plants in the milky endosperm of young maize seeds in 1964. It was named *zeatin* (Letham 1968) (Fig. 7.11).

NATURAL OCCURRENCE OF KININS

Cytokinins required for cell division also regulate and correlate a wide range of activities in morphogenesis (Table 7.1). Young roots, immature fruits and seeds, and the nurse tissues (e.g., liquid endosperm) are rich in kinins. Immature fruits of maize, horse chestnut, banana, apple, and coconut (milky endosperm) are especially rich sources. Since kinins are evidently not translocated to these tissues, they are probably the sites of synthesis.

In general, kinins occur naturally as conjugates of sugar and phosphate ions (Leopold and Kriedemann 1975). Zeatin riboside is the principal kinin in the roots of cocklebur (*Xanthium*), while the principal one found in leaves of disbudded bean plants was zeatin glucoside (Wareing et al. 1977). Eight different kininlike substances have been isolated from maize kernel extract; zeatin was biologically the most active (Wareing and Phillips 1975). The purine base is the common chemical moiety in both natural and synthetic kinins (Fig. 7.11).

A

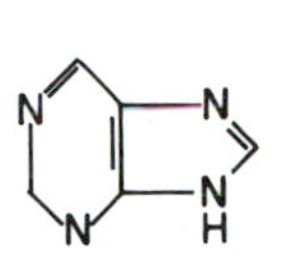

Purine

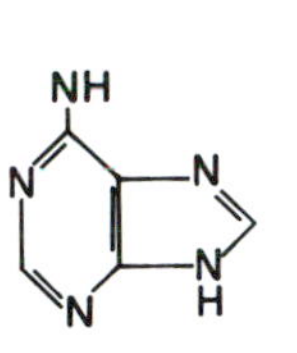

Adenine

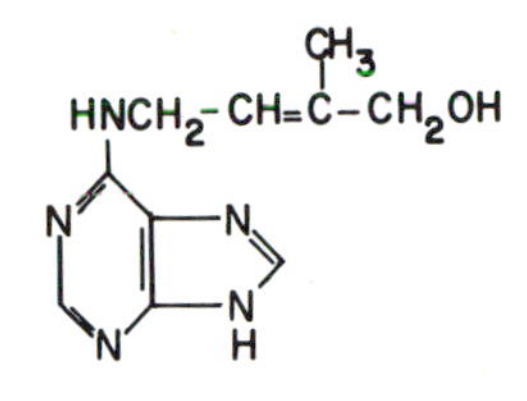

B

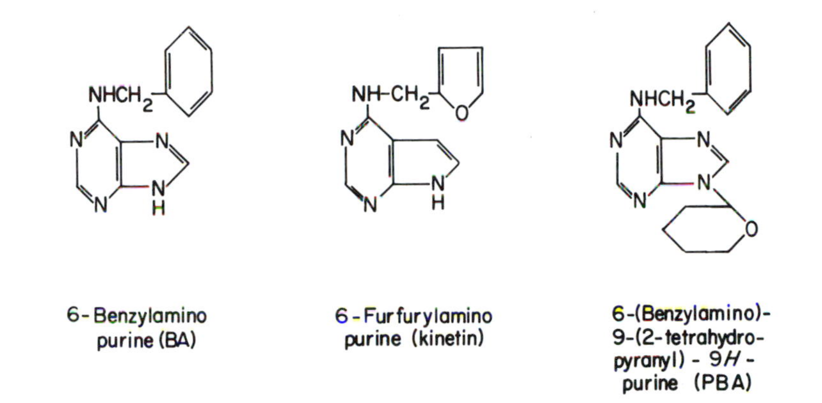

Fig. 7.11. *A*. Structural formulas of purine and adenine bases and the natural cytokinin zeatin; *B*. three commonly used synthetic cytokinins.

KININ METABOLISM

The natural kinins appeared to be synthesized by fixing a side chain, usually a 5-carbon chain, to an adenine molecule. The 5-carbon chain is believed to be derived from isoprene, the basic unit in GA, chlorophyll, xanthophyll, and abscisic acid.

There is evidence that the kinins in leaf and bud tissue are immobile. However, it is well known that kinins produced in roots are translocated throughout the plant via the transpiration stream (Wareing et al. 1977). Buds were stronger sinks for kinins than leaves (Phillips 1965).

KININ ASSAY

Due the difficulty of chemical analysis of the minute concentrations that occur naturally, quantitative analysis of kinins has, until recently, been restricted to biological methods. Presently gas chromatography can be used effectively to separate kinins.

Five bioassays, which reflect kinin activities by increasing parenchymous tissue cell mass, have been described (Weaver 1972); probably the one most commonly used is the tobacco tissue culture assay. This test has advanced the state of knowledge of cytokinins equivalent to the knowledge of auxins revealed by the *Avena* curvature test.

RESPONSES TO KININS

Kinins elicit a broad range of responses, but they act synergistically with auxins and usually with other hormones (Table 7.1; Figs. 7.12, 7.13). Adventitious bud formation was believed to be triggered by the synergism of auxin and cytokinin (Heide 1972). Cytokinin movement from the roots probably has a marked influence on stimulation of new growth from quiescent axillary buds subject to apical dominance. Evidently lateral buds are deficient in kinins, since an exogenous source stimulated their growth (Phillips 1965; Schaeffer and Abdul-Baki 1973). The requirement of kinins for the formation of adventitious buds was manifested on root pieces of field bindweed (*Convolvulus arvensis*) (Torrey 1958). On stem cuttings kinetin, even more than GA, severely inhibited formation of root initials (Fig. 7.14), whereas auxin stimulated it (Fernqvist 1966).

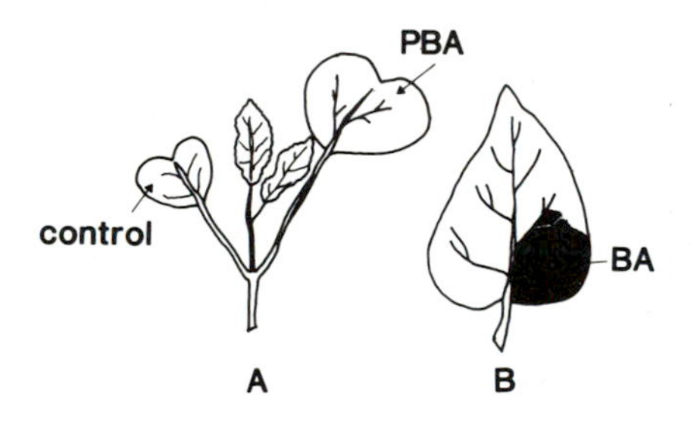

Fig. 7.12. Effects of cytokinins on growth and development. *A*. The right cotyledon of radish was painted with a PBA at 100 mg • l^{-1}, causing increased cotyledon growth. *B*. The right half of leaf *B* was spotted with BA. The green (darkened) area represents attraction of nutrients and delay of senescence. (From Weaver 1972)

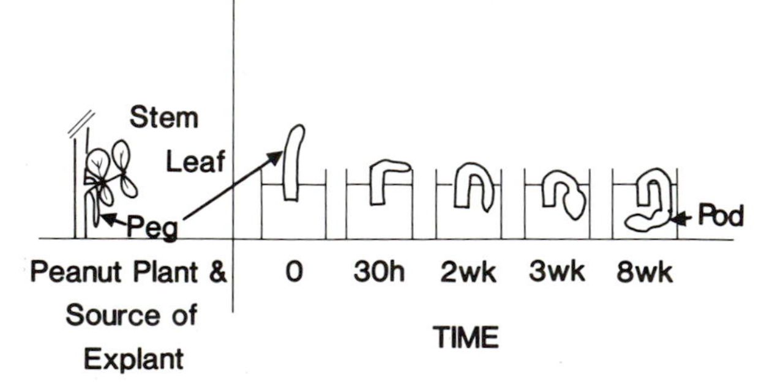

Fig. 7.13. Development of plantlets in the dark from young explants (pegs) of peanut grown in a medium containing kinetin and NAA at 0.5 ppm. The 180° curvature and the diageotropic growth of the pod occurred only when both hormones were in the medium. (From M. Ziev and E. Zamski 1975)

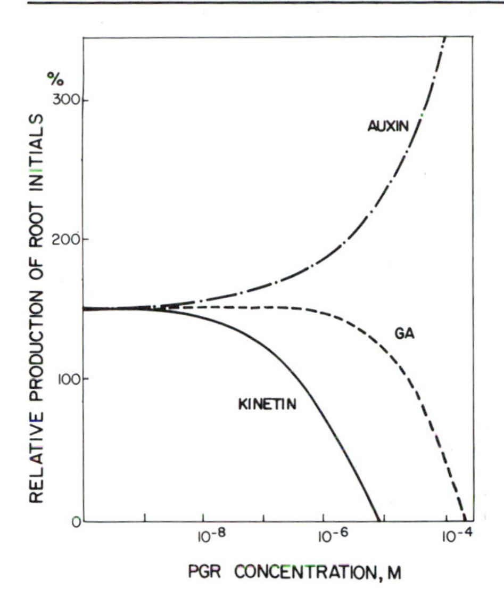

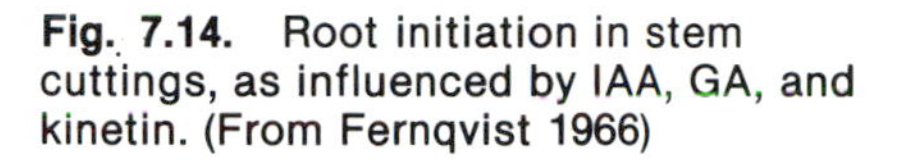
Fig. 7.14. Root initiation in stem cuttings, as influenced by IAA, GA, and kinetin. (From Fernqvist 1966)

GA and kinin promoted germination of seeds of certain species, such as lettuce and white clover (Carr 1972). Seeds of witchweed, a parasitic weed, germinate only if a stimulus is received from the host plant. The stimulus was reported to be kinin (Worsham et al. 1959). The role of kinin in retention of chlorophyll, incorporation of amino acids, and protein retention in leaves, all of which indicate delayed senescence, is of particular interest to physiologists (Quinlan and Weaver 1969) (Fig. 7.12). Searching for more effective means of supplying exogenous kinin sources or, alternatively, of improvement of root delivery to prolong senescence and photosynthetic output could become a fertile area for future research.

AGRICULTURAL USES OF KININS

Weaver (1972) suggested numerous potential uses of kinins, including increasing fruit-set in grapes and improving size and shape of the 'Delicious' apple. Lettuce seeds may acquire a secondary dormancy in soils with high temperatures (25°C); germination is increased if treated with kinetin, which is more effective than benzyladenine (BA). In general, kinin use in crop production is still in the potential stage (Thomas 1976).

The discovery of cytokinin has added a new dimension to plant breeding. Cloning of haploid plants from pollen, diploid plants from somatic cells, embryos from fertilized eggs, and plants from excised tissues provides new tools for achieving breeding objectives that were previously unobtainable. The finding that varying kinin concentration in the culture medium at certain critical stages in the explant growth cycle can induce nonnuclear, transmittable genetic changes opens the way for exciting genetic-engineering possibilities. Using this technique, new potato genotypes have been isolated that produced a significant yield advantage in field trials (Shepard et al. 1980).

Growth Inhibitors

Most growth substances generally stimulate growth and correlate growth and development in morphogenesis. A diverse group of other substances involved in growth correlation generally inhibit growth and are termed *growth inhibitors*. The most common inhibitors are aromatic compounds, such as phenols and lactones, but alkaloids, certain alcohols, organic and fatty acids, and even metallic ions can act as inhibitors (Addicott and Lyon 1969; Abeles 1972).

For convenience of discussion, growth inhibitors (Figs. 7.15, 7.16) are usually classified into three groups:

1. Phytohormones. Terpenoids, such as ABA. ABA-glucose (glucoside), a bound form, also has ABA activity (Abeles 1972).

2. Other natural inhibitors, including phenolic and benzoic acid derivatives and lactones. Unlike the ABA hormone, these apparently are metabolic by-products usually present in large quantities. They may play important inhibitory-correlative roles in growth and development, such as seed dormancy in certain species (Wilkins 1969).

3. Synthetics. A large number of synthetic compounds exhibit growth inhibitory activity. Many have been labeled for agricultural uses. The quaternary salts of ammonium (Amo-1618) and Phosfon-D are growth retardants (Fig. 7.16). Another important synthetic is succinic acid-2,2-dimethylhydrazide (SADH or daminozide). Chlormequat chloride (CCC) is available commercially and widely used in flax and other seed crops to reduce lodging and recently to coordinate barley and wheat tiller growth rate. The morphactins (flurecol and chloroflurecol) are recent additions to the list of growth retardants. The chloro-acid is the most active form of the two morphactins (Fig. 7.16).

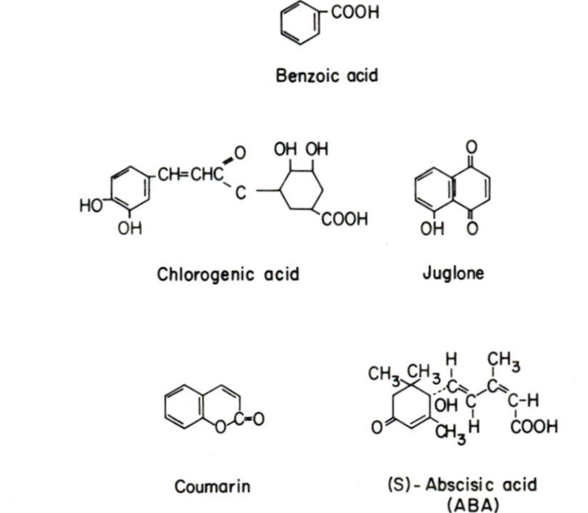

Fig. 7.15. Structural formulas of growth inhibitor hormones.

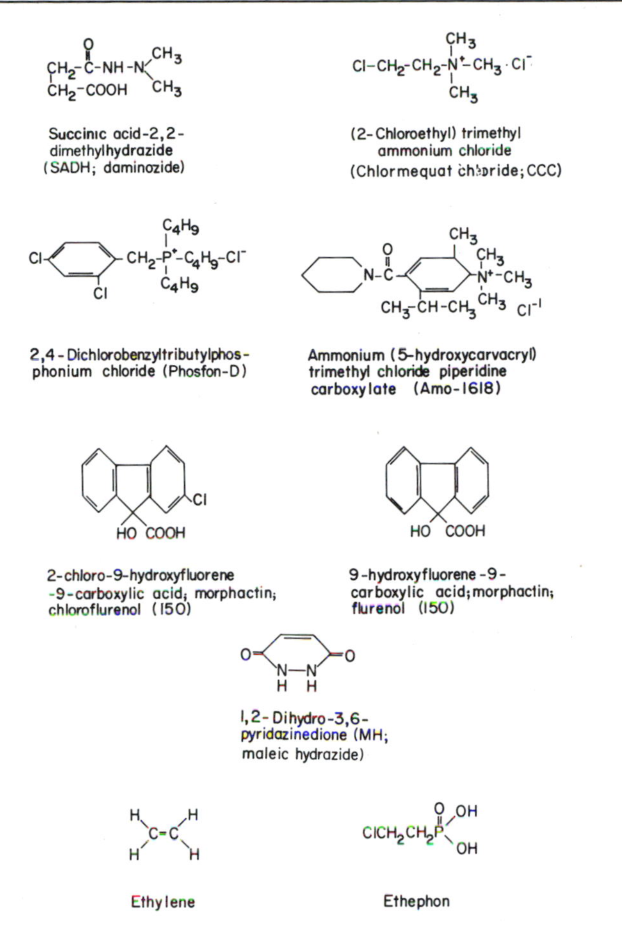

Fig. 7.16. Synthetic growth inhibitors including ethylene and ethephon, a slow-releasing ethylene compound.

NATURAL OCCURRENCE OF GROWTH INHIBITORS

A highly active growth inhibitor was isolated from cotton fruits in the early 1960s (Ohkuma et al. 1963) and named abscisin II. A similar compound was isolated in England from sycamore leaves and named dormin (Cornforth et al. 1965). The two substances were subsequently found to be chemically and biologically identical; it was agreed to name the substance *abscisic acid,* or ABA.

The hormone ABA has been isolated from tubers, buds, pollen, fruits, embryos, endosperms, and seed coats of herbaceous and woody annuals and perennials of some 40 to 50 species (Walton 1980). It is reasonable to conclude that ABA, like IAA, is ubiquitous in higher plants. ABA is normally located in the chloroplast, but after environmental stress it is released to the other organelles (Fenton et al. 1982) and is active in stomata control.

ABA analogs are likewise distributed widely, but they are not as active biologically as ABA (Walton 1980). Phaseic acid is present in the seeds of bean (*Phaseolus multiflorus*), and theospirone is a natural inhibitor and a flavor component in tea leaves.

Secondary substances such as alkaloids, phenolics, and lactones often occur in concentrations sufficient to be stored food reserves (i.e., much greater than hormone level). Juglone, a lactone (Fig. 7.15) is present in high concentrations, especially in the mesocarp (husks) and roots of black walnut (*Juglan nigra L.*). Buffalo gourd (*Curcurbita foetidissima*), a weedy vine adapted to semiarid areas, stores large quantities of carbohydrates in conjunction with an unidentified toxic substance in the fleshy taproots. A water extract containing the chemical was lethal to young tomato and lettuce plants and completely prevented germination of radish and lettuce seeds (Gardner and Reeves 1980). It seems probable that this growth inhibitor may reduce competition from other species in the rhizosphere (*allelopathy*), with resultant survival benefit for buffalo gourd in the ecosystem. Secondary substances probably are allelopathic in numerous species and are operative in natural plant succession (Rice 1974). They may also provide the donor protection from insect and herbivore feeders and perhaps from disease organisms, since many secondary compounds are known to inhibit microbes.

METABOLISM OF GROWTH INHIBITORS

ABA is a terpenoid, as are GA, cytokinin, chlorophyll, carotene, and xanthophyll. Like these, synthesis is evidently through the mevalonic acid and isoprene pathway (Fig. 7.8). Synthesis was also found to result from oxidation of some xanthophylls, such as violaxanthin (Milborrow 1974). Light enhances the occurrence of the most effective form, the *cis-trans* form of ABA. The site of synthesis appears to be in the plastids, especially chloroplasts.

Deactivation of ABA may be by (1) enzymatic conversion to 2-trans ABA (nonactive form), (2) oxidation to phaseic acid, or (3) conjugation with sugars to produce glycosides, primarily glucosides. As with other hormones, bound forms have little or no activity. Free ABA is translocated readily throughout the plant like IAA, but apparently at a greater rate.

Synthesis of phenols was by the shikimic acid pathway using phenylalanine or tyrosine (Leopold and Kriedemann 1975). Cinnamic acid is apparently a precursor to certain benzoic inhibitors. Coumarin, a lactone, is derived from phenylpropane (1-aminocyclopropane-1-carboxylic acid).

RESPONSES TO GROWTH INHIBITORS

Natural or synthetic inhibitors suppress growth and development, as re-

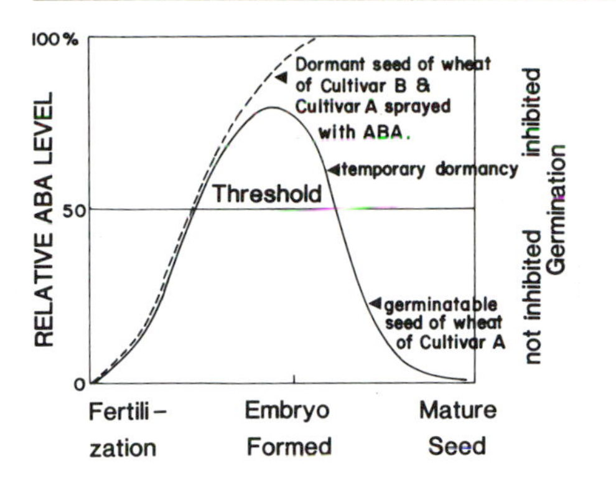

Fig. 7.17. Schematic representation of ABA accumulation in an embryo in relation to germinability during seed ontogeny.

vealed by the standard straight growth test. They also play important correlative roles in morphogenesis and survival. Without dormancy or suspension of active growth, seeds and buds might sprout or resume growth only to be killed by intolerable periods of heat, cold, or desiccation. The dormancy mechanism allows seeds and buds to delay new growth in a resting state and to resume growth only when ABA levels decline concurrently with favorable conditions for completion of the growth cycle (Fig. 7.17). Deciduous species lose their leaves by abscission, creating a winter dormancy cued by natural short days of fall. Potato buds are dormant at tuber maturity due to presence of ABA and do not sprout even though surrounding soil and climatic conditions are favorable (Addicott and Lyon 1969).

Induction of such dormancy mechanisms is controlled by natural inhibitors, principally ABA. Dormancy is generally lost during winter as a result of stratification (cold treatment) or sometimes merely by passage of sufficient time. Production of GA is promoted by stratification, which may be the causative factor for regrowth because the ABA effect is masked by a high GA-ABA ratio (see Chap. 9). Most crop cultivars have been selected against seed dormancy.

Fruits also develop an abscission layer with injury or age; they fall due to ABA accumulation and induced abscission (Fig. 7.18). Diseased lupine pods

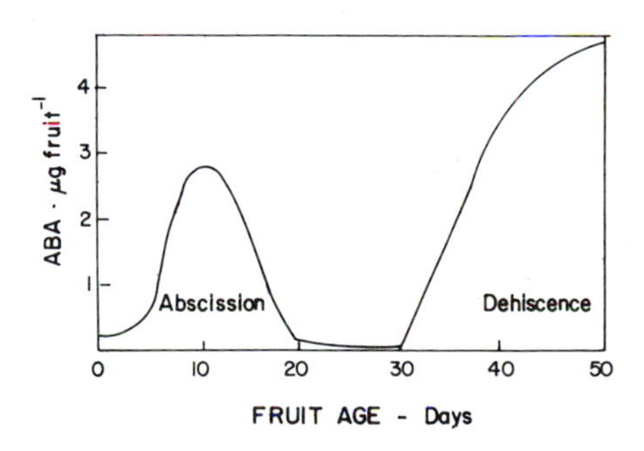

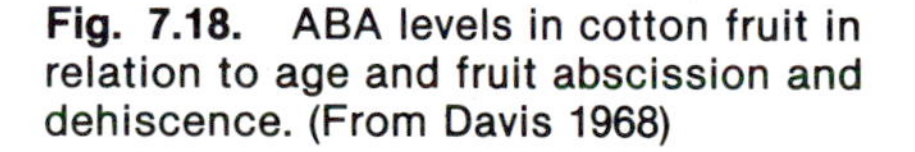
Fig. 7.18. ABA levels in cotton fruit in relation to age and fruit abscission and dehiscence. (From Davis 1968)

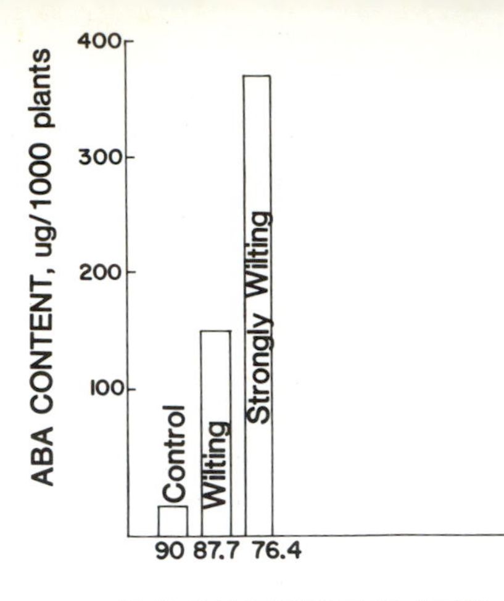

Fig. 7.19. ABA content in wilting pea seedlings in relation to tissue water content. (From Milborrow 1967)

had an ABA content 2.5 times that of the controls (Walton 1980), which resulted in abscission of diseased fruits. The accumulation of ABA in lateral buds points to ABA involvement in apical dominance. It is also highly involved with senescence and dehiscence of fruits such as those of cotton. Distribution of ABA in cotton fruits is biomodal; the first peak occurs soon after fertilization, when it may initiate reduction of fruit load by inducing formation of an abscission layer, and a second peak occurs at senescence (ageing) and dehiscence (splitting) (Fig. 7.18). Treatment with GA can overcome many of these ABA effects.

The association of ABA accumulation and stomata closure as leaves are stressed for moisture (Fig. 7.19) is currently of interest to physiologists. The formation of ABA during stomata closure seems to validate the theory of ABA as the triggering mechanism in stomatal control (Dörffling 1972). It is released from the chloroplast into the epidermis cells during moisture stress.

At the cellular level, ABA can inhibit proton extrusion and potassium uptake (Cocucci and Cocucci 1977). It strongly inhibits activity of enzymes, such as the hydrolytic enzymes induced by GA in barley endosperm (Addicott and Lyon 1969). It also inhibited flowering of a long-day plant given short days (Evans 1966).

Another natural inhibitor of common occurrence is chlorogenic acid. This PGR is believed to be an antiseptic to bacteria in plant wounds.

AGRICULTURAL USES OF GROWTH INHIBITORS

A number of synthetic inhibitors (Fig. 7.16) including growth retardants have been labeled and are available for commercial use. The major effect is to shorten internode length and plant height and generally to reduce lodging, especially in cereal grain crops and flax. Leaf area, light interception, and product yield are usually not reduced by treatment. Leaf area of sugar beet plants was reduced by 25 to 40% as a result of smaller leaf size from spraying with a solution of PP333 at a concentration of 4000 $\mu g \cdot ml^{-1}$ (Jaggard et al.

1982). Daminozide (SADH), chlormequat (CCC), Phosfon-D, and morphactins are effective retardants (Fig. 7.16). An unlabeled compound, BTS 44584, reduced height of 'Williams' soybean (maturity group III) by 20 cm when applied at 1.1 kg • ha^{-1} at the V_4 stage but did not increase yields or decrease lodging (Gardner 1980). In fact lodging was often aggravated by higher rates of this PGR. Morphactins have an advantage in that high dosage rates are not harmful to the plant and extend the period of dwarfing activity (Schneider 1970, 1972). In addition to dwarfing, they induce axillary bud growth and delay senescence. Growth retardants also cause a darker green leaf coloration, ostensibly by increasing chlorophyll content.

Daminozide, marketed as Kylar (Fig. 7.16), is used on more than 140,000 ha runner-type peanuts annually in the southeastern United States. The objective is not to reduce lodging but to curtail late vegetative growth of the vines in order to channel more assimilate to seeds. Less vine also facilitates harvesting. Pod yields of a late, viny-type cultivar 'Dixie Runner' were increased significantly by Kylar application at pod initiation (N'Diaye 1980).

Regim-8 (2,3,5-triiodobenzoic acid, or TIBA), an inhibitor to IAA transport (Galston 1947), caused more upright leaves (Christmas tree effect) and presumably a more optically efficient canopy in soybean (Fig. 7.20). Canopy

Fig. 7.20. Canopy of untreated plants of 'Hawkeye' soybean (*above*) and plants treated with TIBA (*below*). Note the change in canopy architecture (Greer and Anderson 1965).

structure was altered and pod and seed yields were increased (Greer and Anderson 1965). Data of other investigators do not support the canopy improvement theory; they have suggested instead improved partitioning to fruits resulting from reduced vegetative growth (Tanner and Ahmed 1974). The commercial use of Regim-8 has been discontinued, probably because of an unfavorable cost-benefit ratio as well as inconsistent yield benefits.

In the United Kingdom, CCC is used extensively on barley and wheat to regulate tiller growth rate. The primary and secondary tillers are arrested temporarily to prevent dominance over higher order tillers, which distributes growth and yield more evenly among tillers and increases total yield.

Defoliants are used as harvest aids in mechanical harvesting of cotton. Endothall is used commercially, and Harvade, labeled recently, is effective. Probably the most successful growth inhibitor is maleic hydrazide (MH), used for *sucker* (axillary shoot) control in tobacco. MH is used on essentially all the tobacco hectarage in the United States. If tobacco is not treated after topping or removal of the inflorescence, axillary buds freed from apical dominance rapidly produce shoots that shade and withdraw nutrients from the harvestable leaves, lowering the market quality of the leaves.

Ethylene

The emission of ethylene from ripening *climacteric* fruits (those with a climax in ripening, such as apple, avocado, banana, and stone fruits) has been observed for many years (Pratt and Goeschl 1969). The mixing of ripe with green fruits promotes more uniform ripening because ethylene, a gas, moves freely by diffusion from ripe to green tissues. The old saying "a rotten apple spoils the barrel" has a rational basis.

A number of compounds may volatilize from plant tissue and have ethylenelike action, but ethylene showed 60 to 100 times the activity of one such compound, propylene (Pratt and Goeschl 1969). Ethylene is a small gaseous molecule (Fig. 7.16). The small molecule size and the gaseous state make it unique both chemically and physiologically among the phytohormones (Abeles 1973). Ethylene diffusion is passive in plant tissues, and since it escapes as a gas, neither detoxification nor transport systems are required. In contrast, other hormones have elaborate transport and detoxification mechanisms.

NATURAL OCCURRENCE OF ETHYLENE

In addition to the large concentrations in ripe climacteric fruits (Fig. 7.21), ethylene was found present, to some extent, throughout the plant, including leaves, stems, roots, flowers, fruits, and seeds (Abeles 1973). It has a wide range of activity, accelerating some processes and delaying others (Table 7.1). Ethylene production was highly correlated with auxin supply (Leopold 1972); an application of 2,4-D can increase ethylene content in tissue by 50-fold (Burg and Burg 1966). In fact ethylene production may be the causative

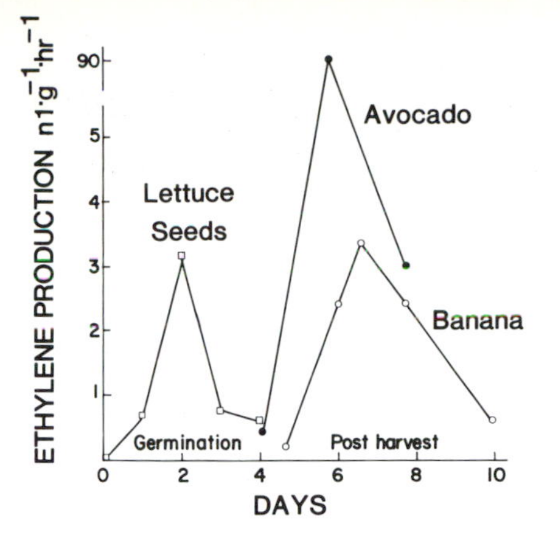

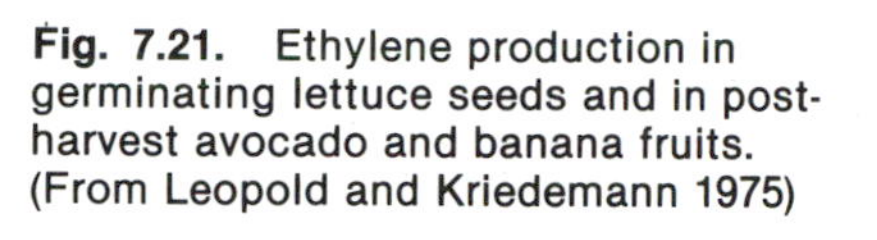
Fig. 7.21. Ethylene production in germinating lettuce seeds and in post-harvest avocado and banana fruits. (From Leopold and Kriedemann 1975)

factor in many responses attributed to 2,4-D.

The highest concentrations of ethylene in climacteric fruits are associated with high respiration and CO_2 release rates (Fig. 7.21). High production of ethylene has also been observed in stressed tissues and in young seedlings. Concentrations of ethylene in fruits and other tissues vary depending on the environment, but nonliving tissues are free of ethylene.

ETHYLENE METABOLISM

The metabolic precursor of ethylene has been somewhat elusive. A number of plausible candidates have been proposed, including pyruvic acid (pyruvate), acetate, formate, acrylate, linolinate, ethanol, and propanol (Abeles 1972). The most widely accepted is methionine, which is hydrolyzed as follows:

$$\overset{5}{CH_3}\text{-S-}\overset{4}{CH_2}\text{-}\overset{3}{CH_2}\text{-}\overset{2}{CH}\text{-}\underset{\displaystyle NH_2}{\underset{|}{\overset{1}{COOH}}} \xrightarrow{\text{enzyme}}$$

$$CO_2 + \underset{\text{formate}}{CHOOH} + \underset{\text{ethylene}}{CH_2{=}CH_2} + CH_3\text{-S-R} + NH_3$$

Acceptance of methionine as the precursor is troublesome, since the natural concentration is hardly adequate to support ethylene generation to the extent it occurs in climacteric fruits, which may have an ethylene concentration 3000-fold that of nonclimacteric fruits (Abeles 1972).

RESPONSES TO ETHYLENE

Ethylene activity is not restricted to postharvest physiological responses. Its activity is known to be involved in a host of responses, ranging from germination to senescence (Burg 1962; Abeles 1972; Leopold 1972).

A sharp elevation of ethylene content occurs during ripening of climacteric fruits (Fig. 7.21). Nonclimacteric fruits (e.g., orange, maize, peanut) do not show the familiar ethylene burst. A sharp peak of ethylene production is present in 2- to 3-day-old seedlings during germination (Fig. 7.21). Ethylene pulses in young seedlings are believed to result in greater stem diameter,

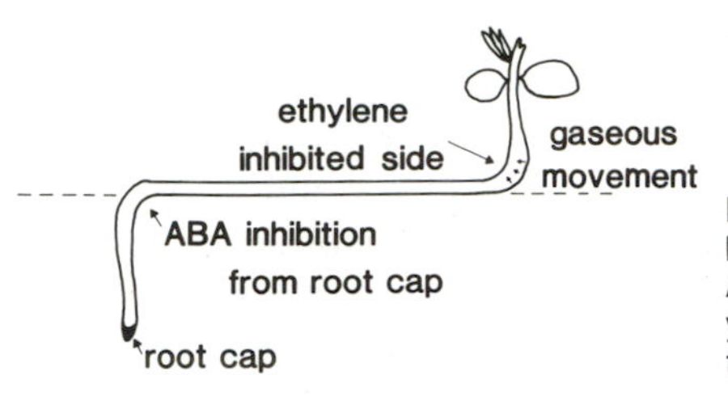

Fig. 7.22. Geotropic response of a horizontally placed seedling indicating ABA translocation from root cap, ethylene production, and upper-side inhibition resulting in bending.

stronger plants, and enhanced survival. The plumule hook of dicot seedlings is an ethylene response (Burg et al. 1971); the hook straightens with exposure to red light. Ethylene effects on seedling growth are referred to as the triple response (Pratt and Goeschl 1969): (1) reduced elongation, (2) increased diameter, and (3) *ageotropic* growth prior to exposure to light. These responses appear to enhance emergence and survival of dicots, particularly the epigeous germination types.

Many plant responses formerly attributed to auxin are now attributed to ethylene, such as geo- and phototropism. According to the ethylene theory, ethylene is generated on the lower side of a stem placed horizontally, due to auxin movement to the lower side in response to gravity. Ethylene diffuses upward as a gas and inhibits growth of the upper side; hence the turned-up response (Fig. 7.22). The inhibition of growth curvature in the presence of CO_2, an ethylene inhibitor, is strong evidence for the ethylene theory (Wheeler and Salisbury 1980). High concentrations of ethylene induced horizontal growth of stems. Increasing concentrations of ethylene in the rhizosphere inhibited root growth, but the inhibition can be corrected by increasing CO_2 at moderate concentrations (Radin and Loomis 1969). This may explain the often reported root growth stimulation by CO_2 enrichment.

Ethylene production has been demonstrated to be associated with the rapid senescence in diseased tissues (Ketring and Melouk 1980). Diseased leaves also abscised. Sprays with $AGNO_3$, an antiethylene-action agent, increased leaf retention on 'Tamnut 75' peanut.

Physical stresses or obstructions to growth have been observed to cause sharp increases in ethylene in the affected tissues. Soil obstruction and resultant ethylene production may explain the *diageotropic* (horizontal) growth of peanut gynophores (pegs), which are positively geotropic until soil penetration. However, ethylene association with this growth habit in the peanut gynophore has not been verified. Ethylene increases germinability of dormant seeds (e.g., peanut) and stimulates germination in witchweed seeds (Eplee 1975). It is necessary to treat seeds of certain peanut varieties with ethephon (an ethylene slow-release compound, 2-chloroethylphosphonic acid) to obtain good germination.

AGRICULTURAL USES OF ETHYLENE

Use of ethylene in agriculture has been limited, in part because of the impracticality of field treatment with a gas. However, a commercial liquid product, ethephon, is now available that releases ethylene slowly to plants. Use on walnut has hastened senescence and dehiscence of hulls, providing earlier harvest and improved nut quality (Fig. 7.23). Ethephon has been used effectively to suppress growth of tobacco seedlings in the starting bed (Kasperbauer and Hamilton 1978). During rainy seasons, seedling growth was too rapid to maintain good transplant material and growth was suppressed for about 10 days by an application of ethephon. Seed treatment with ethephon can break seed dormancy and improve germination.

Elevated levels of ethylene in the atmosphere can cause physiological disorders to plants, such as russet spotting of head lettuce (Morris et al. 1978). In these studies on russet spotting in California the source of the ethylene pollutant was found to be the exhaust gases of internal combustion engines on forklift machines. Ripening fruits that were stored with the lettuce in vacuum coolers were also a source.

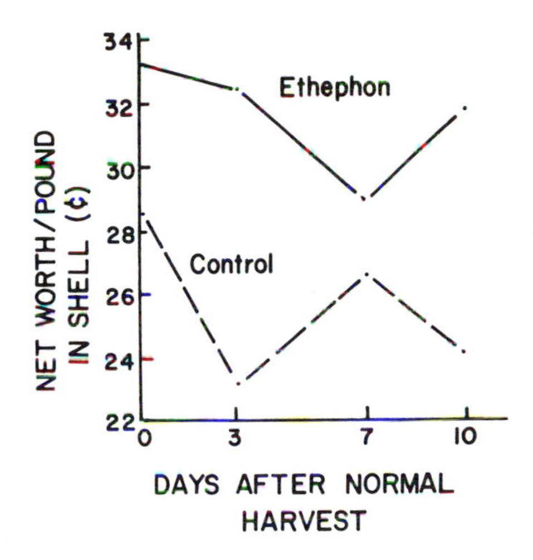

Fig. 7.23. Influence of ethephon on nut quality of walnut by ethephon inducing early dehiscence. (From Sibbett et al. 1978, by permission)

Summary

Chemical substances in minute concentrations, referred to as phytohormones, regulate plant growth and development and coordinate morphogenesis. Plants may be genetically deficient in a specific hormone (endogenous source) and respond to an external application (exogenous source). Plant growth regulators (PGRs) are classified into five groups: (1) auxins; (2) gibberellins; (3) cytokinins, or kinins; (4) inhibitors; and (5) ethylene. There is evidence for a flowering hormone, but it has not yet been isolated or identified. Other natural substances with hormone activity (e.g., triacontanol) that do not fit the five categories have been isolated and identified. The organ of synthesis of growth hormones is generally not the same as the organ of response. Translocation between the two is required except for ethylene, which moves by gaseous diffusion. Young leaves and apical buds are particularly high in auxin, whereas young roots are high in cytokinins and gibberellins. Fruits and seeds generally are rich in all growth hormones.

Numerous synthetic analogs of PGRs have been produced and marketed for commercial use as herbicides, particularly auxins (e.g., 2,4-D, 2,4,5-T, picloram, and certain benzoic acid derivatives). A number of growth inhibitors or retardants have been produced synthetically for agricultural uses (e.g., chlormequat [CCC], daminozide [SADH] and triiodobenzoic acid [TIBA]). Ethephon releases ethylene slowly and is used commercially. Except in a few cases, most modern agronomic crop cultivars evidently have been selected for endogenous levels of hormones high enough to show little or no beneficial response to exogenous sources of auxins, gibberellins, cytokinins, inhibitors, or ethylene. This is not the situation for many horticultural crops with long breeding cycles.

Plant organs respond differently to varying concentrations of PGRs. Shoots are promoted by auxins over a wide concentration range, whereas roots are inhibited except in a short range. Internodes of certain dwarf plant types elongate to normal height if treated with GA over a wide range. Generally hormones act synergistically to induce a response rather than acting alone.

Indoleacetic acid (IAA), GA_3, abscisic acid (ABA), and ethylene are common and widely distributed plant hormones. Zeatin, isolated from maize endosperm, appears to be the most common kinin. Biological assays, such as the *Avena* coleoptile, barley aleurone, and tobacco tissue culture tests, have been developed to quantify the presence of auxins, gibberellins, and kinins, respectively. Most growth hormones affect a wide range of responses so that a number of bioassays are available. Electrochemical assay methods are becoming increasingly important. Hormones are present in the plant in bound or free forms, affecting their availability.

PGRs generally elicit the following characteristic responses (1) auxins stimulate growth by cell elongation and cause apical dominance, (2) gibberellins promote growth of intercalary meristems in the internodes and leaves, (3) cytokinins stimulate growth by cell division, (4) inhibitors retard elongation and induce abscission and senescence, and (5) ethylene promotes ripening in fruits and horizontal growth. Generally PGRs act synergistically, rather than alone, to cause responses.

References

Abeles, F. B. 1972. Annu. Rev. Plant Physiol. 23:259–92.
______. 1073. Ethylene in Plant Biology. New York: Academic Press.
Addicott, F. T., and J. L. Lyon. 1969. Annu. Rev. Plant Physiol. 20:139–64.
Ali, A. A., and R. A. Fletcher. 1971. Can. J. Bot. 49:1727–31.

Allen, O. N. 1973. In Forages, 3d ed., ed. M. E. Heath et al. Ames: Iowa State University Press.
Audus, L. J. 1972. Plant Growth Substances. London: Leonard Hill.
Aung, L. H., A. A. De Hertogh, and G. Staby. 1969. Plant Physiol. 44:403–6.
Bailey, K. M., I. D. J. Phillips, and D. Pitt. 1976. J. Exp. Bot. 27:324–36.
Brian, P. W. 1958. Nature 181:1122–23.
Brian, P. W., and H. G. Henning. 1961. Nature 183:74.
Burg, S. P. 1962. Annu. Rev. Plant Physiol. 13:265–302.
Burg, S. P., and E. A. Burg. 1966. Science 152:1269.
Burg, S. P., A. Apelbaum, W. Eisinger, and B. G. Kang. 1971. Hortic. Sci. 6:359–64.
Carr, D. J., ed. 1972. The Plant Growth Substances. 1970. Berlin: Springer-Verlag.
Cathy, H. M. 1964. Annu. Rev. Plant Physiol. 15:271–302.
Chlor, M. A. 1969. Nature 214:1263–64.
Cocucci, S., and M. Cocucci. 1977. Plant Sci. Lett. 10:85–95.
Collins, G. B., W. E. Vian, and G. C. Phillips. 1978. Crop Sci. 18:286–88.
Cornforth, J. W., B. V. Milborrow, G. Ryback, and P. F. Wareing. 1965. Nature 204:1269–70.
Davis, L. A. 1968. Ph.D. diss., University of California, Davis.
Dörffling, K. 1972. In Hormonal Regulation in Plant Growth and Development, ed. H. Kaldewey and Y. Vardar. Weinheim: Verlag Chemie.
Eplee, R. E. 1975. Weed Sci. 23:433–36.
Evans, L. T. 1966. Science 151:107–8.
Fenton, R., T. A. Mansfield, and R. G. Jarvis. 1982. In Chemical Manipulation of Crop Growth and Development, ed. J. S. McLaren. London: Butterworth.
Fernqvist, I. 1966. Lantbrukshogskol. Ann. 32:109–244.
Galston, A. W. 1947. Am. J. Bot. 34:356–60.
Gardner, F. P. 1980. Western Ill. Univ. Annu. Rep., unpublished.
Gardner, F. P., and J. W. Reeves. 1980. Abstr. Ill. State Acad. Sci.
Greer, H. A. L., and I. C. Anderson. 1965. Crop Sci. 5:229–32.
Hart, R. C., and G. E. Carlson. 1967. USDA-ARS, CR-55-67.
Hedden, P., J. MacMillan, and B. O. Phinney. 1978. Annu. Rev. Plant Physiol. 29:149–92.
Heide, O. M. 1972. In Hormonal Regulation in Plant Growth and Development, ed. H. Kaldewey and Y. Vardar. Weinheim: Verlag Chemie.
Jaggard, K. W., D. K. Lawrence, and P. V. Briscoe. 1982. In Chemical Manipulation of Crop Growth and Development, ed. J. S. McLaren. London: Butterworth.
Jones. E. R. H., H. B. Henbest, G. F. Smith, and J. A. Bently. 1952. Nature 169:485.
Kasperbauer, M. J., and J. L. Hamilton. 1978. Agron. J. 70:363-66.
Ketring, D. L., and H. A. Melouk. 1980. Proc. Am. Peanut Res. Educ. Soc. 12:64.
Krishnamoorthy, H. N., ed. 1975. Gibberellins and Plant Growth. New York: Wiley.
Lang, A. 1970. Annu. Rev. Plant Physiol. 21:537–70.
Leopold, A. C. 1964. Plant Growth and Development. New York: McGraw-Hill.
______. 1972. In Hormonal Regulation in Plant Growth and Development, ed. H. Kaldewey and Y. Vardar. Weinheim: Verlag Chemie.
Leopold, A. C., and P. E. Kriedemann. 1975. Plant Growth and Development. 2d ed. New York: McGraw-Hill.
Leshem, Y. 1973. The Molecular and Hormonal Basis of Plant Growth Regulation. New York: Pergamon.
Letham, D. S. 1968. In Biochemistry and Physiology of Plant Growth Substances, ed. F. Wightman and G. Setterfield. Ottawa: Runge.
Loveys, B. R., and P. F. Wareing. 1971. Planta 98:109–16.
Luckwill, L. C. 1976. Outlook Agric. 9:46–51.
Marre, E. 1977. In Plant Growth Regulators, ed. P. E. Pilet. New York: Springer-Verlag.
Masuda, Y. 1977. In Plant Growth Regulators, ed. P. E. Pilet. New York: Springer-Verlag.
Milborrow, B. V. 1967. Planta. 76:93–113.
______. 1974. Annu. Rev. Plant Physiol. 25:259–307.

Miller, C. O. 1961. Annu. Rev. Plant Physiol. 12:395–408.
Mitchell, J. B., and P. C. Marth. 1947. Growth Regulators for Garden, Field and Orchard. Chicago: University of Chicago Press.
Morris, L. L., A. A. Kader J. A. Klaustermeyer, and C. C. Cheyney. 1978. Calif. Agric. 32:12–13.
N'Diaye, O. 1980. Ph.D. diss., University of Florida, Gainesville.
Nelson, P. M., and E. C. Rossman. 1958. Science 127:1500–1501.
Nitch, C. 1950. Am. J. Bot. 37:211–15.
Ohkuma, K., J. L. Lyon, F. T. Addicott, and F. T. Smith. 1963. Science 142:1592–93.
Paleg, L. G. 1965. Annu. Rev. Plant Physiol. 16:291–322.
Phillips, I. D. J. 1965. Annu. Rev. Plant Physiol. 16:341–67.
Phinney, B. O. 1956. Proc. Natl. Acad. Sci. 42:185–89.
Phinney, B. O., and C. A. West. 1960. Annu. Rev. Plant Physiol. 11:411–36.
Pratt, H. K., and J. D. Goeschl. 1969. Annu. Rev. Plant Physiol. 20:541–84.
Probsting, W. M., P. J. Davis, and G. A. Marx. 1978. Planta 141:231–38.
Quinlan, J. D., and R. J. Weaver. 1969. Plant Physiol. 44:1247–52.
Radin, J. W., and R. S. Loomis. 1969. Plant Physiol. 44:1584–89.
Rice, E. L. 1974. Allelopathy. New York: Academic Press.
Schaeffer, G. W., and A. A. Abdul-Baki. 1973. Bull. Torrey Bot. Club 100:143–46.
Schneider, G. 1970. Annu. Rev. Plant Physiol. 21:499–536.
______. 1972. In Hormonal Regulation of Plant Growth and Development, ed. H. Kaldewey and Y. Vardar. Weinheim: Verlag Chemie.
Scott, P. C., and A. C. Leopold. 1967. Plant Physiol. 42:1021–22.
Shepard, J. F., D. Bidney, and E. Shanin. 1980. Science 208:17–24.
Sibbett, G. S., G. C. Martin, U. C. Davis, and T. Draper. 1978. Calif. Agric. 32:12–13.
Steward, F. C. 1964. Plants at Work. Reading, Mass.: Addison-Wesley.
Tanner, J. W., and S. Ahmed. 1974. Crop Sci. 14:371–74.
Thimann, K. V. 1937. Am. J. Bot. 24:407–12.
______. 1963. Annu. Rev. Plant Physiol. 14:1–18.
______. 1972. In Plant Physiology: A Treatise, vol. 1B, ed. F. C. Steward. New York: Academic Press.
Thomas, T. H. 1976. Outlook Agric. 9:62–68.
Thomas, T. H., P. F. Wareing, and P. M. Robinson. 1965. Nature 205:1270–72.
Torrey, J. G. 1958. Plant Physiol. 33:358–63.
Wain, R. L., and C. H. Faucett. 1969. In Plant Physiology: A Treatise, vol. 1B, ed. F. C. Steward. New York: Academic Press.
Walton, D. C. 1980. Annu. Rev. Plant Physiol. 31:453–89.
Wareing, P. F. 1976. Outlook Agric. 9:42–45.
Wareing, P. F., R. Horgan, I. E. Henson, and W. Davis. 1977. In Plant Growth Regulators, ed. P. E. Pilet. New York: Springer-Verlag.
Wareing, P. F., and I. D. J. Phillips. 1978. The Control of Growth and Differentiation in Plants. 2d ed. New York: Pergamon.
Weaver, R. J. 1972. Plant Growth Substances in Agriculture. San Francisco: W. H. Freeman.
Went, F. W., and K. V. Thimann. 1937. Phytohormones. New York: Macmillan.
Wheeler, R. M., and F. B. Salisbury. 1980. Science 209:1126–27.
Wilkins, M. B., ed. 1969. Physiology of Plant Growth and Development. New York: McGraw-Hill.
______. 1977. In Plant Growth Regulators, ed. P. E. Pilet. New York: Springer-Verlag.
Wittwer, S. H. 1958. Econ. Bot. 12:213–55.
Worsham, A. D., D. E. Moreland, and D. Klingman. 1959. Science 130:1654–56.
Yang, S. F. 1967. Arch. Biochem. Biophys. 122:481–87.
Ziev, M., and E. Zamski. 1975. Ann. Bot. n.s. 39:579–83.

8 Growth and Development

PLANT GROWTH and development are essential processes of life and propagation of a species. They are continuous during the life cycle, depending on availability of meristems, assimilate, hormones and other growth substances, and a supportive environment.

Empirically plant growth can be expressed as a function of genotype × environment = f(internal growth factors × external growth factors). Certain traits in a plant are primarily influenced by genotype, others by the environment; the degree of each depends on the particular trait. DNA codes the sequencing of amino acids into specific proteins and enzymes, establishing the genetic potential for growth, development, and complete morphogenesis. Interaction of genotype with the environment gives expression to the genetic potential.

In modern crop production the object is to maximize growth rates and yield through both genetic and environmental manipulation. Genotypes can be changed by plant breeding and selection, often with dramatic results. The *microclimate* (environment near the plant surface) can be altered in many ways, such as site selection, tillage, irrigation, drainage, fertilization, pest control, and numerous cultural strategies (e.g., planting date, stand density, and spatial arrangement). Most or all of these are commonly used by a grower in modern agriculture and the list could be expanded.

Definition of Growth

Growth is easier to describe than to define. In a restricted sense it is *cell division* (increase in number) and *cell enlargement* (increase in size). Both processes require protein synthesis and are irreversible. The latter process involves hydration and vacuolation. The process of *differentiation* (cell specialization) is frequently considered a part of growth. Plant development requires both growth and differentiation (Fig. 8.1).

While some define plant growth as the processes of cell division and elongation, agronomists generally define growth as increase in dry matter. This definition includes the process of differentiation, which contributes

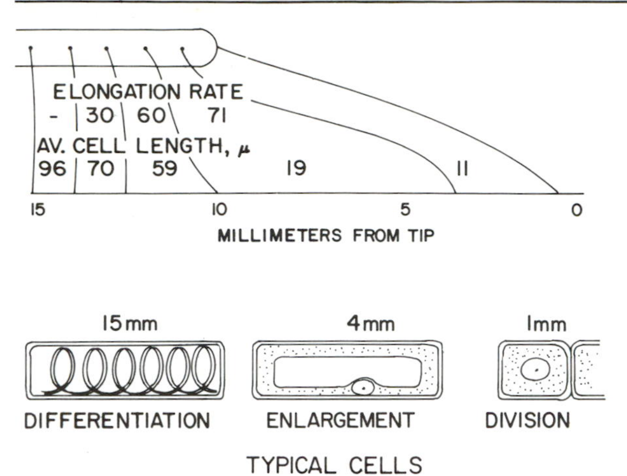

Fig. 8.1. Growth and differentiation regions and representative cell types of a maize root tip. (From Baldovinos 1953)

greatly to dry matter accumulation. In the final analysis, plant development and morphogenesis result from all three: growth by cell division, enlargement, and differentiation.

Dry weight accumulation is commonly used as a parameter characterizing growth because it usually has the greatest economic importance. Any of a number of somewhat related parameters, such as height, volume, and leaf area, also can be used. Fresh weight is less useful because it fluctuates, depending on the moisture status of the plant. Vegetable, flower, and fruit producers, however, are more interested in fresh weight (combined with quality factors) than in dry weight.

Growth Factors

Factors affecting growth, which can be broadly categorized as external (environmental) and internal (genetic), are grouped as follows:

External factors

1. Climatic: light, temperature, water, day length, wind, and gases (CO_2, O_2, N_2, SO_2, nitrogen [N] oxides, Fl, Cl, and O_3). These gases are often atmospheric pollutants (except for the first three) and can be sufficiently concentrated to inhibit growth.
2. Edaphic (soil): texture, structure, organic matter, cation exchange capacity (CEC), pH, base saturation, and nutrient availability. A total of sixteen elements are required by plants (see Chap. 5).
3. Biological: weeds, insects, disease organisms, nematodes, various types of herbivores, and soil microorganisms, such as N_2-fixing and denitrifying bacteria and mycorrhiza (symbiotic fungal association with plant roots)

INTERNAL FACTORS

1. Resistance to climatic, edaphic, and biological stresses
2. Photosynthetic rate
3. Respiration
4. Partitioning of assimilate and N
5. Chlorophyll, carotene, and other pigment contents
6. Type and location of meristems
7. Capacity to store food reserves
8. Enzyme activity
9. Direct gene effects (e.g., heterosis, epistasis)
10. Differentiation

Factors under genetic control that contribute to yield are numerous, so this is only a partial list.

Limitation of Growth Factors

Plant response to nutrient limitation was one of the earliest subjects of scientific plant investigation. The classic works of Liebig, Sachs, Blackman, Mitscherlich, and others were the basis for the formulation of several theories on growth factor limitation and plant response. Some of these concepts have been elevated, erroneously, to the status of laws. Plant responses and the myriad possible interactions are too extensive and complex to be so predictable. However, a knowledge of the theories can lead to a better understanding of plant response and can aid in planning management strategies (Fig. 8.2).

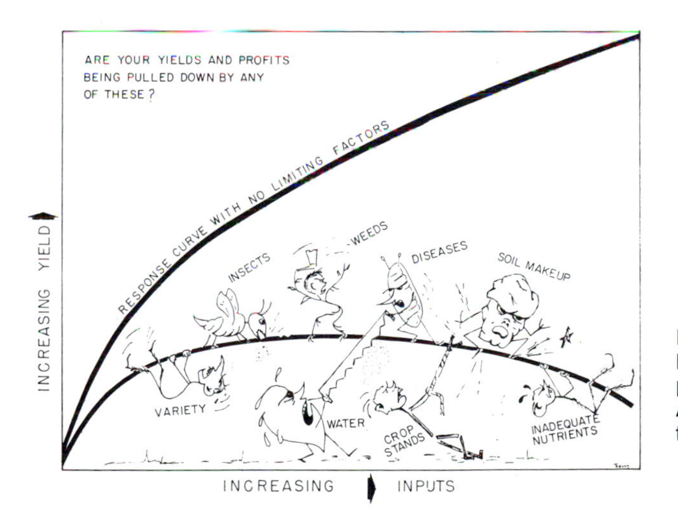

Fig. 8.2. Some of the limiting factors in crop production. (Courtesy of American Potash Institute)

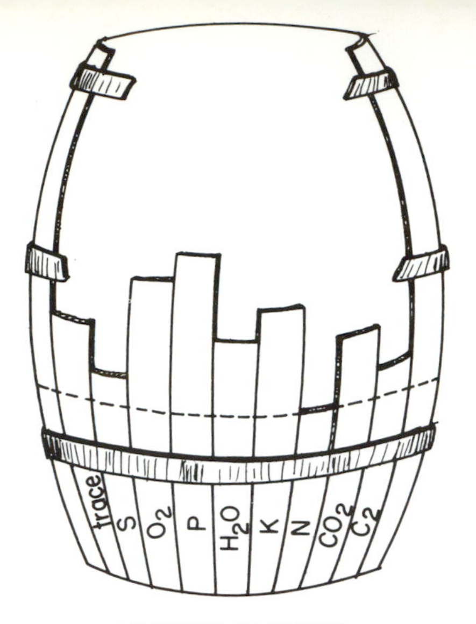

Fig. 8.3. Law of the minimum, illustrated by a barrel with staves of different heights. Nitrogen, the lowest stave, establishes the maximum capacity of the barrel and, by analogy, the maximum possible growth.

LIEBIG: LAW OF THE MINIMUM

The "law of the minimum," proposed by Justus von Liebig (1862), is probably the best known of the limiting factor theories. He stated it as follows: "A deficiency or absence of one necessary constituent, all others being present, renders the soil barren for crops for which that nutrient is needed." It is sometimes referred to as the "barrel concept." If a barrel has staves of different heights, the lowest one establishes the capacity of the barrel (Fig. 8.3). Accordingly, the growth factor in lowest supply (whether climatic, edaphic, biological, or genetic) sets the capacity for yield.

BLACKMAN: OPTIMA AND LIMITING FACTORS

F. F. Blackman's theory of "optima and limiting factors" (1905) was stated as follows: "When a process is conditioned as to its rapidity by a number of separate factors, the rate of the process is limited by the pace of the slowest factor." Light and carbon dioxide are required for photosynthesis. The Blackman theory suggests that there is an abrupt cessation of the process, the "Blackman response," if one of these factors becomes limiting (Fig. 8.4). However, nature seems to abhor angles and this type of response is seldom

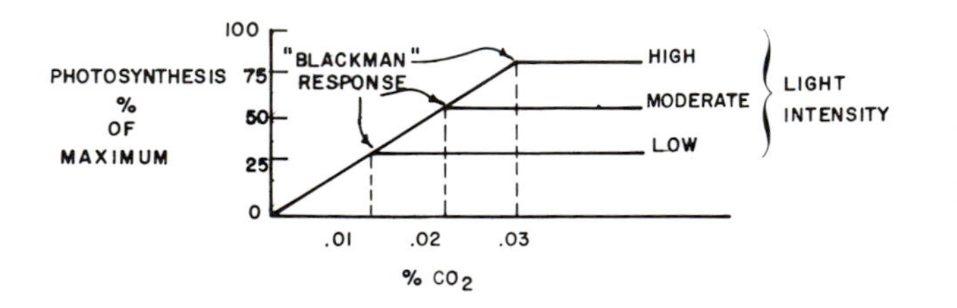

Fig. 8.4. Assimilation of CO_2 and its interaction with light intensity (Blackman 1905).

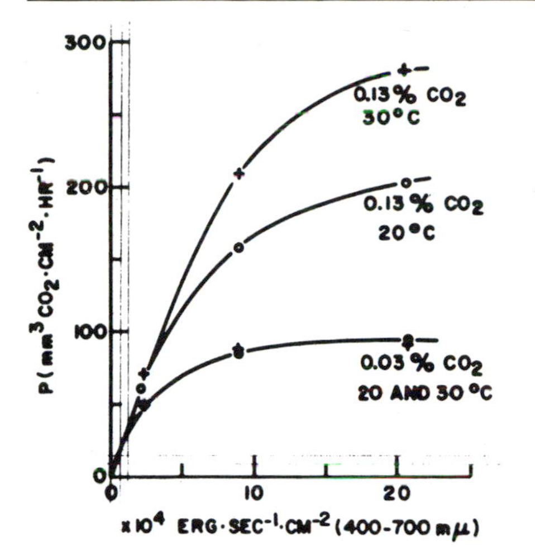

Fig. 8.5. Photosynthesis of a cucumber leaf in relation to light intensity, temperature, and CO_2 concentration (Gaastra 1963).

found. Rather, response lines to factors limiting photosynthesis are curvilinear and approach a maximum limit asymptotically (Fig. 8.5).

MITSCHERLICH: LAW OF DIMINISHING RETURNS

A German soil scientist, E. A. Mitscherlich (1909), developed an equation that related growth to the supply of growth factors. He observed that when plants had adequate amounts of all but one limiting element the growth response was proportional to the limitation element. Plant growth increased with additional increments of a limiting factor, but not in direct proportion (Fig. 8.6). The Mitscherlich "law of diminishing returns" states, "The increase in any crop produced by a unit increment of a deficient factor is proportional to the decrement of that factor from the maximum." The response is curvilinear instead of linear, as Blackman had suggested.

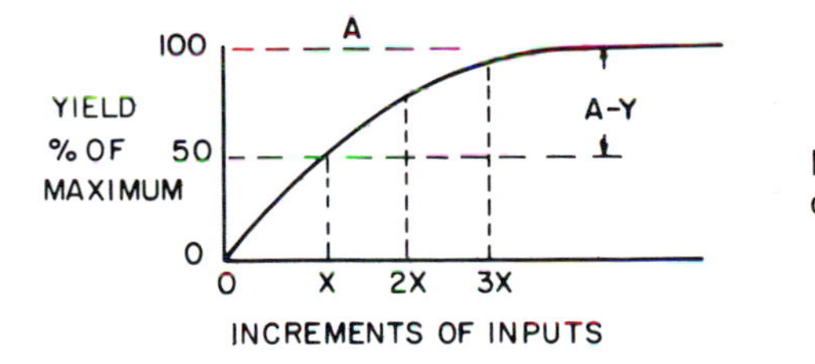

Fig. 8.6. Growth response curve described by the Mitscherlich equation.

The Mitscherlich equation is as follows:

$$dy/dx = C(A - Y)$$

where d is the increment of change, dy is the amount of increase in yield (y)

resulting from an increment of the growth factor (dx), A is the maximum possible yield obtained by an unlimited supply of the growth factor in question, Y is the yield obtained with any given quantity of the factor x, and C is a proportionality constant that depends on the nature of the growth factor. The growth increase in y is greatest for the first increment of x (Table 8.1); the amount of increased yield (y) becomes progressively smaller with each added increment of x, theoretically about one-half the response from the previous increment.

TABLE 8.1. Amount of nutrient or factor required to give one-half yield (Baule unit), as applied in the Mitscherlich equation

Baule Units Added	Predicted Yield % of Maximum
0	0.00
1	50.00
2	75.00
3	87.50
4	93.75
∞	100.00

When yields are expressed on a relative basis ($A = 100$), $C = 0.301$. If $A = 100\%$, the equation may be written as follows:

$$\log (100 - Y) = \log 100 - 0.301x$$

The amount of a nutrient or other factor necessary to give one-half the maximum yield is referred to as a *Baule unit*. From this hypothesis, as is seen in the equation, yields can be predicted from additional Baule units (Table 8.1).

Willcox (1937) proposed that the value of C in the equation is a constant for all crops, but this idea has not been generally accepted.

MACY: CRITICAL PERCENTAGES

Macy (1936) added a new dimension to these concepts by suggesting a relationship between the sufficiency of nutrients and plant response in terms of both yield and nutrient concentration of plant tissues. Macy proposed a critical percentage for each nutrient in each kind of plant (Fig. 8.7). In the tissue-minimum-percentage range, an added increment of a nutrient increases the yield but not the nutrient percentage. In the poverty-adjustment range, added increments of a nutrient increase both yield and nutrient percentage. In the luxury-consumption range, added increments of a nutrient have little effect on yield but increase the composition percentage of the nutrient. He suggested that Liebig's law holds in the tissue-minimum-percentage range because there is not enough of a nutrient to allow much plant growth. Liebig's law holds again in the luxury-consumption range because, although there is a large sup-

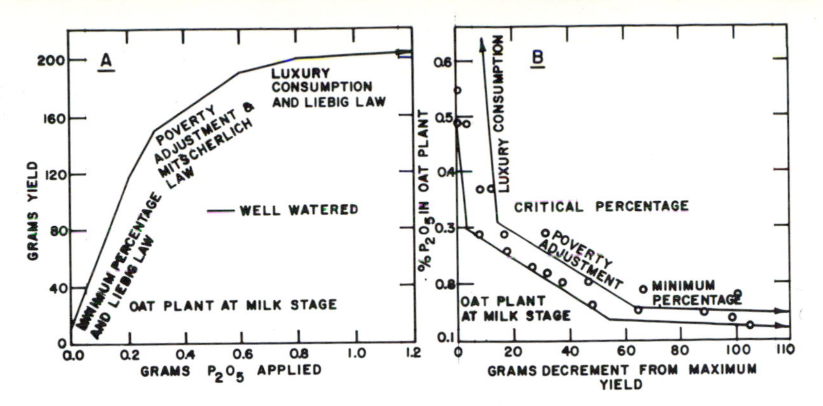

Fig. 8.7. Macy's interpretation of the Liebig and Mitscherlich concepts, which he correlated with the minimum percentage, poverty adjustment, and luxury consumption levels of a nutrient. The point between poverty adjustment and luxury consumption was the critical percentage (Macy 1936).

ply of one nutrient, some other nutrient becomes limiting and stops growth. Mitscherlich's law of diminishing returns holds during the poverty-adjustment range because the response is curvilinear (i.e., representing diminishing returns) to added increments.

Limiting Factor Qualifications

Several reasons are apparent as to why the above are only theories and not accepted as laws.

1. Biological reactions are complex and may proceed by more than a single pathway. Compound A can go to compound B and then to C, or directly to C, all reversibly. Slowing one pathway or lowering one reactant does not necessarily slow or stop the reaction.
2. Factors substitute for factors. Sodium, for example, can partially substitute for potassium in certain species.
3. Factors modify or affect other factors. Phosphorus decreases zinc uptake, and potassium decreases magnesium uptake, for example. Increasing the irradiance increases temperature and decreases water availability.
4. Plants affect factors, and factors in turn affect plants. For example, additional N increases plant growth and leaf area, which in turn reduces light to lower leaves. Soil temperature is decreased and humidity increased by shading from additional growth.
5. More than one factor may be limiting simultaneously.

Meristems

Growth by cell division and enlargement occurs in specialized tissues known as *meristems,* which are found at a number of locations in the plant (Fig. 8.8). The number of meristems of a plant is large, but in terms of total

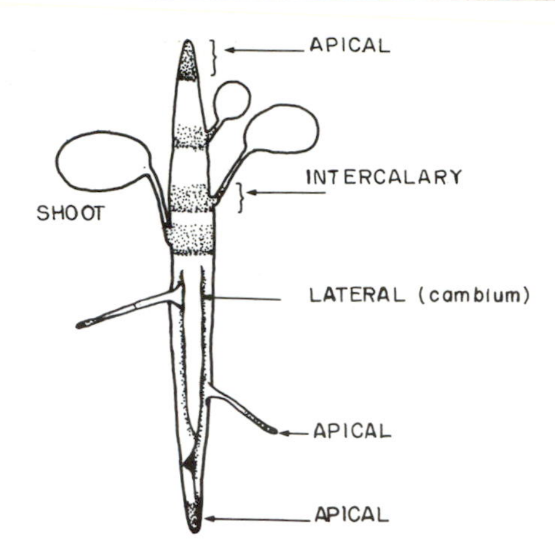

Fig. 8.8. Plant meristems. (From Janick 1963)

mass, meristematic tissue is minor. Meristems may compete strongly with each other for organic and mineral nutrients. In fact much of the art of crop production depends on the management of meristem competition (e.g., increasing the number of tillers, branches, inflorescences, and leaf area) by promoting meristematic development in certain potential apical buds (e.g., axillary buds) and *intercalary* meristems (those between differentiated tissues). It is usually desirable to promote more and larger leaves from apical and intercalary meristems and often desirable to promote more tillers or branches from apical meristems arising from dormant or quiescent buds in leaf axils.

Lateral meristems generate the new cells that expand the breadth or diameter of an organ. The *vascular cambium* is a specialized lateral meristem from which secondary xylem and phloem are derived. Another type of lateral meristem is located at the margin of young, expanding leaves.

Apical meristems generate new cells in the tip of roots or shoots, resulting in increased height or length (Fig. 8.8). The relationship between apical and lateral meristem development to growth hormones is discussed in Chapter 7.

A specialized, intercalary meristem is located between two previously differentiated tissues of certain organs (Fig. 8.8), such as between node and internode or between leaf blade and sheath. The pulvinus in a grass stem contains an intercalary meristem. Stems contain intercalary meristems at internode bases. If a stem lodges, growth from the intercalary meristems by cell elongation occurs at differential rates around the stem circumference, resulting in regaining the upright position. Intercalary meristems at the bases of leaf blades and sheaths function to extend leaf length.

In discussing meristems, it is useful to differentiate between diffuse and massed meristems. Diffuse meristems have a low cell number and/or cell activity and also require an external source of hormones for growth. Cambium and recently fertilized eggs are examples; if a rudimentary organ, such as a fertilized egg, does not soon develop into a massed meristem with high cell activity and generate its own hormone supply, it aborts. An apical bud contains a massed meristem and has a self-generating hormone supply. Very often crop management practices are designed to control formation of massed meristems; for example, maize is planted at a moderate density to favor the development of one or two ear shoots (from massed meristems), no less and no more. Wheat is seeded at a moderate rate to favor the production of tillers, but not so many as to cause much head abortion in the pseudostems prior to heading. Therefore massed meristems have sufficient cell numbers and activity to assure adequate hormone production for perpetual cell division and the directed flow of carbohydrates and other nutrients for morphogenesis.

Growth Correlations

Plants acquire a characteristic shape or form by correlated growth of component parts. Component parts also have a characteristic shape or form that is repeatable in time and space. A favorable environment can enhance growth quantitatively, but the geometry of the parts and the whole is relatively constant.

ALLOMETRY

The relationship between the growth rates of individual parts of an organ or organism is referred to as *allometry.* The relationship between two variants (X and Y) may be expressed as $Y = bx^K$, where x and y are physical parameters, and b and K are constants, K being the allometric constant. The quantity K can be calculated from the equation $\log y = \log b + K \log x$. It may be obtained by plotting y against x on a double logarithm scale, which produces a straight line, the slope of which is K. It also may be determined by linear regression analysis of the data set y and x. If the length and breadth of an organ, such as a leaf, expand at the same rate, the slope of the regression line (the coefficient of allometry, or K) is 1.0; the growth rates of the two parameters are perfectly correlated. Hammond (1941) showed that the allometry of normal and okra-type leaf of cotton was highly heritable and controlled by a single gene. Allometry coefficients of top and root relationships are based on dry weights rather than dimensions and usually exhibit a lower K. The *harvest index* (proportion of seed weight to whole plant weight) has a relatively high coefficient of allometry and is a stable parameter in time and space. Although allometry usually deals with physical parameters of the plant, it logically follows that physiological processes are correlated. Allometry calculations of various correlations can provide useful approximations, but it can be shown mathematically that they are not exact.

SHOOT-ROOT RATIO

The allometry of top growth to root growth, usually expressed as the shoot-root (S-R) ratio, has physiological significance, since it can reflect one

TABLE 8.2. Distribution of dry weight and ^{14}C between shoots and roots in rice plants receiving three rates of nitrogen fertilization

Nitrogen Added (g/plant)	Plant Part[a] and Ratio	Dry Wt. (g/plant)	$^{14}C_2$ after 4 Days[b] (%)
0 (low)	Shoot	1.86	59
	Root	0.76	41
	S-R ratio (2.45)		
	Total	2.62	
3 (medium)	Shoot	7.41	86
	Root	2.11	15
	S-R ratio (3.51)		
	Total	9.52	
6 (high)	Shoot	8.40	89
	Root	2.33	11
	S-R ratio (3.60)		
	Total	10.73	

Source: From Murata 1969.

[a]Shoots refer to all above-ground parts, roots to below-ground parts.

[b]$^{14}CO_2$ applied, 50 μcurie/plant, for 1 hr, at maximum tiller number stage; counts of ^{14}C as counts • min^{-1} • $plant^{-1}$ × 10^3 made after 4 hr $^{14}CO_2$ treatment.

type of tolerance to drought stress. Although the S-R ratio is under genetic control, it is also strongly influenced by environment. Murata (1969) showed that N fertilization had a pronounced influence on the S-R ratio of rice (Table 8.2). Under a high-N regime approximately 90% of the photosynthate was partitioned to the shoot, compared with only 50% to the shoot under low N. New shoot growth, stimulated by N, was a stronger assimilate sink than were roots.

A deficiency of water, while curtailing both top and root growth, had a relatively greater effect on top growth (Loomis 1953). Tops are favored differentially when N and water are plentiful; roots are favored when these factors are limited, as reflected by S-R ratios. Roots have the first access to water, N, and other edaphic factors. Tops have the first access to light, CO_2, or climatic factors. (Water and other factors affecting top and root growth are discussed in detail in Chaps. 10 and 11.)

APICAL AND LATERAL GROWTH

Plants assume a characteristic form or geometry largely due to the extent of growth from apical and lateral buds. Growth from lateral buds can profoundly change plant shape and appearance. Lateral growth as new shoots normally arises from buds in leaf axils, frequently from the compacted nodal section of the basal stem referred to as the *crown*. New shoots can also emerge adventitiously from any position. The end result is that plants tend to fill the space available to them, an advantage in natural survival and productivity. Light is a primary factor controlling growth from lateral buds.

VEGETATIVE AND REPRODUCTIVE GROWTH

Reproductive growth in annual plants appears to make virtually absolute demands on assimilate. In annuals vegetative growth is generally terminated by reproduction. Leaves, stems, and other vegetative parts not only fail to compete for current assimilate during ripening of fruits but to some extent may sacrifice previously accumulated carbon and minerals via mobilization and redistribution. This process accelerates senescence and eventually results in the death of the plant.

Perennials appear to make only a partial commitment to reproduction; that is, shoots that bear fruits may remain healthy, or even if they die, new vegetative shoots are generated from axillary buds to replace them at senescence of the fruiting shoots. Perennials such as apple and citrus trees do not appear to be greatly affected by the presence of ripening fruits. The shoots of perennial herbaceous grasses and legumes that bear fruits usually senesce like annual plants, but new shoots arise from crown buds, which results in perennation. (Vegetative and reproductive growth are discussed in Chaps. 11 and 12.)

GROWTH AND DIFFERENTIATION

Plant development is a combination of a host of complex processes of growth and differentiation that leads to an accumulation of dry matter. Differentiation processes have three requisites: (1) available assimilate in excess of most metabolic uses, (2) a favorable temperature, and (3) a proper enzyme system to mediate the differentiation process. If these requisites were met, one or more of three differentiation responses occurred: (1) cell wall thickening, (2) deposit of cell inclusions, and (3) hardening of the protoplasm (Loomis 1953). The latter is important in preventing protoplasmic damage from natural stresses, such as cold, heat, or drought. For example, well-hardened nursery stock or transplants can be set out with more success than nonhardened or tender stock.

The first essential to differentiation processes is availability of carbohydrate, assuming the necessary enzyme system is in place. Assimilate in excess of growth requirements normally results from factors that curtail growth without curtailing photosynthesis. Factors that limit growth more than they limit photosynthesis, such as water or N deficiencies, result in surplus photosynthate to drive the differentiation process, given a favorable temperature and necessary enzymes. Cell wall thickening, secondary-product accumulation (e.g., alkaloids and starches), and protoplasm hardening may occur, depending on enzymes and temperature. These chemical changes can result in changes in anatomy and morphology.

The production of quality crop products often requires production strategies that achieve an appropriate balance between growth and differentiation. Growth is essential but generally should not be favored (for example, with water and N) as to preclude differentiation.

Cereal crops grown under high-water and high-N regimes, especially with low irradiance (as in thick stands), have thin walls in the stems and tend to lodge. Limiting these factors will cause the reverse. Thin cell walls are desirable in the petioles of celery for tenderness, so the object is to promote petiole growth with adequate supplies of water and N and to reduce differentiation by these treatments and by shading the petioles to reduce assimilate.

Management strategies for cell inclusions are essentially the same. Sugar beet accumulates sugar poorly if N and water are excessive (Fig. 8.9). While N promotes biomass yield, the percentage of sucrose is negatively correlated with soil N. High rates of N decrease the yield of sucrose per land unit. Cool nights are also necessary for sugar accumulation. After production of good assimilatory and sugar-storage systems in the vegetative growth, which requires most of the growing season, then high radiation, cool temperatures, and less-than-optimum N and water levels are needed (as are typical of northern latitudes). Production of quality sweet melons requires similar management strategies, except for the cool temperatures. Sandy soils are used for commercial production of melons in humid areas in order to control the levels of water and N during ripening. Melons produced on heavy soils in humid areas will usually be larger but lack sweetness (cell inclusions).

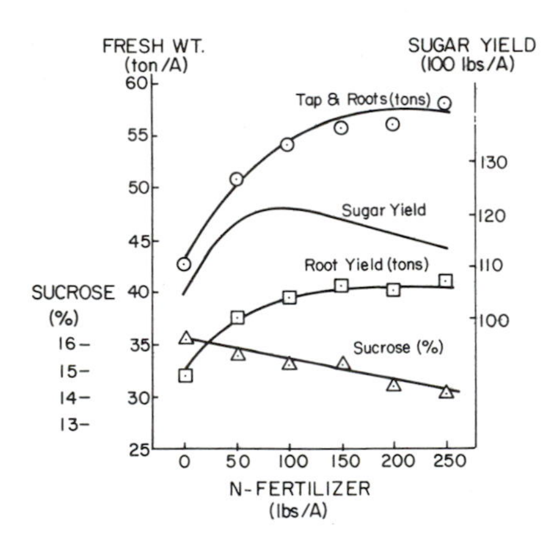

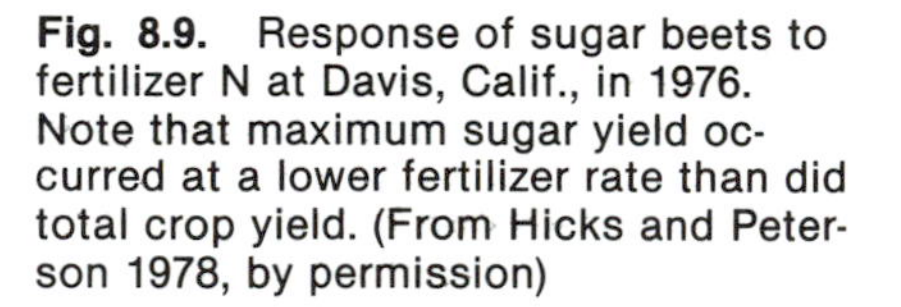
Fig. 8.9. Response of sugar beets to fertilizer N at Davis, Calif., in 1976. Note that maximum sugar yield occurred at a lower fertilizer rate than did total crop yield. (From Hicks and Peterson 1978, by permission)

Alfalfa accumulates starches in fleshy taproots in sunny, cool fall days. High irradiance during the day and cool nights favor excess assimilate, since photosynthesis output exceeds growth and maintenance respiration requirements. The result was accumulation of carbohydrate food reserves in the taproots and hardening of the protoplasm for overwintering (Escalada and Smith 1972).

HARVEST INDEX

Harvest index reflects the proportion of assimilate distribution between economic and total biomass (Donald and Hamblin 1976). Harvest index is similar to the partitioning coefficient. (The harvest index is discussed in greater detail in Chap. 3.) It is important to stress, as has Stoy (1969), that translocation to the metabolic sinks (e.g., roots, new shoots, developing fruits) is extremely complex and that the mechanism or attracting force that directs or regulates partitioning to the metabolic sinks is not known.

Growth Dynamics

The pattern of growth over a generation is typically characterized by a growth function referred to as the *sigmoid curve*. The time frame may vary from less than days to years, depending on the organism or organ; but the sigmoid accumulation pattern typifies all organisms, organs, tissues, and even cell constituents. If plant mass (dry matter), volume, leaf area, height, or accumulation of chemical substances are plotted against time, a line fitted to the data will normally be sigmoidal (Fig. 8.10). The S-shaped curve results from differential rates of growth during the life cycle. For example, seedling growth is slow and usually negative in dry matter gain for a short period of 1 or 2 wk. This phase is followed by a period of exponential growth rate (phase *a* in Fig. 8.10); the exponential phase is relatively short in crop canopies. The linear phase (*b*) follows for a relatively long period, during which dry matter increases are at a constant rate. In crop stands the linear phase is an expression of the crop growth rate (CGR). The CGR of stems, and other parts to a lesser degree, may become negative as the stem loses weight with the onset of grain formation, due to mobilization and redistribution of labile food reserves to the seeds. This principle of redistribution of assimilate from vegetative to repro-

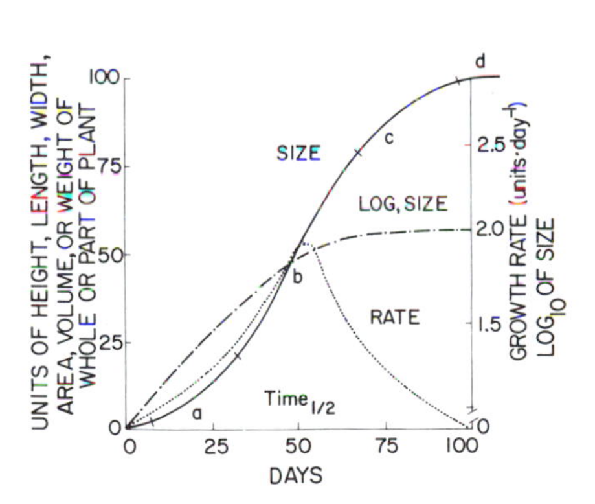

Fig. 8.10. Generalized lines or curves for size (weight, height, length, width, area, volume), log of size, and growth rate, each plotted against time. Growth phases are indicated in the size curve as follows: exponential or logarithmic (*a*), linear (*b*), dampened exponential (*c*), and steady state (physiological maturity) (*d*). Note that the growth rate peaks at about $t_{1/2}$.

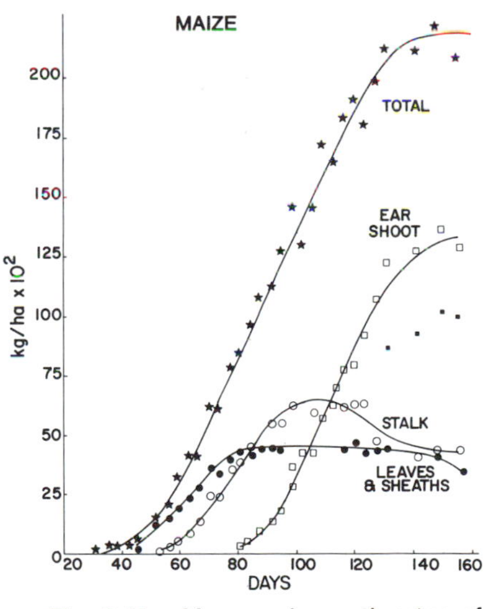

Fig. 8.11. Measured growth rates of maize plants and component parts. (From Duncan 1981. Data from Ohio Agricultural Experiment Station, Wooster)

ductive structures is illustrated in Figure 8.11. The linear rate is followed by a phase of declining rate (*c*); the increases in growth become progressively less with time, until a steady state (*d*) is reached. This steady state phase is referred to as *physiological maturity.* At this stage gains, for example, in dry matter, are in balance with losses.

Quantitative aspects of growth can often best be visualized by studying single-celled organisms, such as yeast, or small plants, such as duckweed. Assuming no environmental limitation, the number of cells or the number of duckweeds can be predicted as follows: $N_1 = N_0 \times 2$ after the first generation, where N_1 = the number of single cells or duckweeds and N_0 = the initial number of cells or duckweeds. Thus at the end of the first generation two cells or duckweeds have taken the place of one. The number at the end of the second generation is $N_2 = N_0 \times 2 \times 2 = N_0 \times 2^2$, and the number at the end of the third generation, starting with a single cell or duckweed, is $N_3 = N_0 \times 2^3 = 8$. The number at the end of *n* generations would be $N_n = N_0 \times 2^n$.

If the interval of time between generations is relatively constant, as is usually true, the equation can be written $N = N_0e^{kt}$, where *e* is the natural logarithm base, *k* is a constant for growth rate, and *t* is time. The function reflects an exponential or logarithmic, rather than linear, rate of growth, often referred to as the compound interest equation. It has been applied under certain conditions to growth of multicellular or higher plants by substituting weight for cell number as follows: $W = W_0e^{rt}$, where *W* is weight and the *r* is the constant for growth rate ("interest rate").

With either single-celled or higher plants, growth at the compounded (exponential) rate cannot persist once there is competition (e.g., for space, substrate, or nutrients). In field crops the exponential phase may persist only a few days, especially in dense stands. Even in spaced plants, internal competition or some other limitation ends the exponential rate and initiates the linear one (Fig. 8.10). Once the canopy closes, growth rate is linear until senescence, which finally slows the rate to zero or to a steady state.

Many agronomists attach little significance to the exponential phase, since it is at a juvenile state and short-lived. The duration and slope (rate) of the linear phase (Fig. 8.10, *b–c*) best describes yield. Linear growth is expressed by the equation $a + bx$, where *a* is the intercept of the *y* axis and *b* the slope (rate) per unit of *x*.

Growth Analysis

Frequently, investigators need to know more than the end result, the final yield of dry matter. Events along the way may have had a marked influence on the final outcome. One approach to the analysis of yield-influencing factors and plant development as net photosynthate accumulation is naturally integrated over time has come to be known as *growth analysis.*

The basic concept and physiological implications in growth analysis are relatively simple and have been explained in the early classic approaches by V. H. Blackman (1919), Briggs et al. (1920), and Fisher (1920). Growth analysis came to be used extensively in the British Commonwealth countries, including the classic work of Watson (1947, 1952). In recent years growth analysis has been used by plant physiologists and agronomists in the United States (Radford 1967). Two extensive treatments of the subject have been published (Evans 1972; Hunt 1978).

Only two measurements, made at frequent intervals, are required for growth analysis: leaf area and dry weight. Other quantities in the analysis are derived by calculation (Table 8.3). The commonest approach, classic growth analysis, involves measurements made at fairly long intervals (1–2 wk) on a relatively large number of plants. The second approach involves measurements at more frequent intervals (2–3 days) on a smaller number of plants. Both approaches provide mean values of the quantitative changes that occur over any particular time interval. The second approach, using more frequent harvests, has been suggested to make better use of the materials and time of the researcher (Hunt 1978).

Dry weight is determined by standard procedures. Leaf area (one side only) can be determined a number of ways. Currently the commonest is by a photoelectric device that reads leaf area directly as individual leaves are fed into it. Another common method is using linear regression analysis: area $= a + b\ (l \times w)$, where $b =$ slope, $l =$ leaf length, and $w =$ leaf width. From regression analysis on 60 leaflets, the author obtained the equation for leaf area determination of bush bean: $a = .624 + .583\ (l \times w)$. Equations for most crop plants have been published (Sepaskhah 1977). Other methods of leaf area determination include tracing fresh leaves on grid, blueprint, or photocopy paper to determine area-weight ratio. The leaf weight experimentally determined can then be converted to leaf area by calculation. Other quantities for the growth analysis can be calculated (Table 8.3).

Growth analysis can be made of individual plants or of plant communities. Analysis of individual plant growth, generally made at the early stages, includes the following: (1) relative and absolute growth rates, (2) unit leaf rate or net assimilation rate, (3) leaf area ratio, (4) specific leaf area, and (5) specific leaf weight and allometry in growth (e.g., S-R ratio) (Table 8.3).

Agronomists generally analyze plant community growth, since it represents the accumulation of economic yield. Quantities used in growth analysis of plant communities include (1) leaf area index, (2) leaf area duration, (3) crop growth rate of total biomass (usually the above-ground parts) and of the economic biomass (e.g., seeds, tubers), and (4) net assimilation rate. The partitioning coefficient or index can be calculated as a ratio between the economic biomass and the total biomass. A complete growth analysis evaluates both the individual plant and the plant community. The formulas, symbols, and other information for calculating the quantities derived in growth analysis are given in Table 8.3.

TABLE 8.3. Growth analysis quantities derived from plant weight and leaf area

Derived Quantity	Symbol	Instantaneous Value[a]	Formula for Mean Value over Time Interval $(T_2 - T_1)$[b]	Unit
Relative growth rate	RGR	$1/W \cdot dw/dt$	$\overline{RGR} = (\ln W_2 - \ln W_1)/(T_2 - T_1)$	$W \cdot W^{-1} \cdot T^{-1}$
Leaf area ratio	LAR	L_A/W	$\overline{LAR} = (L_{A_2}/W_2 + L_{A_1} + W_1)/2$	$A \cdot W^{-1}$
Specific leaf area	SLA	L_A/L_W	$\overline{SLA} = (L_{A_2}/W_2 + L_{A_1}/W_1)/2$	$A \cdot W^{-1}$
Specific leaf weight	SLW	L_W/L_A	$\overline{SLW} = (L_{W_2}/L_{A_2}) + (L_{W_1}/L_{A_1})/2$	$W \cdot A^{-1}$
Net assimilation rate	NAR	$1/L_A \cdot dw/dt$	$\overline{NAR} = (W_2 - W_1)/(T_2 - T_1) \cdot (\ln L_{A_2} - \ln L_{A_1})/(L_{A_2} - L_{A_1})$	$W \cdot A^{-1} \cdot T^{-1}$
Leaf area index	LAI	L_A/P	$\overline{LAI} = (L_{A_2} + L_{A_1})/2 \cdot (1/G_A)$	dimensionless
Crop growth	CGR	$1/P \cdot dw/dt$	$\overline{CGR} = 1/G_A \cdot (W_2 - W_1)/(T_2 - T_1)$	$W \cdot A^{-1} \cdot T^{-1}$
Leaf area duration (leaf area basis)	LAD	None	$\overline{LAD} = (L_{A_2} + L_{A_1})(T_2 - T_1)/2$	$A \cdot T$
Leaf area duration (leaf area index)	LAID	None	$\overline{LAID} = (L_{A_1} + L_{A_2})(T_2 - T_1)/2$	T
Biomass duration	BMD	None	$\overline{BMD} = [(W_2 + W_1)/2] \cdot (T_2 - T_1)$	$W \cdot T$

[a] L_A = leaf area, L_W = leaf weight, G_A = ground area, T = time, W = weight, A = area.
[b] Approximate except for relative growth rates.

RELATIVE GROWTH RATE

Relative growth rate (RGR) expresses the dry weight increase in a time interval in relation to the initial weight. In practical situations, the mean relative growth rate ($\overline{RGR}$) is calculated from measurements taken at times t_1 and t_2. The equation for calculating the RGR is derived from the compound interest equation discussed earlier, $W = W_0e^{rt}$, where W is the weight at any given time, W_0 is the initial weight, e is the natural logarithm base (2.71828), r is the relative growth rate, and t is the length of the time period. The $\overline{RGR}$ does not imply a constant growth rate during a particular t_1 to t_2 time frame; it can vary from instantaneous values of RGR. RGR is the slope of the line when $\log_e W$ is plotted against t.

The example in Table 8.4 shows that plants *A* and *B* had the same RGR, although *B* gained 10 g and *A* gained 5 g. This is because *B* was twice as large at the start. The RGR of crop plants generally begins slowly just after germination, peaks rapidly soon afterward, and then falls off. Species vary in RGR. For example, during a 5-wk period from germination Grime and Hunt (1975) observed a wide distribution of RGRs among woody and herbaceous species under favorable conditions. The maximum $\overline{RGR}$s ranged from a low of 0.22 g • wk^{-1} for Sitka spruce to a high of 2.70 g • wk^{-1} for *Poa annus* (annual bluegrass).

TABLE 8.4. Relative growth rate of two hypothetical plants of different sizes

	Plant	
Quantity	A	B
W_1 (g)	5	10
W_2 (g)	10	20
Gain in 1 week (g)	5	10
$\log_e W^2 - \log_e W^1$	2.30–1.60	3.00–2.30
$t_2 - t_1$ (wk)	1	1
RGR g • g^{-1} • wk^{-1}	.70	.70

LEAF AREA RATIO

Leaf area ratio (LAR) expresses the ratio between the area of leaf lamina or photosynthesizing tissue and the total respiring plant tissues or total plant biomass (Table 8.3). The LAR reflects the leafiness of a plant, but mean values ($\overline{LAR}$) were not precise (Hunt 1978). Plants like sunflower and beet had a high LAR, compared with plants like pine, and also had a 10-fold greater RGR (Jarvis and Jarvis 1964). Such differences suggest, all other factors being equal, a strong competitive position for sunflower in the juvenile stages.

NET ASSIMILATION RATE

Net assimilation rate (NAR), or unit leaf rate, is the net gain of assimilate, mostly photosynthetic, per unit of leaf area and time. It also includes gain in minerals, but this is not a large fraction since minerals constitute only 5% or less of the total weight. The equation to calculate mean values ($\overline{NAR}$) (Table 8.3) assumes that the relationship between plant weight and leaf area is linear; this assumption may hold for early phases of ontogeny but not for latter phases, as growth rate of leaf area may exceed that of dry matter or vice versa. NAR is not constant with time but shows an ontogenetic downward drift with plant age. The age drift was accelerated by an unfavorable environment (Hunt 1978), and dry matter gain per unit leaf surface decreases as new leaves are added due to mutual shading. Increased competition for nutrients and other factors are probably also important as age and size increase.

LEAF AREA INDEX

Crop production is the practical means of trapping solar energy and converting it into food and other usable materials. Crop production strategies are usually designed to maximize light interception by achieving complete ground cover through manipulating plant density and spatial arrangement and promoting rapid leaf expansion. Bare ground does not trap and convert light energy.

Leaf area index (LAI) expresses the ratio of leaf surface (one side only) to the ground area occupied by the crop. Mean values ($\overline{LAI}$) are calculated in Table 8.3. A LAI of 1, which is one unit of leaf surface area per unit of land surface, theoretically could intercept all incident light; but it seldom does, due to leaf shape, thinness (light is transmitted), inclination, and vertical distribution variations. A LAI of 3–5 is usually necessary for maximum dry matter production of most cultivated crops. Forage crops, such as grasses, with erectophile (upright) leaf orientation may require a LAI of 8–10 under favorable conditions to maximize light interception. A higher LAI is also required where total biomass, not economic yield, is the objective (e.g., forage crops). In this case assimilate in excess of growth and maintenance respiration in amounts sufficient to produce seeds or tubers are not required or desired.

The LAI and its seasonal distribution varies considerably with species (see Fig. 8.13). Values required for maximum production increase with the level of solar radiation.

CROP GROWTH RATE

Crop growth rate, the gain in weight of a community of plants on a unit of land in a unit of time, is used extensively in growth analysis of field crops. See Table 8.3 for a calculation of the mean value ($\overline{CGR}$). A ($\overline{CGR}$) of 20 g • m^{-2} • day^{-1} (200 kg • ha^{-1} • day^{-1}) is considered respectable for most crops, particularly C_3 types. A $\overline{CGR}$ of 30 g • m^{-2} • day^{-1} (300 kg • ha^{-1} • day^{-1}; for grain, 6 bu • a^{-1} • day^{-1}) is obtainable from C_4 types, such as maize. The CGR of economic yield, such as grain or tubers, is of equal or greater interest. When the total and economic dry weights are plotted against time, the slope of the regression line of the linear phase (slope = CGR) is usually similar for high-yielding cultivars. The ratio of CGR_{econ} to CGR_{total} produces a useful quantity, referred to as the *partitioning coefficient* or *index* (Duncan et al. 1978). The partitioning coefficient of a crop expresses efficiency in conversion of assimilate to economic yield. Modern cultivars of peanut partitioned at about 75%, some of the older cultivars at only 40 to 50% (Duncan et al. 1978). The nut yields from new peanut cultivars, such as 'Florunner,' are about twice those from 'Dixie Runner' and other older cultivars with low partitioning coefficients.

LEAF AREA DURATION

Leaf area duration (LAD) expresses the magnitude and persistence of leaf area or leafiness during the period of crop growth. It reflects the extent or seasonal integral of light interception and in wheat has been shown to correlate highly with yield. Mean leaf area duration ($\overline{LAD}$) is calculated from the leaf area from individual plants (Table 8.3). In field crops the primary interest is the relationship of leaf area to land surface, or LAI, and $\overline{LAD}$ can also be calculated on this basis (Table 8.3).

If LAI is plotted against time, it produces a function that indicates the crop's assimilatory capacity during the period in question (Fig. 8.12). Using four crops—barley, potato, wheat, and sugar beet—Watson (1947) found that the $\overline{NAR}$s during periods of rapid growth were quite similar but that the $\overline{LAD}$s for the four crops differed significantly (Fig. 8.13).

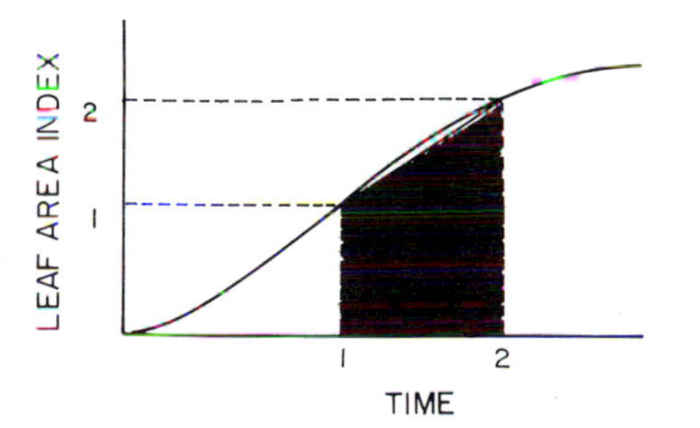

Fig. 8.12. Leaf area duration (shaded area) determined from a plot of leaf area index against time. (From Hunt, *Plant Growth Analysis,* Institute of Biological Studies 96, © 1978, by permission of Edward Arnold, Ltd.)

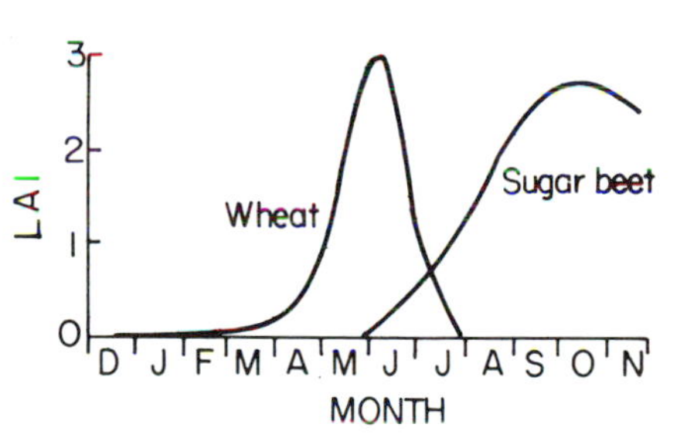

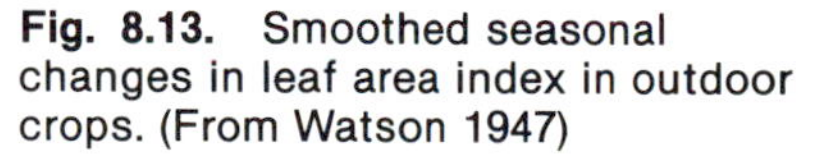

Fig. 8.13. Smoothed seasonal changes in leaf area index in outdoor crops. (From Watson 1947)

BIOMASS DURATION

Biomass duration (BMD) (Table 8.3) is analogous to LAD. If the area under the time curve for biomass production is calculated as explained for LAD (Fig. 8.12), the value for biomass persistence with time is obtained. This quantity may be less useful when used alone than in the calculation of maintenance respiration losses over time, a function of live weight and temperature.

Those and other derived quantities can assist in better understanding crop responses. They can be used to construct models of plant responses to measurable parameters. McCollum (1978) applied growth analysis to the data obtained from potato grown at different soil phosphorus (P) levels (Fig. 8.14). Growth curves for plants with adequate P were sigmoid and exhibited four postemergence ontogenetic phases: (1) pretuber vegetative (emergence to 28 days); (2) tuber initiation, bulking, and rapid leaf growth (28–50 days); (3) continued tuber growth and leaf loss, and (4) death of haulms (50–80 days). Plants with low P never entered phase 3. Low-P plants developed only 50% of the LAI, had a reduced NAR in the early growth phases and had a reduced LAD, compared with high-P plants (Fig. 8.14).

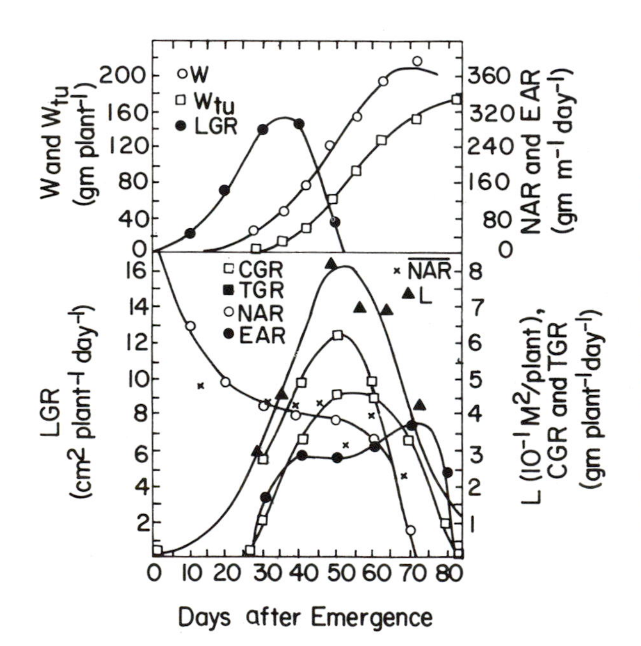

Fig. 8.14. Seasonal trends in parameters of potato growth (e.g., 'Pungo,' 40,000 plants/ha) under high-phosphorus regimes: total dry weight (*W*), tuber dry weight (W_{tu}), crop growth weight (*CGR*), tuber growth rate (*TGR*), leaf area (*L*), leaf growth rate (*LGR*), net assimilation rate (*NAR*), and economic assimilation rate (*EAR*). (From McCollum 1953, by permission)

Summary

Growth and development are continuous processes leading to morphogenesis characteristic of the species. Both processes are controlled by genotype and environment, the degree of influence depending on the particular plant characteristic.

Growth can be defined as cell division and enlargement, but the most widely used definition is the gain in dry matter, which includes differentiation. Growth is the result of the interaction of numerous internal growth-influencing factors (those under genetic control) with the climatic, edaphic, and biological elements of the environment.

A limitation of a growth factor results in a reduction of growth and development. Several theories related to effects of limitation have been postulated, beginning with Liebig in 1862.

The loci of growth are in the apical, lateral, and intercalary meristems. Apical growth tends to produce length, lateral to produce breadth. Stem and leaf length occur primarily in intercalary meristems, which require an added source of growth hormones and have a low cell number or activity; apical or massed meristems, on the other hand, have a high cell number and activity and self-generated hormones. Meristems play a major role in growth correlations, including allometry, which is the growth correlation between two parts (e.g., root and shoot). The growth relationship between parts of the same plant is usually high, that is, there is a high allometric coefficient. The allometric coefficient between shoot and root, the S-R ratio, can vary widely depending on the soil environment, particularly water and N levels. The coefficient between biomass and harvested part such as tubers or grain (harvest index) is quite stable for a given cultivar.

Growth is favored particularly by water and N; differentiation (e.g., cell wall thickening, cell inclusions, protoplasmic hardening) is favored by an excess of photosynthate over growth needs, a favorable temperature, and a proper enzyme system to mediate differentiation. Control of water and N is necessary to obtain thicker cell walls, sugar accumulation (sugar beet), and protoplasmic hardening.

The sigmoid growth curve typifies the pattern of whole plant growth and its constituent parts, including leaf area, over time. The linear phase of the curve, representative of the crop growth rate, is of greatest interest, since rate and duration express crop yield. Rates of 200 kg • ha^{-1} are common; rates of 300 or higher are possible with C_4 in plants such as maize. Other quantities of the classic growth analysis include leaf area index, relative growth rate, and leaf area ratio. All quantities, including crop growth rate, are calculated from measurements at frequent intervals (weekly or biweekly) of leaf area and dry weight. Growth analysis is useful in obtaining a better understanding of yield-influencing development during the crop growth cycle.

References

Baldovinos, G. 1953. In Growth and Differentiation in Plants, ed. W. E. Loomis. Ames: Iowa State College Press.
Blackman, F. F. 1905. Ann. Bot. 19:281–95.
Blackman, G. E., and G. L. Wilson. 1951. Ann. Bot. n.s. 15:373–409.
Blackman, V. H. 1919. Ann. Bot. 33:353–60.
Briggs, G. E., F. Kidd, and C. West. 1920. Ann. Appl. Biol. 7:103–23.
Daynard, T. B., J. W. Tanner, and D. J. Hume. 1969. Crop Sci. 9:831–34.
Donald, C. M., and J. Hamblin. 1976. Adv. Agron. 28:361–405.
Duncan, W. G. 1981. Personal communication.
Duncan, W. G., D. E. McCloud, R. L. McGraw, and K. J. Boote. 1978. Crop Sci. 18:1015–20.
Escalada, J. A., and D. Smith. 1972. Crop Sci. 12:745–49.
Evans, C. 1972. The Quantitative Analysis of Plant Growth. Berkeley and Los Angeles: University of California Press.
Fisher, R. A. 1920. Ann. Appl. Biol. 7:367–72.
Gaastra, P. 1963. In Environmental Control of Plant Growth, ed. L. T. Evans. New York: Academic Press.
Grime, J. P., and R. Hunt. 1975. J. Ecol. 63:393–422.
Hammond, D. 1941. Am. J. Bot. 28:124–38.
Hicks, F. J., and G. R. Peterson. 1978. Calif. Agric. 32:8–9.
Hunt, R. 1978. Plant Growth Analysis. London: Edward Arnold.
Janick, J. 1963. Horticultural Science. San Francisco: W. H. Freeman.
Jarvis, P. G., and M. J. Jarvis. 1964. Physiol. Plant 17:654–66.
Loomis, W. E. 1953. Growth and Differentiation in Plants. Ames: Iowa State College Press.
Macy, P. 1936. Plant Physiol. 11:749–64.
McCollum, R. E. 1978. Agron. J. 70:58–67.
Mitscherlich, E. A. 1909. Jahrb. Landwirtsch. Schweiz 38:537–52.
Murata, Y. 1969. In Physiological Aspects of Crop Yield, ed. J. D. Eastin et al. Madison, Wis.: American Society of Agronomy.
Radford, P. J. 1967. Crop Sci. 7:171–75.
Sepaskhah, A. R. 1977. Agron. J. 69:783–85.
Stoy, V. 1969. In Physiological Aspects of Crop Yield, ed. J. D. Eastin et al. Madison, Wis.: American Society of Agronomy.
von Liebig, J. 1862. Die Chemie in ihre Anwendung auf Agrikultur und Physiologie. Braunschweig.
Watson, D. J. 1947. Ann. Bot. n.s. 11:41–76.
______. 1952. Adv. Agron. 4:101–45.
Willcox, O. W. 1937. ABC of Agrobiology. New York: Norton.

9 Seeds and Germination

SEEDS have always been vital to human existence. Prehistoric humans gathered and preserved seeds for food and for propagation. The rise of ancient civilizations can be linked to cereal grain production: wheat and barley in the Middle East, rice in Southeast Asia, and maize in the Americas. The Romans worshiped Ceres, the goddess of grain, and maize played a prominent role in the ancient American religious rituals. Seeds are currently a major source of food, drink, and numerous drugs.

The seed is the living link between parents and progeny and the primary means of dissemination. It must often survive extremely hostile environments (freezing, fire, flooding, animal ingestion) while waiting for conditions favorable to germination and growth.

Biologically a *seed* is a ripe, fertilized ovule; agriculturally the definition is somewhat more inclusive. In many species, including those of the grass family, the seed is a single-ovule fruit that is dry and is *nondehiscent* (nonsplitting of the fruit wall). In certain other species the seed is a two-seeded fruit that is dry and nondehiscent. In beet the seed is an aggregate of dry fruits, each with single ovules (a seed ball). Variation in seeds among thousands of cultivated and wild species is indeed great, physically (size, shape, and color), biochemically, and physiologically.

Seed Development

The seed has its origin with *microsporogenesis* and *megagametogenesis,* the formation of pollen grains (male gametophytes) and embryo sac (female gametophyte), respectively (Fig. 9.1). Microspore mother cells in the anther and the megaspore mother cell in the embryo sac further divide, first by meiosis, producing haploid daughter cells, and subsequently by mitosis to multiply the number of haploid nuclei. The final result is pollen cells or grains with two nuclei each and the embryo sac with two nuclei. Subsequently, a nucleus in the embryo sac cell divides to form the egg cell and a nucleus that divides again to form the polar nuclei of the ovule.

At fertilization, one of the two pollen nuclei fuses with the egg cell of the

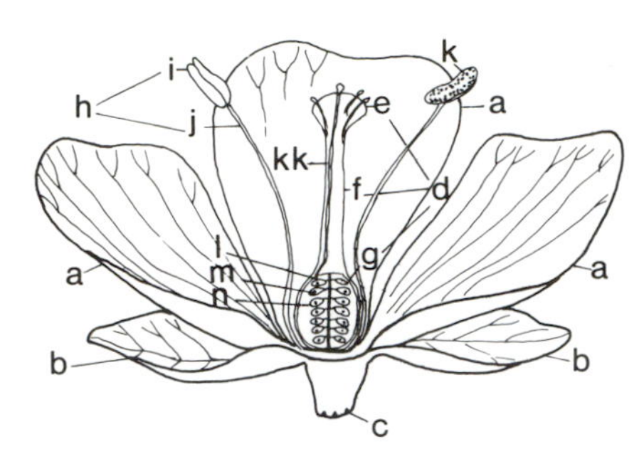

Fig. 9.1. Typical flower: petals (corolla) (*a*); sepals (calyx) (*b*); receptacle (*c*); pistil (*d*), composed of stigma (*e*), style (*f*), ovary (*g*); stamen (*h*), composed of anther (*i*) and filament (*j*); pollen (*k*); pollen tube (*kk*); sperm nucleus (*l*); egg cell (*m*); and ovule (*n*). (Courtesy of USDA)

embryo sac to form the embryo, thus restoring the diploid complement of chromosomes (2N). The second sperm nucleus fuses with the polar nuclei to form the endosperm (3N).

In monocotyledonous plants, the endosperm is a discrete, major structural seed unit (Fig. 9.2D). The monocot endosperm is composed of undifferentiated parenchyma cells encased in a protein-rich, thin outer layer or envelope of living cells, the *aleurone.* A detailed drawing of the structural units and subunits of a *caryopsis* (grass seed) is diagramed in Figure 9.3

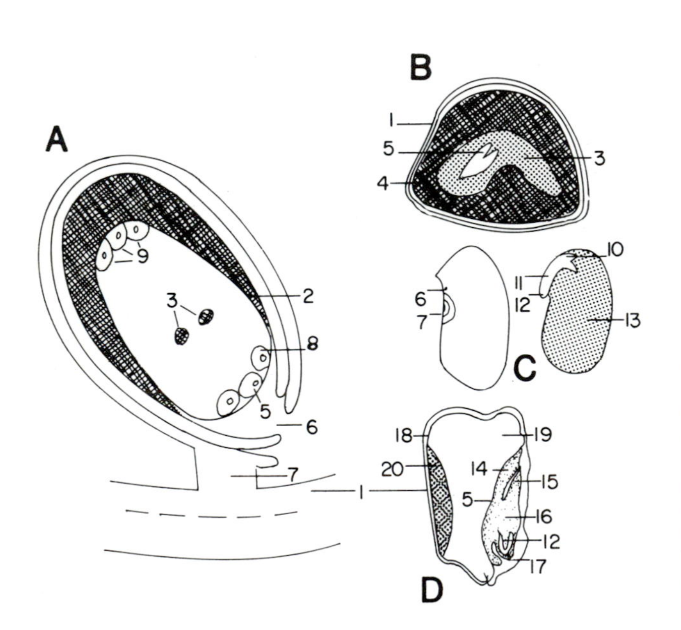

Fig. 9.2. *A*. Typical ovule. *B*. Beet seed. *C*. Bean seed. *D*. Maize seed. Each seed has embryo, endosperm, and seed coat, which are derived from egg, polar nuclei, and integuments, respectively. Reproductive structures: pericarp (*1*); outer and inner integuments (testa, or seed coat) (*2*); polar nuclei, or endosperm (*3*); nucellus, or perisperm (*4*); egg, or embryo (*5*); opening to embryo sac, or micropyle (*6*); funiculus, or hilum (*7*); synergid cell (*8*); antipodal cell (*9*); true leaf (*10*); hypocotyl (*11*); radicle (*12*); cotyledon, or seed leaf of bean embryo (*13*); scutellum, or cotyledon (*14*); coleoptile (*15*); first internode, or mesocotyl (*16*); coleorhiza (*17*); aleurone (*18*); starchy endosperm (*19*); and flinty endosperm (*20*).

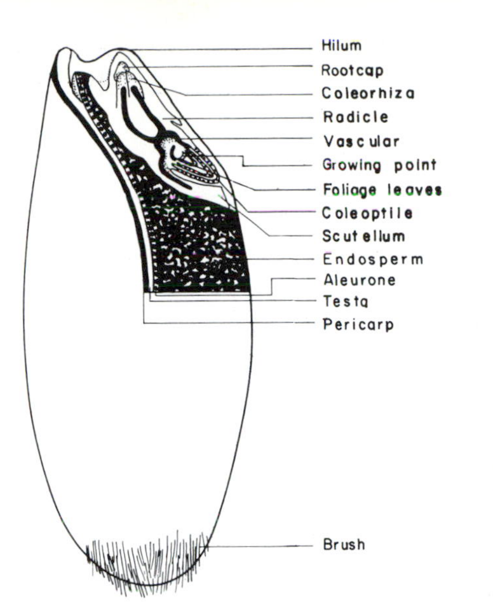

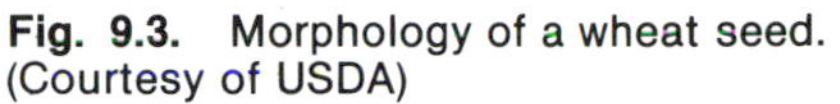
Fig. 9.3. Morphology of a wheat seed. (Courtesy of USDA)

In dicots, the endosperm is partially or completely absorbed by the embryo, specifically by the *cotyledons,* or seed leaves. The *seed coat,* or *testa,* is derived from the outer integument(s) of the ovary, which is maternal tissue. The *hilum* is the *funicular* (vascular connection) scar. It aids the two-way passage of water and dissolved oxygen (O_2), both essential for germination. Water and dissolved gases also enter the *micropyle,* a microscopic scar from the pollen tube entry into the integuments. Often the hilum is equipped with a plug to permit water loss but not uptake (Leopold and Kriedemann 1975).

A mature seed has four components significant both physiologically and ecologically for survival: (1) the seed coat, a protective envelope; (2) the embryo, an embryonic plant or sporophyte; (3) stored food and mineral reserves that nourish the new sporophyte until independence; and (4) enzymes and hormones required for the digestion of food reserves and the synthesis of new tissues in the seedling during germination. These features also provide the seed with protective mechanisms to withstand harsh environments while in a *quiescent* state (resting in a dry state). If quiescent or dormant the seed is inactive but alive, a state of existence that prevails until conditions are suitable for germination. The moisture content and metabolic rate of seeds during quiescence may be only one-tenth or less that in plant tissues.

ONTOGENY

For the first 8 to 10 days dry weight in a wheat seed, which is characteristic of cereal grains, was found to consist primarily of the seed coat (*testa,* or ovule wall, and *pericarp,* or ovary wall) and a small embryo (Fig. 9.4) Jennings and Morton (1963). During the next 2 wk, dry weight increased linearly, due to rapid endosperm starch deposition. At the end of the seed-filling period, a steady state (*physiological maturity*) is reached, at which time growth increments are in balance with metabolic loss increments. The period of seed

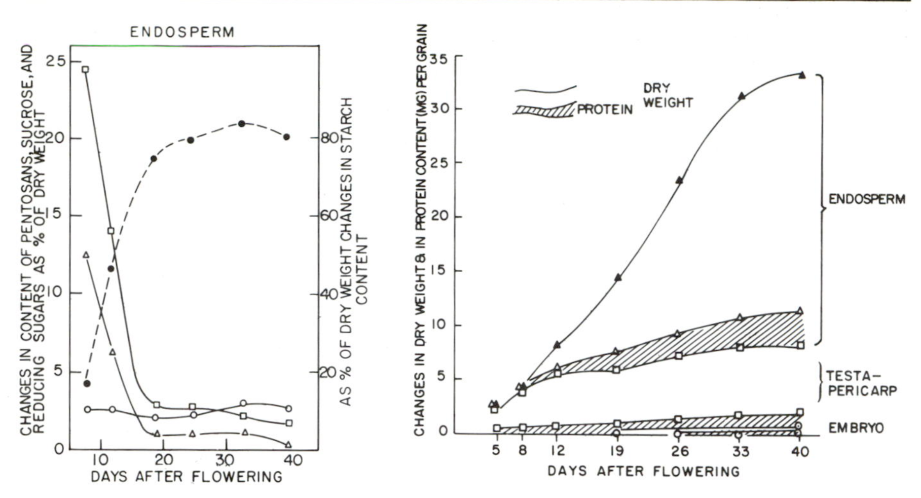

Fig. 9.4. (*Left*) Changes in endosperm of wheat seeds after flowering: pentasans (○), sucrose (□), reducing sugars (△), and starch (●) as a percentage of the dry weight and as a percentage of the starch; (*right*) changes in dry weight, endosperm, and protein during maturation of a wheat seed (Jennings and Morton 1963).

growth for crop plants varies from about 20 to 40 days, depending on genotype and environment, especially temperature.

EMBRYO

The seed embryo consists of the *embryo axis* with the *hypocotyl* (portion of the embryo axis just below the cotyledonary node); one or two cotyledons are at one end, the radicle at the other end (Fig. 9.2). The cotyledons of legume seeds absorb the endosperm and make up 90% or more of the total seed weight. Grass seeds have a single, comparatively small cotyledon (*scutellum*), which functions during germination more for the absorption of hydrolyzed substances from the separate endosperm than for storage, except for oil storage. Seeds of certain dicots have both cotyledon and endosperm storage.

FOOD STORAGE STRUCTURES

Species vary as to the primary structure for food and mineral storage (Table 9.1). Bewley and Black (1978) classified seeds relative to storage structure (Fig. 9.2) into three types: (1) endosperm (grasses, castor bean, tomato, buckwheat), (2) embryo (legumes, lettuce),and (3) perisperm, from nucellus (beet, yucca, coffee).

Any one of the above structures can be the primary, but usually not exclusive, storage structure of a seed. For example, carbohydrates and protein are stored primarily in the endosperm and fats in the embryo in maize seeds.

TABLE 9.1. Types of food reserves and the major storage structures in some crop seeds

Species	Average Composition (% dry wt)			Major Storage Structure
	Protein	Fat	Nitrogen-free extract[a] (major component)	
Maize (*Zea mays*)	11	5	75 (starch)	Endosperm
Sweet corn (*Zea mays*)	12	9	70 (starch)	Endosperm
Oats (*Avena sativa*)	13	8	66 (starch)	Endosperm
Wheat (*Triticum aestivum*)	12	2	75 (starch)	Endosperm
Rye (*Secale cereale*)	12	2	76 (starch)	Endosperm
Barley (*Hordeum vulgare*)	12	3	76 (starch)	Endosperm
Broad bean (*Vicia faba*)	23	1	56 (starch)	Cotyledon
Flax (*Linum usitatissimum*)	24	36	24 (starch)	Cotyledon
Field pea (*Pisum arvense*)	24	6	56 (starch)	Cotyledon
Garden pea (*Pisum sativum*)	25	6	52 (starch)	Cotyledon
Peanut (*Arachis hypogaea*)	31	48	12 (starch)	Cotyledon
Soybean (*Glycine max*)	37	17	26 (starch)	Cotyledon
Cotton (*Gossypium* spp)	39	33	15	Cotyledon
Rape (*Brassica napus*)	21	48	19 (starch)	Cotyledon
Watermelon (*Citrullis vulgaris*)	38	48	5	Cotyledon
Brazil nut (*Bertholletia excelsa*)	18	68	6	Radicle, hypocotyl
Oil palm (*Elaeis guineensis*)	9	49	28	Endosperm
Ivory nut (*Phytelephas macrocarpa*)	5	1	79 (galactomannan)	Endosperm
Date (*Phoenix dactylifera*)	6	9	58 (galactomannan)	Endosperm
Castor bean (*Ricinus communis*)	18	64	trace	Endosperm
Pine (*Pinus pinea*)	35	48	6	Megagametophyte

Source: Bewley and Black, *Physiology and Biochemistry in Relation to Germination,* vol. 1, *Development, Germination, and Growth,* © 1978, by permission of Springer-Verlag.

[a]Nitrogen-free extract consists of material that is not protein, fat, fiber (including cellulose) or ash (mineral nutrients). Thus starch, free sugars, and dextrins are the usual components.

The aleurone, an outer layer of the endosperm, is dense, living,and high in protein (Fig. 9.3). The chemical and physiological significance of the aleurone exceeds that of storage.

UNIQUE SEEDS

Members of the grass family and some dicots do not produce true seeds. In grasses a single-ovule fruit ripens and dries, without dehiscence, to form a dry fruit (caryopsis), or seed. The pericarp (ovary wall) and testa (ovule wall) fuse, to form the seed coat (Fig. 9.3). However, in sunflower, dock, and dandelion the pericarp and testa are not fused, so that a dry, nondehiscent fruit (*achene*) is the seed. The ripe ovule or true seed is loosely attached to the pericarp and is easily separated, as demonstrated by seed-eating birds. The seed (*schizocarp*) of carrot and cocklebur is a dry, nondehiscent fruit with two ripe ovules. The seed (*seed ball*) of beet is a dry aggregate of fruits produced by the pericarps of individual flowers fusing at the base. Modern cultivars of beet have a single seed (*monogerm*) similar to an achene.

Some species, particularly grasses and citrus, produce seeds asexually, or without fertilization (a process called *apomixis*) from diploid, maternal tissue. Seeds produced by apomixis are vegetatively produced and, consequently, are true breeding, like vegetative propagules from root or stem pieces. Many Kentucky bluegrass (*Poa pratensis*) cultivars produce seeds by apomixis.

Chemical Composition

Aside from propagation, seeds are also important for food, feed, and as raw material for a vast array of products. Seeds store proteins, amino acids, and other substances often quite disparate chemically from those in the vegetative tissues. For example, seeds are usually high in fats, whereas vegetative parts are low.

Seeds can be classified as carbohydrate or lipid, based on the principal type of food reserve stored. Seeds in either category can be rich in protein as well (Table 9.1).

Changing the chemical composition of seeds is frequently a primary objective in crop breeding. Dudley and Lambert (1969) reported changes made in maize seeds over 65 generations. Oil and protein contents were 4.77 and 10.9%, respectively, at the beginning of the selection period. After 65 generations, oil for the low and high lines ranged from 1.0 to 15.2%. Protein for the low and high lines ranged from 4.97 to 19.57%. Chemical composition of soybean seeds is presented in Table 9.2. The chemical composition of seeds is genetically controlled but is also influenced by environment. Irrigation (Stone and Tucker 1969), fertilization (Early and DeTurk 1948) and other cultural practices (Osler and Cartter 1954) influenced the chemical composition, including oil and protein contents of seeds of different species.

CARBOHYDRATES

Carbohydrates and lipids are the dominant seed energy reserve in most cultivated and wild plants (Bewley and Black 1978). Seeds of cereal and pulse crops store starch (carbohydrate). The pulses are also high in protein. The seeds of many species (e.g., soybean, peanut, sunflower, rape, and cotton) are high in oil and protein. Seeds of some species may also contain significant amounts of simple sugars.

Starch. Starch is the most common carbohydrate, or polysaccharide, stored in seeds. Two glucosan starches, *amylose* and *amylopectin,* are common. Both are long-chain polymers of glucose molecules with an α,1-4 linkage. Amylose is a straight chain of 300 to 400 glucose molecules (Fig. 9.5). Amylopectin has side chains of glucose with a β,1-6 linkage to the main molecule. Amylopectin may contain more than a thousand glucose units; consequently it has a higher molecular weight and different chemical and physical properties than amylose. In the iodine test for starch, amylopectin stains red and amylose stains blue. Amylopectin is more viscous upon wetting. The cooked product from waxy corn starch, amylopectin (tapioca), is more gelatinous, making it desirable for certain food uses.

TABLE 9.2. Chemical composition of soybean seeds

AMINO ACID COMPOSITION OF SOYBEAN PROTEIN[a]

Essential Amino Acids	Meal	Nonessential Amino Acids	Meal
Lysine	6.9	Arginine	8.4
Methionine	1.6	Histidine	2.6
Cystine	1.6	Tyrosine	3.9
Tryptophan	1.3	Serine	5.6
Threonine	4.3	Glutamic acid	21.0
Isoleucine	5.1	Aspartic acid	12.0
Leucine	7.7	Glycine	4.5
Phenylalanine	5.0	Alanine	4.5
Valine	5.4	Proline	6.3
		Ammonia	2.1

[a]Grams amino acid per 16 g nitrogen.

ULTRACENTRIFUGE FRACTIONS OF WATER-EXTRACTABLE SOYBEAN PROTEINS

Fraction	Percentage of Total	Components	Molecular Weight
2 S	22	Trypsin inhibitors	8,000–21,500
		Cytochrome *c*	12,000
7 S	37	Hemagglutinins	110,000
		Lipoxygenases	102,000
		β-Amylase	61,700
		7 S globulin	180,000–210,000
11 S	31	11 S globulin	350,000
15 S	11	. . .	600,000

FATTY ACID COMPOSITION OF SOYBEAN OIL

Fatty acid	Percentage
Myristic	0.1
Palmitic	11.0
Palmitoleic	0.1
Stearic	4.0
Oleic	23.4
Linoleic	53.2
Linolenic	7.8
Arachidic	0.3
Behenic	0.1

Source: Orthoefer, in *Soybean Physiology, Agronomy, and Utilization,* © 1978, by permission of Academic Press, Inc.

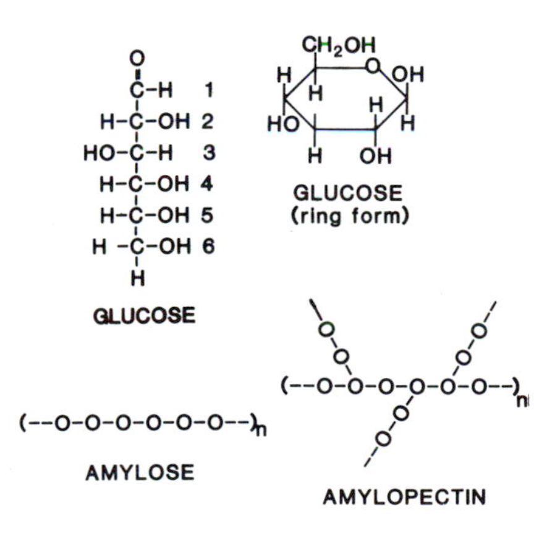

Fig. 9.5. Glucose sugar: straight chain glucose, ring form (natural) glucose, and two glucosan starches, amylose and amylopectin. Cornstarch is about 25% amylose and 75% amylopectin.

Amylose is 100% digestible by α-amylase, amylopectin approximately 50% digestible. Of the two types, amylopectin is dominant in most starchy seeds. Standard cultivars of dent corn (maize) contain approximately 72% amylopectin and 28% amylose. Maize cultivars containing virtually 100% amylopectin (waxy) or 100% amylose (starchy) have been selected and are available commercially. The endosperm starch of sweet corn (maize) has a high sugar content.

Hydrolysis of glucosan starches yields *glucose* (a monosaccharide) and *maltose* (a disaccharide), both soluble and easily converted to sucrose for transport to the root and shoot meristems.

Inulin, a relatively small starch molecule composed of fructose sugar molecules, is the principal food reserve in barley and certain other temperate grasses. Fructosan is partially water soluble, whereas the glucosan starches are insoluble in water.

Other Polysaccharides. *Pentosans,* polymers of 5-carbon sugar molecules, are usually found on or in the coats of certain seeds. Pentosans imbibe water strongly, an adaptive characteristic in dissemination.

Seeds of some legumes are rich in *mannans,* long-chain polymers of mannose sugar. Seeds of alfalfa and honey locust contain *galactomannan,* which consists of a mannan and a galactose (6-carbon) sugar side chain. Glucose and arabinose are also found as side chains on mannans.

Although poorly defined chemically, *hemicellulose* is an important food reserve in seeds (Bewley and Black 1978). The mannans, xylans, and galactans (polymers of mannose, xylose, and galactose simple sugars, respectively) are classified as hemicellulose. Seeds of guar (*Cyanopsis tetragonolobus*) contain over 20% galactomannan, which is used in pharmaceuticals and is the basis for the industrial market for the crop.

The *mucilages* are a complex family of carbohydrates composed principally of polyuromides and galactomides. Mucilages may serve as food reserves, but they also function as a seed coat covering, becoming sticky when wet. This sticky characteristic aids in seed dispersal by animals. These mucilages are used by the seed industry to remove certain weed seeds from legumes with small seeds, for example, removing buckhorn plantain (*Plantago lanceolata*) seeds from alfalfa seeds; the weed seeds become sticky when wetted and adhere to a velvet roller, while the alfalfa seed passes.

Pectins, carbohydrates that are long-chain polymers of galacturonic acid, are the bonding between cell walls (*middle lamella*) of seeds. Pectins are composed principally of pectic acid and propectin and their calcium and magnesium salts.

Other carbohydrates often found in seeds include stachyose (a tetrasaccharide), raffinose (a trisaccharide), sucrose (a disaccharide), and reducing sugars such as glucose (a monosaccharide). Sugar crops accumulate sucrose in the root or stem, but not in the seed. Soybean seeds contain appreciable amounts of reducing (monosaccharide) sugars (Smith and Circle 1972; Orthoefer 1978).

LIPIDS

By definition, lipids are compounds soluble in ether, benzene, and chloroform but insoluble in water (Bloor 1928). *Lipid* is the generic term for fats and oils; *oils* are liquid at normal temperatures, whereas *fats* are solid. Oils, the principal energy storage in numerous species, are frequently found to some extent in starchy seeds.

Generally lipids are esters of the trihydric alcohol *glycerol* and three fatty acids:

$$
\begin{array}{l}
H_2{-}C{-}O{-}R_1 \\
\quad\;\; | \\
\;H{-}C{-}O{-}R_2 \\
\quad\;\; | \\
H_2{-}C{-}O{-}R_3
\end{array}
$$

where R_1, R_2, and R_3 are fatty acids (discussed below).

The *degree of unsaturation* (i.e., the ratio of single to double bonds between carbon atoms) and the number of carbon atoms determine the kind and properties of the fatty acids. *Oleic, linoleic,* and *linolenic* acids—which have 18 carbons with 1, 2, and 3 double bonds, respectively—are the principal fatty acids in oil seeds. The dominant fatty acid depends on plant species (Table 9.3), linoleic being primary in soybean, for example. Certain crops are grown for their characteristic seed fatty-acid composition, for example, crambe for erucic acid and castor bean for lesquerolic acid (Table 9.3).

TABLE 9.3. Fatty acids in seeds of crop plants

Name	Carbon: Unsaturated Bonds	Plant Species
Caprilic	C_7:0	...
Capric	C_9:0	...
Lauric	C_{12}:0	Coconut, palm
Palmitic	C_{16}:0	Cotton
Stearic	C_{18}:0	Hubbard squash
Oleic	C_{18}:1	...
Linoleic	C_{18}:2	Soybean
Linolenic	C_{18}:3	Flax
Erucic	C_{22}:1	Rape, crambe
Vernolic	C_{18}:1 + 1 epoxy group $-c-\overset{O}{c-c}-c-$	Vernonia
Lesquerolic	C_{18}:2 + 1 hydroxy group $-c-\underset{OH}{\underset{\vert}{c}}-c-$	Castor bean

Seeds selected for high oil content also tend to be high in protein; selection for the one objective is likely to achieve the other.

Palm oil is extracted from the fleshy fruit of the palm tree rather than from its seeds. Unlike soybean the palm oil industry does not produce a high-protein meal as a by-product. The yield of oil from soybean is considerably less than palm on a land area basis, but soybean has maintained a healthy competitive position in the market because of the value of soybean meal.

Waxes, esters of fatty acids and a monohydric alcohol, are found particularly in seed coats and are solids at room temperature.

Phospholipids are important for membrane metabolism and storage, functioning as an energy and phosphorus reserve for seedling growth. *Phospholipids* are esters of fatty acids and alcohol but in addition contain a phosphate group and the nitrogen (N) of choline (Fig. 9.6). Lecithin, a phospholipid widely distributed in nature, is highly important for commercial uses. *Soybean lecithin* is a generic name used by the industry for a mixture of three phospholipids: lecithin, cephalin, and phytin (Smith and Circle 1972). Cephalin is important in soybean and other oil seeds. The principal fatty acids of lecithin and cephalin are linoleic, oleic, palmitic, and hexadecanoic acids.

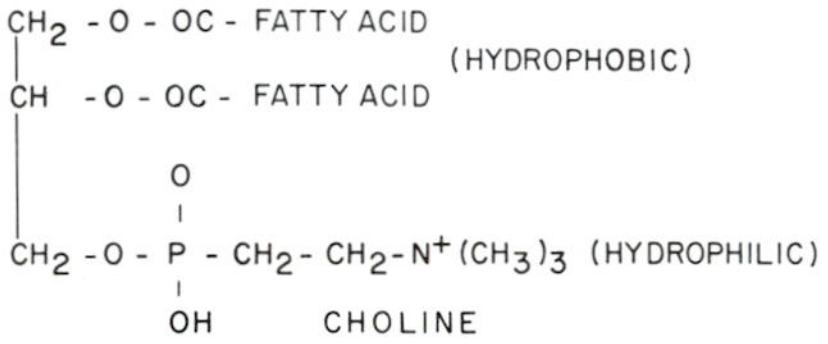

CH_2 - O - OC - FATTY ACID
(HYDROPHOBIC)
CH - O - OC - FATTY ACID

O

CH_2 - O - P - CH_2 - CH_2 - $N^+(CH_3)_3$ (HYDROPHILIC)

OH CHOLINE

Fig. 9.6. Lecithin phospholipid.

During germination, the lipids are hydrolyzed into component fatty acids and glycerol. These metabolites are mobile and readily transported to the embryo axis, where they undergo further oxidation via the Krebs cycle or, alternatively, the pentose phosphate pathway shunt.

Strong alkalis are used commercially to hydrolyze fats, a process called *saponification.* Oils are converted to fats commercially by saturating the double bonds of the fatty acids with hydrogen, a process called *hydrogenation.*

PROTEINS

Proteins, the N reserves of seeds for seedling growth, are polymers of amino acids connected by peptide bonds (Fig. 9.7). Twenty amino acids that form protein occur in nature; a few or all 20 may be sequenced with varying frequencies to form different protein molecules. Hence protein molecules are extremely large, complex, high in molecular weight (40,000 or more), and they can vary chemically nearly infinitely. Sequencing of amino acids in biological systems is coded by the polynucleotides DNA and RNA. Protein complexity is increased by hydrogen (H) bonding, a weak cross-linkage between H and O_2

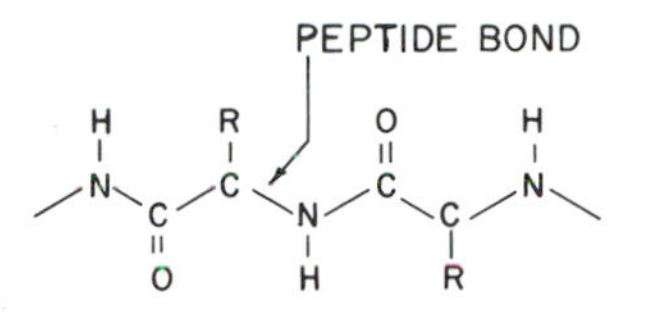

Fig. 9.7. Peptide bond, a low-energy input bond between carbon and nitrogen.

atoms in the molecules, and also by sulfhydryl bonding. Physiologically proteins are the matrix of life in seed and other living cells.

As previously stated, the amino acid composition of seed storage proteins is different from that in leaves or vegetative tissues. Seed proteins are usually deficient in one or more of three essential (i.e., required in the diet of monogastric animals) amino acids—lysine, tryptophan, and methionine—depending on plant species and cultivar. Therefore, if used as the single protein source, seed protein is lower in biological or nutritional value for monogastric (simple-stomached) animals, including man, than is animal protein.

Based on solubility and method of separation, Osborne (1924) divided proteins into four groups:

1. *Albumins,* which are water soluble at a neutral or slightly acid pH and coagulated by heat. Enzymes and egg white are predominantly albumins.
2. *Globulins,* which are water and salt solution soluble and not readily heat coagulated. Legume seeds are generally rich in globulins (e.g., glycinin in soybean).
3. *Glutelins,* which are insoluble in water but soluble in salt solutions and in strong acid or alkali solutions. The bread wheats are high in glutenin, a glutelin protein. Glutenin imparts the capacity of wheat dough to stretch, or rise.
4. *Prolamins,* which are soluble in 70 to 90% alcohol. The cereal grains are high in prolamins (e.g., the protein zein in maize seeds). While the prolamins are good reserves of N for seedling growth, they are low in biological or nutritional quality for monogastric animals.

In general, cereals are high in prolamin and glutelin proteins. Oat is an exception, with a protein composition of 80% globulin (Mayer and Poljakoff-Mayber 1963). Unlike cereals, the pulses are rich in globulins and albumins, suggesting better nutritional quality.

Important prolamins in cereals include zein in maize, gliadin in wheat, and hordenin in barley (Bewley and Black 1978). Some important glutelins in cereals include zecanin in maize, glutenin in wheat, hordenin in barley, and oryzenin in rice. Some important globulins in various legume seeds are legumin, vicilin, glycinin, vignin, and arachin.

Cereal prolamins, zein, in particular, are deficient in *lysine* and *tryptophan.* Thus the quality of maize seeds as a sole protein source is particularly low for monogastrics.

The amino acid imbalance in pulse and soybean seeds, in particular, is due to a methionine deficiency. Maize and soybean seeds complement each other in a diet, providing a protein source fairly well balanced in lysine and methionine. In germination, proteins are hydrolyzed into amino acids, transported, and resynthesized in the embryo axis into protein balanced in amino acid composition. Therefore seed sprouts provide protein of excellent quality and are widely used in human nutrition (e.g., alfalfa and bean sprouts).

Stored proteins in seeds also exist as *lectins,* which are glycoproteins (sugar-protein polymers). Over 1% of soybean protein is lectin (Daubert 1950).

OTHER COMPOUNDS IN SEEDS

The seed must contain adequate minerals to support the seedling until it is autotrophic. The mineral composition of seeds is similar to that of somatic tissue except for a higher content of phosphorus and certain other minerals in organic (chelate) form. *Phytin* (phytate) is the primary source of phosphorus, and it also contains organic-salt complexes of calcium, magnesium, manganese, and potassium (Copeland 1967). These minerals are liberated at germination by the enzyme *phytase.* Phytin is concentrated in the aleurone in grass seeds and in the cotyledons in dicots. Species and cultivars vary in phytin content.

Alkaloids are cyclic nitrogenous storage compounds found in seeds and vegetative parts of plants. Alkaloids cause strong flavors and odors and may be toxic to other plants and animals. The cup of hemlock that Socrates drank contained the alkaloid coniine. Other important and well-known alkaloids include nicotine, caffein, morphine, strychnine, and theobromine (in tea). Gramine alkaloid, found in the vegetation of certain forage grasses, causes reduced palatability and thus reduced forage intake and possible impairment of animal health (Martin and Heath 1973). Seed alkaloids tend to function primarily as germination inhibitors. Alkaloids may be allelochemicals in the ecology of natural vegetation, possibly protecting the young seedling from competition.

Seeds of certain species contain phenolic compounds (e.g., tannins, chlorogenic acid, coumarin, furelic acid, and caffeic acid). These compounds are also classified as lactones. The lactones can inhibit germination, that is, serve as a dormancy mechanism.

Seeds are a rich source of certain vitamins, especially the B complex, and free amino acids, sugars, and nucleic acids are present in low concentrations. Seeds also contain growth regulators—auxins, gibberellins, cytokinins, and growth inhibitors—which have a vital function in germination and seedling growth. Interestingly the first natural cytokinin, zeatin, was isolated from maize seeds.

Germination

A definition of the term *germination* depends on perspective. A seed analyst may accept a morphological change, such as protrusion of the radicle, but to a grower germination means seedling emergence. Technically germination is the resumption of active growth that results in rupture of the seed coat and emergence of the seedling (Amen 1963). Germination includes the following physiological and morphological events (Toole and Hendricks 1956): (1) imbibition and absorption of water, (2) hydration of tissues, (3) absorption of O_2, (4) activation of enzymes and digestion, (5) transport of the hydrolized molecules to the embryo axis, (6) increase in respiration and assimilation, (7) initiation of cell division and enlargement, and (8) embryo emergence.

In lettuce seeds imbibition peaked in 2 hr, whereas respiration began after 2 hr and peaked first at 8 hr (Mayer and Poljakoff-Mayber 1967) (Fig. 9.8).

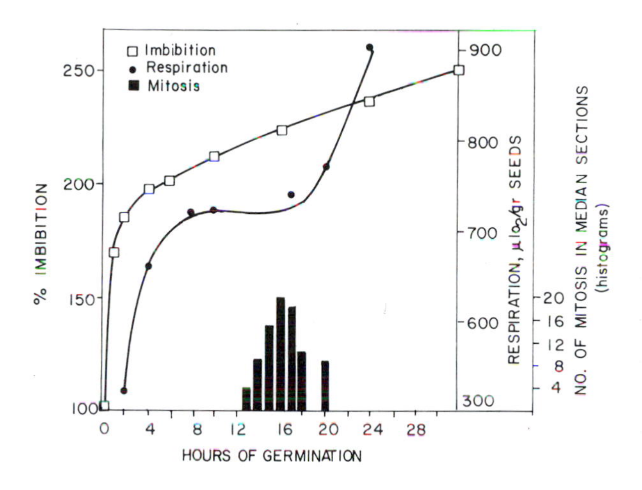

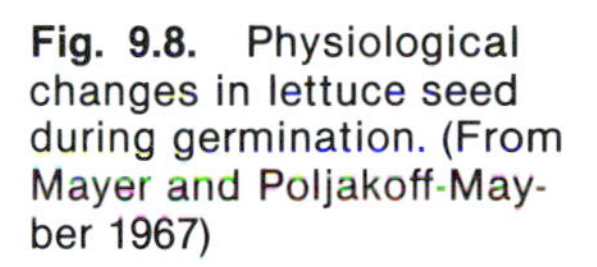

Fig. 9.8. Physiological changes in lettuce seed during germination. (From Mayer and Poljakoff-Mayber 1967)

After the first peak a second respiration peak began at approximately 16 hr and peaked at 24 hr or later. The two peaks were interpreted to be associated with chemical hydrolysis and synthesis, respectively. Mitosis was apparent at 12 hr and peaked at 16 hr. The ontogeny of germination suggested two distinct metabolic phases: enzymatic hydrolysis of stored reserves and synthesis of new tissue from hydrolyzed compounds (i.e., from the liberated sugars, amino acids, fatty acids, and minerals). In the growth of the embryo axis, the initial growth rate of the radicle is more rapid than that of the plumule, and it is generally the first to emerge from the ruptured seed coat. Dry weight of the shoot overtakes that of roots in a few days. Overall weight of seed-seedling declines for about 10 days due to respiration loss (Fig. 9.9). A growth sequence with root growth preceding top growth appears to have survival advantage for a seedling.

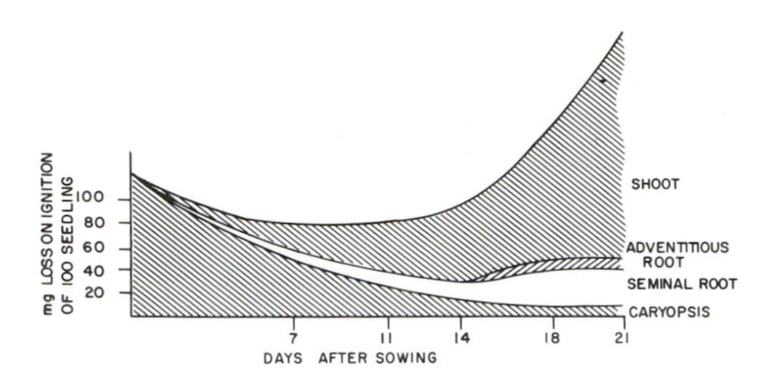

Fig. 9.9. Dry weight distribution in the seed, root, and shoot of barley during 21 days after sowing (Anslow 1962).

Phytohormones initiate and mediate essential germination processes. There are several known activities of growth hormones (Fig. 9.10).

1. Gibberellins activate the hydrolytic enzymes of digestion.
2. Cytokinins stimulate cell division, resulting in radicle and plumule emergence. The initial expansion off the coleorhiza (emerging root tip) is primarily by cell enlargement.
3. Auxins promote growth by enlargement of coleorhiza, radicle, and plumule and by activation of geotropism (i.e., correct orientation of root and shoot growth, irrespective of seed orientation).

METABOLISM OF STORED FOODS

Germination and seedling emergence have a high demand for energy via respiration of seed food reserves. The energy in chemical bonds of carbohydrates, fats, and proteins is released by digestion and oxidative phosphorylation, which produces energy-rich nucleotides, such as adenosine triphosphate (ATP), in the *mitochondria* (the loci for respiration). As ATP is converted to adenosine diphosphate (ADP), energy is released for biological activities as follows:

$$\text{carbohydrate, fat, or protein} \xrightarrow{(\text{ADP} + \text{Pi}) \curvearrowright \text{ATP}} \text{degradation products} \xrightarrow[\text{ATP} \curvearrowright (\text{ADP} + \text{Pi})]{} \text{biosynthesis}$$

Starches are hydrolyzed by α- and β-amylase, mediated by gibberellins, to maltose (disaccharide) and glucose sugar. Van Overbeek (1968) clearly illustrated the role of growth hormones in hydrolysis and seedling emergence (Fig. 9.10). Some glucose is converted by the enzyme *invertase* to sucrose, the sugar commonly transported in plants. Glucose is metabolized by (1) *glycolysis,* which forms two molecules of pyruvic acid and ATP, and (2) oxidation by the *Krebs* or *tricarboxylic acid* cycle, which can completely oxidize the acid intermediates into CO_2, H_2O, and ATP, or, alternatively, by the pentose phosphate pathway shunt.

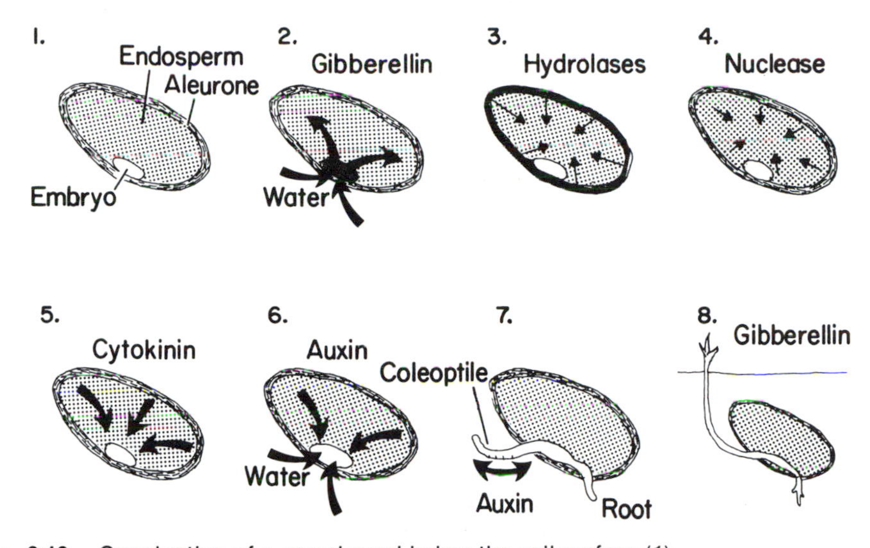

Fig. 9.10. Germination of a cereal seed below the soil surface (*1*) is regulated by a number of hormones working in sequence. First the absorption of water from the soil causes the embryo to produce a small amount of gibberellin (*2*). The gibberellin then diffuses into a layer of aleurone cells that surrounds the endosperm's food storage cells, causing them to form enzymes (*3*) that in turn lead the endosperm cells to disintegrate and liquefy (*4*). Cytokinins and auxins formed in this process (*5, 6*) then promote the growth of the embryo by making its cells divide and enlarge. If the shoot is pointing down into the soil, the auxins tend to migrate to the lower side of the seedling, causing it to grow faster and hence turning the growing point of the shoot upward toward the surface of the soil (*7*). Once the shoot has broken into the sunlight, the plant begins to produce its own food by photosynthesis (*8*). (From Van Overbeek 1968)

Fats are hydrolyzed by lipase into glycerol and fatty acids. Fatty acids are further degraded by peroxidase and aldehydrogenase in α-oxidation, which removes successive carbon atoms to yield CO_2 and stored energy (NADPH). A more common degradation of fatty acids is in β-oxidase, which splits the fatty acid into two-carbon units (*acetyl coenzyme A*) and ATP. Acetyl coenzyme A may enter the Krebs cycle for further oxidation and production of ATP.

Protease breaks peptide bonds in protein molecules, yielding amino acids. The fates of amino acids are as follows: (1) resynthesis into new proteins in growth; (2) *transamination,* the transfer of an amino group from amino acid to an organic acid; or (3) *deamination,* hydrolysis of amino acids into organic acids and ammonia. Organic acid residues enter the Krebs cycle for further oxidation.

Phosphorus is liberated from phytin (inositol hexaphosphate) by the enzyme phytase. To a lesser extent phospholipids may also be hydrolyzed, freeing phosphorus. Phosphorus in plant tissue exists primarily as a constituent of the nucleotides (ADP, ATP, NAD, and NADP), nucleic acids, phospholipids, phosphoproteins, and phosphorylated sugars.

RESPIRATORY QUOTIENT

Germinating seeds respire rapidly, but respiration is hardly detectable in quiescent seeds. Aerobic respiration consumes O_2 and liberates carbon dioxide (CO_2). The ratio of the volume of CO_2 released to the volume of O_2 consumed is referred to as the respiratory quotient (R.Q.) (R.Q. = CO_2/O_2). The R.Q. indicates the type of substrate respired and its completeness of degradation. The R.Q. for glucose is 1.0; hence the R.Q. for starch is approximately 1.0 (Copeland 1967). The R.Q. for fats is 0.7, due to the greater O_2 requirement of unsaturated fatty acids.

GERMINABILITY AND VIABILITY

Mature seeds are viable before or at separation from the mother plant, but they may not be germinable (capable of rapid germination under favorable conditions). Seeds of some species are dormant and become germinable only after exposure to a special set of conditions. Crop seeds are viable and quiescent (i.e., alive but do not germinate due to the lack of a supportive environment for germination, such as not enough water or the wrong temperature) and generally germinable when separated from the mother plant. Seeds of many, or probably most, wild species and certain forage crop species remain dormant (i.e., do not germinate if given a supportive environment for germination) despite conditions suitable for germination. Hence germinability and viability may differ by as much as 100% for different seed populations. Germination does not occur until dormancy is lost, although seeds may be completely viable and germinable. Reliable tests are available to test both germinability (standard germination test) and viability (tetrazolium test). In general viability decreases and germinability increases with age, due to the natural breakdown of dormancy factors in seeds.

Germination Requirements

ENVIRONMENT

Water, nonrestrictive temperature, and suitable atmosphere are required for germination of nondormant, or after-ripened, seeds (Fig. 9.11). Generally conditions favoring growth of seedlings also favor germination. Seeds of different species, however, have varying degrees of genetically or environmentally imposed dormancy. Germination does not occur until what is referred to as *after-ripening,* (the loss of such dormancy, through exposure to certain environmental conditions for a sufficient length of time) (Meyer and Anderson 1949). After-ripening of dormant seed is achieved by a special set of environmental conditions (or in many species, generally by ageing up to many years).

In dry storage most crop seeds after-ripen (become germinable) during maturation, or soon after, so germination of the next crop is not a problem. But the lack of sufficient dormancy for normal quiescence in crop plants can be a problem with some cultivars if humidity is high during maturation; this can result in germination of seeds in the panicle, or head, prior to harvest (see Fig. 9.12). Spanish-type peanut cultivars often lack sufficient dormancy to prevent growth from the subterranean fruits (pods) prior to digging. Therefore, in humid environments dormancy sufficient to assure quiescence until harvest is completed is highly desirable.

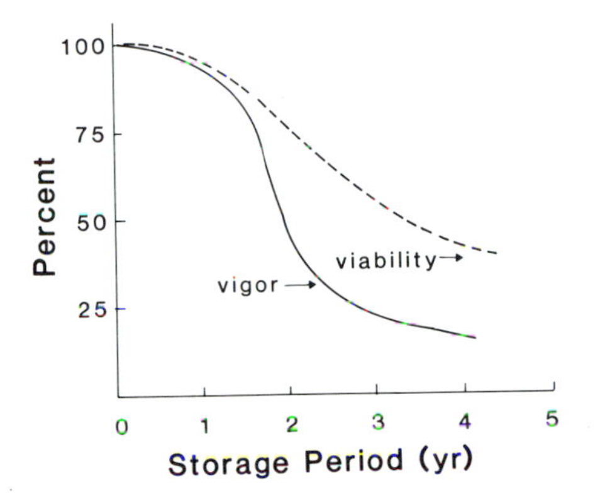

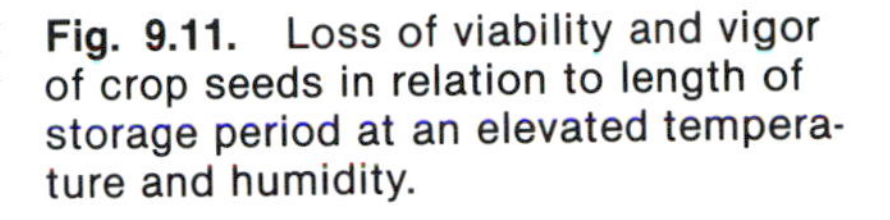
Fig. 9.11. Loss of viability and vigor of crop seeds in relation to length of storage period at an elevated temperature and humidity.

Water. Imbibition of water is the first process of germination. Both live and dead seeds imbibe water and swell; the amount imbibed is related to the chemical composition of the seed. Proteins, mucilages, and pectins are more colloidal and hydrophylic and imbibe more water than starches. Cereal grains, such as maize, imbibe water to approximately one-third seed weight, soybean seeds to one-half seed weight. A soil moisture level of field capacity is generally optimum for germination. Germination proceeds at slower rates as soil moisture nears the wilting point. Less-than-optimum water content usually results in partial imbibition and slowed or arrested germination. Seeds may be wetted and dried repeatedly during the process of germination but usually not without a loss of viability, the extent depending on species and number of wetting and drying cycles. Composition of the medium, particularly the solute content, affects water availability. Ryan (1973) found that as the osmotic concentration was increased, water availability decreased, but specific ions, especially sodium and magnesium, affected germination more than water availability.

Temperature. Except for imbibition, germination involves numerous enzymatically controlled processes of catabolism and anabolism and hence is highly responsive to temperature. The *cardinal temperatures* (maximum, optimum, and minimum) for germination of most crop seeds are essentially those of normal vegetative growth (Table 9.4). The optimum temperature is that one giving the highest germination percentage in the shortest period of time. Non-after-ripened seeds with a partial, or relative, dormancy (Borriss 1949) germinated in a narrow range of temperatures, like 5°–15°C for low-temperature species (Amen 1968). After-ripened seeds (as are found in the cultivars of most crops) do not have such a narrow germination range. Cardinal temperatures for germination of different crop seeds overlap, but germination rate of all is slower at low temperatures.

TABLE 9.4. Temperature ranges in which germination occurs for different seeds

Seeds	Temperature (°C)		
	Minimum	Optimum	Maximum
Corn	8–10	32–35	40–44
Rice	10–12	30–37	40–42
Wheat	3–5	15–31	30–43
Barley	3–5	19–27	30–40
Rye	3–5	25–31	30–40
Oat	3–5	25–31	30–40
Buckwheat	3–5	25–31	35–45
Field bindweed (*Convolvulus arvensis*)	0.5–3	20–35	35–40
Tobacco (Florida cigar wrapper)	10	24	30

Source: From Mayer and Poljakoff-Mayber 1963.

Seeds of some species, such as cotton, are very sensitive to chilling during germination, especially during imbibition. Germination of seeds of many grasses and trees are benefited by diurnal temperature alternations, for reasons not fully understood. It appears that seeds of many wild species or those with a short domestication history have a relative dormancy that is responsive to temperature alternation.

Gases. Germination requires high levels of O_2 unless respiration associated with it is by fermentation. Most species respond best to the ambient composition of the air: 20% O_2, 0.03% CO_2, and 80% N. Decreasing the O_2 content below 20% usually decreases germination. Rice seeds can germinate anaerobically but not without the occurrence of abnormal seedlings. While seed germination in most species is favored by an ambient or higher O_2 concentration, seeds of cattail (*typha latifolia*) and bermudagrass (*Cynadon dactylon*) were found to germinate better at an O_2 concentration lower than ambient (Morinaga 1926). A CO_2 concentration higher than ambient favored bermudagrass germination (Mayer and Poljakoff-Mayber 1963). Generally poorly drained soils provide an atmosphere less than optimum for germination.

Light. That seeds of many species require light for germination has been known for nearly a century (Mayer and Poljakoff-Mayber 1963). Kinzel (1926) identified a sensitivity to light in seeds of a large number of species and classified several hundred species according to whether (1) germination was favored by light, (2) germination was favored by dark, or (3) germination was unaffected by light or dark. Seeds in the first category are said to be *photoblastic*. The ecological significance of light in germination is understandable, since many weed seeds germinate only after light exposure at the soil surface, such as that induced by soil disturbance. Exotics in a forest ecosystem germinate only after the land is cut over and disturbed by harvesting operations, exposing these seeds to light.

Germination of crop seeds with relatively long domestication histories (except for tobacco and lettuce) is generally nonphotoblastic. Lettuce seeds germinate readily with light or if after-ripened by a special set of conditions. These facts suggest that light is an after-ripening factor, a trigger mechanism in breaking a certain kind of dormancy. The dormancy mechanism (secondary) in lettuce is acquired by exposure to high air or soil temperature (i.e., acquired after seed maturation). Normally lettuce seeds mature during long days and high temperatures, which may induce the acquired dormancy. This dormancy is broken and germination and growth occur at the onset of cool season.

A group of USDA scientists convincingly demonstrated that the light-response mechanism in seed germination is the same as that controlling other formative processes, such as flowering, pigment formation, stem elongation, and straightening of the hypocotyl hook (Borthwick et al. 1952; Toole and Hendricks 1956). Red (R) light is the active pigment; the process is photoreversible with far-red (FR; infrared) light (Table 9.5). The quantity (energy level), quality (color or wavelength), and duration of light (photoperiod) in the cycle has a marked influence on germination, depending on the species. Generally light at a low energy level (1/50–1/100 full sunlight) is adequate to stimulate germination. Commercial seed germinating machines are equipped to deliver radiation at this level. However, bentgrass (*Agrostis tenuis*) germinated best in full sunlight (1.3 cal • cm^{-2} • min^{-1}). Like flowering, germination in many species is also responsive to photoperiod and, depending on species, is favored by short days or long days or is day neutral (Copeland 1967).

TABLE 9.5. Germination of lettuce seeds at 26°C with dark broken by light exposure

Sequence of Light Exposure[a]	Germination (%)
R	70
R–FR	6
R–FR–R	74
R–FR–R–FR	6
R–FR–R–FR–R	76
R–FR–R–FR–R–FR	7

Source: From Borthwick et al. 1954.

[a]Red (R) light exposure 1 min; far-red (FR) light exposure 4 min.

The effect of light quality on light-sensitive seeds is pronounced, as is shown by the following summary of some early work:

Wavelength (nm)	*Color*	*Response*
<290	Ultraviolet (invisible)	Inhibition
290–400	Ultraviolet (invisible)	No clear-cut effects
420–500	Blue (visible)	Inhibition
560–700	Orange-red (visible)	Promotion
>700	Far-red (invisible)	Inhibition

The most effective wavelengths for promoting and inhibiting seed germination were reported to be red (peak at 660 nm) and infra-(far) red (730 nm), respectively (Flint and McAlister 1937; Borthwick et al. 1954).

Borthwick et al. first reported in 1952 that germination of lettuce seeds was photoreversible by exposure of moist seeds to a few minutes of R or FR radiation (Table 9.5).

After this early work, these investigators found that the pigment *phytochrome* was the light receptor controlling the response. This is a protein that exists in two interconvertible forms, P_r and P_{fr} (also see Chap. 12); P_r appears blue and P_{fr} a faded shade of blue, which becomes evident after exposure to red light. A proposed mechanism for seed germination is as follows (Borthwick et al. 1954; Amen 1968):

$$\underset{\text{reactant}}{P_r \longrightarrow P_{fr}} \longrightarrow \underset{\text{product}}{\text{hormone}} \longrightarrow \underset{\text{enzyme activity}}{\text{hydrolytic}} \longrightarrow \text{germination}$$

P_{fr} is believed to be the biologically active form of the controlling mechanism in germination and in other plant phytochrome responses (Hendricks 1968).

Exogenous Chemicals. A number of chemicals in the medium promote germination of some species (see Table 9.7); they can be regarded as stimulators rather than germination requirements. Certain chemicals, such as gibberellins, can enhance or substitute for the light or cold requirements for after-ripening. Some of the more important chemicals used to stimulate germination follow:

1. Potassium nitrate (KNO_3), which is used routinely in germination tests of many grass seeds and generally on photoblastic seeds (Copeland 1967).
2. Thiourea, or $CS(NH_2)_2$, which is not used extensively but does stimulate germination in certain species. It cannot substitute for light or temperature requirements (Tukey and Carelson 1945).
3. Hydrogen peroxide (H_2O_2), which is effective on the seeds of certain legumes, tomato, and barley (Copeland 1967).
4. Ethylene (C_2H_4), which stimulates germination in certain species (e.g., peanut) and increases the girth of the germinating seedling axis. Peanut seed

dormancy (and the need to remove it with ethylene) varies greatly with cultivar.

5. Gibberellins (GAs), which can substitute, at least partially, for light and cold in photoblastic seeds. GA_3 (gibberellic acid) is most commonly used, but GA_4 and GA_7 have been found to be more effective than GA_3 (Borris 1967).

MATURITY

Even in a favorable environment, germination cannot occur until after a minimum level of morphogenesis in the seed. Generally sufficient development for viability and germinability occurs long before seed maturity. Smooth bromegrass seeds germinated in about 6 days after fertilization (Grabe 1956). Many weed seeds have been observed to be viable and germinable 8 to 10 days after flowering. Seed dormancy generally increases with seed maturity.

Longevity

Duration of viability or seed longevity depends on genotype, dormancy mechanisms, and the storage environment. Seeds of a lupine (*Lupinus articus*) removed from a peat bog in Canada germinated after 10,000 yr (Porsild and Harrington 1967). Indian lotus seeds from a Manchurian lake bed germinated after 1,000 yr (Copeland 1967). Mimosa (*Mimosa glomerata*) seeds from the Herberia of the National Museum of Paris demonstrated germinability after 221 yr (Becquerel 1934). Certain legume seeds have been found to germinate after 100 to 150 yr of dry storage (Ohga 1926). The longevity studies of Beal and those of Duvel, which involved burying containers with seeds in a moist medium and removing them at intervals, were summarized by Klingman and Ashton (1975) (Table 9.6).

TABLE 9.6. Seed longevity

Experiment	Years Buried	No. Germinated[a]
Duvel[a]	1	71
	10	68
	20	57
	38	36[b]
Beal[c]	20	11
	40	9[d]
	70	3[e]
	90	1[f]

Source: From Klingman and Ashton 1975.

[a]Duvel buried 107 species in clay pots at 8, 22, and 42 in.

[b]Examples: jimson weed 91%, moth mullein 48%, velvetleaf 38%, evening primrose 17%, lambsquarters 7%, green foxtail 1%, curley dock 1%.

[c]Beal buried 20 species in inverted uncapped glass bottles.

[d]Redroot pigweed, prostrate pigweed, common ragweed, black mustard, evening primrose, broadleaf plantain, purslane, curley dock, moth mullein.

[e]Moth mullein, evening primrose, curley dock.

[f]Moth mullein 70–80%.

In the Beal experiment, three species germinated after 70 yr and one, moth mullein, after 90 yr. Seeds of most crop plants maintained viability for many years, germinating 70 to 90% after 7 to 10 yr if stored in suitable conditions (Mayer and Poljakoff-Mayber 1963).

Longevity studies have emphasized the importance of favorable storage conditions: low temperature, low humidity, and low O_2 (Ching et al. 1959). Seeds of certain forage crops stored at a high temperature (38°C) germinated well after 6 yr if the seed moisture content was maintained at or below 6%. At 22°C viability was lost in 3 mo if the seed moisture content was elevated to 16%. A general rule is that the sum of values of the ambient humidity (%) and temperature (°F) of the storage environment should not exceed 100. For example, 40°F and 40% relative humidity (totaling 80) should constitute good seed storage conditions, but lower temperatures and/or humidities should be superior.

Seedling Vigor

Seedling *vigor* (growth rate) declines rapidly with length of seed storage. Short periods of storage under adverse conditions may affect vigor more than viability (Copeland 1967) (Fig. 9.11). As can be noted, vigor may be lost more rapidly than viability; that is, the half-life of vigor of seeds in storage in this case was about 2 yr, compared with about 4 yr for the viability half-life. One explanation is that loss of vigor is due to a breakdown of the structures protective against microorganisms that can weaken or destroy the seed or seedling. Storage time or adverse conditions may cause seed membranes to lose selectivity and metabolites during germination (Abdul-Baki and Anderson 1970), further inviting microbial attack. Loss of vigor is probably more complex than just physical deterioration. For example, mitochondria of soybean seedlings from new seeds were found to differ significantly in respiration from those from old seeds (Abu-Shakra and Ching 1967). The rate of photophosphorylation of the old-seed plants was only 40 to 70% of the new seed plants per unit of O_2 consumed. Seedlings from old seeds had fewer mitochondria per unit weight than those from new seeds.

Whatever the mechanisms, neither vigor nor viability of a seed population is lost instantaneously; both generally decline exponentially with time at a rate that essentially fits a sigmoid curve (Borriss 1949). Vigor cannot be measured reliably by the standard germination test. For this reason a *cold test* was developed that is used extensively in the maize-seed industry and is attracting some interest for certain other crops. This cold test involves placing imbibed seeds in cold (5–10°C), moist, unsterilized organic soil for a period of about 7 days and then transferring the system to a warm temperature (22°C) for completion of germination. These conditions predispose the seed and seedling to diseases or rots by *Pythium* and other damping off organisms. Germination may differ widely from that of the standard germination test and probably better reflects field performance in spring. Acceleration ageing, involving exposure of seeds to a high temperature for a specified period, is another test for viability; still other tests are undergoing further study and refinement.

Tilden (1984) observed that a process referred to as *priming* (slow, controlled water uptake) resulted in *annealing* (healing) of the plasma membrane, reducing electrolyte loss and improving germination and seedling vigor.

Dormancy

Dormancy, a suspended state of growth or rest, is a condition that may persist for an indefinite period despite conditions favoring germination. Technically a seed is dormant at the point of physical or physiological separation from the mother plant. This immediate dormancy ceases, however, with a new set of conditions favoring germination. *Quiescence* is a more descriptive term for the seed rest that results from an environment nonsupportive for germination (e.g., on the maturing plant or in storage). More aptly stated, *dormant* seeds are those that fail to germinate when placed in an environment that supports germination of nondormant members of the seed population (Amen 1963, 1968).

Selection pressures during thousands of years of domestication have virtually eliminated crop plant dormancy. Most crop seeds germinate readily after maturation and desiccation, or drying down. In rainy weather germination of quiescent seeds in the inflorescence of the standing crop is not uncommon (Fig. 9.12). On the other hand, seeds from wild species (e.g., weeds and trees,

Fig. 9.12. Sprouting of rice in the panicle, indicating after-ripening during seed maturation. (Courtesy of N. P. Sarma)

including fruit trees) generally exhibit intense dormancy. Crop plants with a short domestication history often exhibit dormancy to some degree and require special conditions or more storage time for germination (e.g., many hard-seeded forage legumes and many grass species with physiological dormancy, including *Sorghum, Poa,* and *Festuca*). The fact that dormancy is ubiquitous in wild species suggests its ecological significance in species survival. Natural selection pressures during evolution have resulted in plants with dormant seeds and/or dormant buds as an adaptation to periods of environmental adversity such as those found in temperate climates. If germination or bud growth were not synchronized with the incidence of suitable climatic conditions for growth and reproduction, the species could not persist. Dormancy is a major principle in the success of weeds, which survive and find an ecological niche despite the all-out war directed against them. Seeds of many weedy species remain viable and ultimately germinate despite severe stresses from temperature, water, fire, cultivation, and animal and bird ingestion.

TYPES OF DORMANCY

Amen (1968) classified dormancy mechanisms of certain species as follows:

1. Immature embryo: Orchideacae spp.
2. Impermeable seed coats: Leguminoseae (to water), Gramineae (to O_2)
3. Mechanically resistant seed coats: certain species of Gramineae and species with nuts as seeds.
4. Physiological: a wide range of species with seeds that contain growth inhibitors or a supply of growth promoters in the embryo sac, seed coats, or hulls insufficient to initiate the vital processes of germination (Simpson 1978).

The process of becoming germinative, commonly referred to as after-ripening, may be accomplished by maturation on the mother plant, desiccation in storage, or merely ageing in dry storage. On the other hand, after-ripening in some species may require prolonged cold treatment or a more complex set of conditions, such as alternating temperatures, cycles of radiation, presence of salts, leaching, or hull removal. These treatments are effective only on imbibed seeds.

Immature embryos, common in parasitic seed plants, such as witchweed (*Striga lutea*), require a host-provided stimulus. Cytokinin from the host plant has been suggested as the necessary stimulus. In some species embryo maturation can occur in storage or during germination.

A hard seed coat is the principal dormancy mechanism in legume seeds (Figs. 9.13, 9.14). Water impermeability of legume seeds results from two factors: (1) a seed coat with a densely compacted layer of *scleroid* Malpighian cells at right angles to the surface of the testa plus phenolic, or other water-repellent compounds, such as are common in legume seed coats (Evenari 1949;

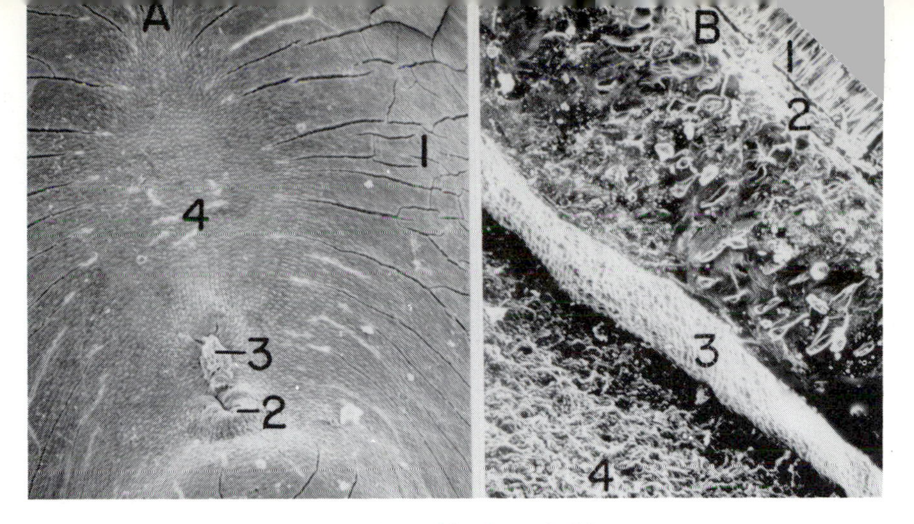

Fig. 9.13. Scanning microscope view of hard seed of *Leucaena*. *A.* Seed structures: seed coat (*1*), micropyle (*2*), funicle (*3*), and pleurogram (*4*). *B.* Cross section: testa, or seed coat (*1*); Malpighian layer (*2*); light line (*3*); and endosperm (*4*). (Courtesy of E. Olvera, S. H. West, and W. G. Blue, University of Florida)

Amen 1963); (2) closure of the natural openings in the seed coat, which include the micropyle, funicle, and *pleurogram* (a depression below the micropyle and funicle). Olvera et al. (1982) concluded that the main factor responsible for the hard seed in *Leucaena* (legume) is pleurogram closure. These structures close as the moisture level outside the seed becomes lower than inside, allowing moisture to leave but not to enter (Fig. 9.14).

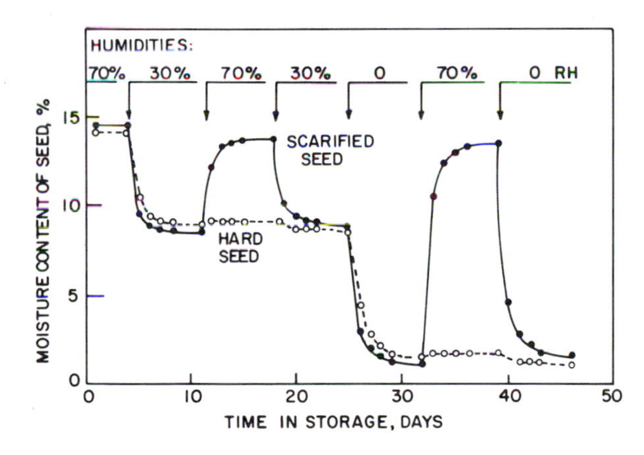

Fig. 9.14. Hard seeds of white clover placed in alternating high and low relative humidities. Seeds lose moisture in lowered humidities but do not gain it back in elevated humidities because of the operation of the hilum valve. Scarified seeds can gain moisture readily. (From Leopold and Kriedemann 1975)

A large number of after-ripening treatments are effective in breaking hard seed dormancy (Table 9.7). Strong acids or alkalis are highly effective but can also damage the seeds. Heat at 100°C for 1.5 minutes, as delivered by a 250-W infrared lamp (Rinker 1954) or hot water, are effective in reducing hard seed

TABLE 9.7. Seed dormancy and after-ripening treatments promoting germination

Species	After-ripening Treatment Promoting Germination	Remarks	Reference
Alfalfa (*Medicago sativa*)	Scarifying: abrasion, heat, acid, electricity, puncturing coat, biotic factors	Legume hard seed impervious to water; may be beneficial agronomically; hard seeds soften naturally in soil	Dexter 1955
Indian ricegrass (*Oryzospsis hymenoides*)	Scarifying with acid	Only effective method; hulls contain inhibitor ABA	McDonald and Khan 1977
Peach (*Perricum malum*)	Stratifying	Embryo dormancy; requires 750 chilling units at 6–10°C or removal of cotyledons	Tukey and Carelson 1945
Cocklebur (*Xanthium pennsylvanium*)	High O_2, kinetin	Upper seed dormant; germinated 100% at 100% O_2 and 30–33°C	Crocker 1906
Wild oat (*Avena fatua*)	High O_2	R light and kinetin removed inhibitors	Khan and Tolbert 1965
		Kinetin cannot substitute for light	Copeland 1967
	R light	Gibberellins can substitute light and temperature in most species	Toole and Hendricks 1956
	GA	GA activated maltase production	Bonner and Varner 1965
	Dehulling (removal of lemma and palea), KNO_3	High O_2 and H_2O_2 and puncture of seed coats effective	Hay 1967; Toole and Hendricks 1956
Lettuce (*Lactuca sativa*)	Dry storage, GA R light, ethylene	Dormancy lost on dry storage of freshly harvested seed of Grand Rapid cultivar; acts synergistically with light and GA	Black and Naylor 1959 Borthwick et al. 1954
	Thiourea	Can substitute for light and certain temperature requirements (cold)	Poljakoff-Mayber et al. 1958; Tukey and Carelson 1945

Kentucky bluegrass (*Poa pratensis*)	Alternating temperatures plus light		Copeland 1967
Bermudagrass (*Cynodon dactylon*)	Low O_2		Morinaga 1926
Bentgrass (*Agrostis palustris*)	Alternating temperature, light	Length of exposure to high temperature important for most cultivars; R light effective during high temperature; light response of two cultivars at marginal temperature only	Toole and Koch 1977
Tall fescue (*Festuca arundinacea*)	Low germinating temperature	Seed held at 35°C for 7 days dormant; light induced germination in lower temperature	Danielson and Toole 1976
Desert grasses	Leaching	Leaching by winter rains	Vegis 1963
New Zealand browntop (*Agrostis tenuis*)	N salts (KNO_3)	Widely used in seed labs; 0.1 to 1.0% solution effective; most light-sensitive seeds affected	Gadd 1955
Wild rice (*Zizania palustris*)	Cold-water storage > 100 days	ABA inhibitor in embryo and pericarp	Albrecht et al. 1979

content. Hot water (100°C) for 5 to 20 sec caused the pleurogram in *Leucaena* to open and a resultant germination of 95 to 100%, depending on variety (Olvera et al. 1982). *Scarification* (mechanical abrasion, acid, or hot water treatment of the seed coat) can remove hilum plugs and increase permeability. Hard seeds in moderate amounts are believed to be of value to forage seeds (Dexter 1955), so scarification is not always advisable. After-ripening is accomplished naturally by freezing and thawing, wetting and drying, animal ingestion, microbial action, and/or enough ageing in storage.

The hulls of many grass seeds and weedy species, such as green needlegrass (Weisner and Kinch 1964) and Indian ricegrass (McDonald and Khan 1977), are impervious to O_2. In green needlegrass the lemma and palea (seed hulls) act as a barrier. Germination peaked at 72% 7 yr after harvest. Laboratory chilling and KNO_3 solution pretreatments induced nearly complete germination. Cocklebur and wild oats are classical examples of seed dormancy that results from an O_2-impermeable seed coat (Crocker 1906; Hay 1967).

The cocklebur seed is actually a spiney, dry, nondehiscent fruit containing two seeds. The lower seed may germinate readily, while the upper one remains dormant for several years because of the low O_2 tension surrounding it (Crocker 1906). Hulls of wild oats also impose a low O_2 tension. Removal of the hulls of the seeds of both these plants greatly improves germination.

Mechanically resistant seed coats can imbibe water readily, unlike hard seeds, but resist swelling and embryo protrusion. Seeds of some grasses and most species with hard-seeded fruits (nuts) as seeds have seed coats mechanically resistant to embryo emergence. A prolonged period of wet storage can weaken the hard covering and accomplish after-ripening. In the case of black walnut (*Juglans nigra*) several weeks of cold (2–5°C) wet storage (*stratification*) are required (Meyers et al. 1979); this species appears to have at least a double dormancy mechanism: a mechanically resistant seed coat and an unripe (physiologically immature) embryo. It is also known that walnut husks contain a strong growth inhibitor, juglone. This substance, probably lost by leaching during wet storage, seems to constitute even a third dormancy mechanism in black walnut. The ecology of nut-bearing species is generally closely associated with small animals burying nuts during fall, which accomplishes seed dispersal, softening of the seed coat, and stratification during the overwintering period.

PHYSIOLOGICAL DORMANCY

Physiological dormancy is often referred to as embryo dormancy (Amen 1968) and has been called deep dormancy (Borriss 1949). A physiologically immature embryo is considered a physiological dormancy. The presence of growth inhibitors, a deficiency of growth-promoting substances, or a lack of proper balance between the two hormones has been postulated as the factor causing embryo dormancy. Abscisic acid (ABA), coumarin, and other inhibitors (Table 9.8) have been shown to induce dormancy, but these factors may be

TABLE 9.8. Naturally occurring germination inhibitors in various plant species

Germination Inhibitor	Species Producing Inhibitor	Part of Plant Containing Inhibitor	Remarks
Amygdalin	Plum (Prunaceae), almond	Seeds, fruit juice	Contains HCN
Ammonia	Sugar beet	Seed balls	Inhibits germination of sugar beet and other seeds
Ethylene	Climacteric fruits	Fruit juice	Volatile inhibitors
Mustard oils	Cruciferae	Seeds	Volatile inhibitors
Organic acids	Apple, citrus fruits	Fruit juice	Direct effect, rather than pH
Unsaturated lactones	Lettuce	Hulls	Coumarin
Aldehydes	Unripe maize, peas, bitter almond	Seeds	Produced in anaerobic conditions
Essential oils	Citrus fruit	Peels	Prevented germination of wheat seeds
Alkaloids	Tobacco, coffee, cocoa	Seeds, other plant parts	Contain nicotine, caffein, and cocaine alkeloids, respectively
Phenols	*Vicia, Rapa*	Seeds	Thymol most active inhibitor

Source: From Evenari 1949.

in the hull, seed coat, aleurone, or embryo. Growth-promoting substances (GAs and cytokinins) release dormancy in a wide variety of species (Table 9.7).

A theoretical model of growth promoter–growth inhibitor balance in germination is illustrated in Fig. 9.15. According to the model, germination can occur when a critical hormone balance is reached, either by elevation of growth-promoting substances or by depression of growth inhibitors. Amen (1963) stated that most dormancy mechanisms can be broken by growth-promoting substances. The fact that GA treatment replaces the light requirement in numerous photoblastic seeds (lettuce, tobacco) and the cold requirement in species requiring stratification (wild oat, many tree species) supports this conclusion. Growth-promoting substances often decline during seed ontogeny, whereas growth inhibitors such as ABA increase, the result being dormancy at seed maturity due to hormone imbalance (Fig. 9.15). Various conditions during postharvest generally cause the reversal of the above process, which explains the loss of light and stratification requirements in certain species during dry storage.

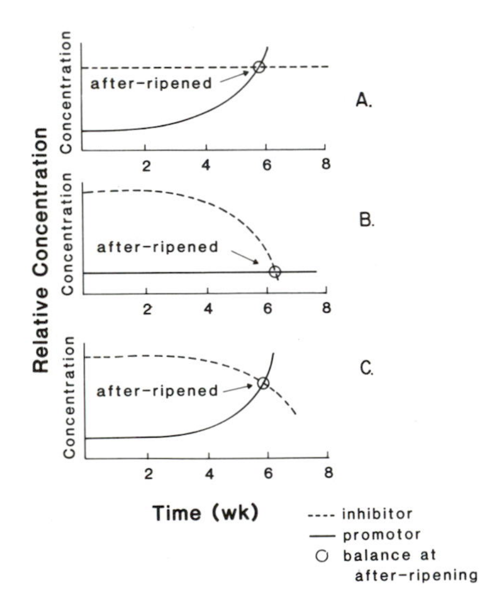

Fig. 9.15. Model for after-ripening in seed as a result of reaching a proper balance of growth-promoting and growth-inhibiting hormones. *A.* Increasing growth-promoting hormones while growth-inhibiting hormones remain static. *B.* Decreasing growth-inhibiting hormones. *C.* Simultaneously increasing growth-promoting and decreasing growth-inhibiting hormones.

Coumarin is a common natural chemical inhibitor in physiological dormancy, but abscisic acid (ABA, or dormin), unsaturated lactones, alkaloids, phenols, ethylene, ammonia, essential oils, hydrocyanic acid, and organic acids also have been reported to cause dormancy (Evenari 1949; Hay 1967) (Table 9.8). Growth inhibitors controlling dormancy may be in the embryo, as in a number of grasses; in the hull, as in lettuce and buckwheat; or in the fruit, as in apple and tomato. Varying degrees of dormancy found in ten varieties of wheat were due to water- or methanol-soluble growth inhibitors, which disappeared after a month or so of dry, warm storage (Ching and Foote 1961). Leaching or hull removal increases germination of certain grass species. In sorghum, dormancy has been associated with brown pericarp fused with the testa (Clark et al. 1968). The dormancy factor was removed by scarification or by hot water treatments. The endosperm and aleurone of certain species also contained dormancy factors (Amen 1968). The embryo and pericarp of wild rice contained inhibitory levels of ABA, which were removed by more than 100 days of cold water (3°C) storage (Albrecht et al. 1979).

From the previous discussion it is seen that seed dormancy is extremely complex, involving several seed structures, growth stimuli of the environment, endogenous growth substances, exogenous chemicals, and all possible interactions of these factors. A schematic model (Fig. 9.16) serves to illustrate the point. If only four factors are operative—for example, seed coat, temperature, endogenous substance A, and exogenous substance B—12 main effects, 72 primary interactions, or 84 possible causative factors may influence dormancy. The after-ripening of freshly harvested lettuce is illustrative (Amen 1968): (1) due to a secondary dormancy (acquired postharvest), germination occurred in a narrow range of low temperatures (15–20°C); (2) germination was rapid at a wide temperature range after an adequate period of dry storage, in which dormancy was lost; (3) imbibed germinative seeds became dormant (secondary dormancy) if exposed to high temperatures (30–35°C); (4) the induced (secondary) dormancy did not break down to allow germination unless returned to a narrow range of cool temperatures; (5) seeds with secondary, or even true, dormancy germinated readily if exposed to R radiation (Borthwick et al. 1954) or GA_3 (Borriss 1967); or (6) germination occurred if seed hulls were removed. Treatment of imbibed lettuce seeds with coumarin induced dormancy in germinative seeds, just as a high temperature caused secondary dormancy (Nutile 1945). Primary dormancy (initially developed during seed ontogeny), *second-*

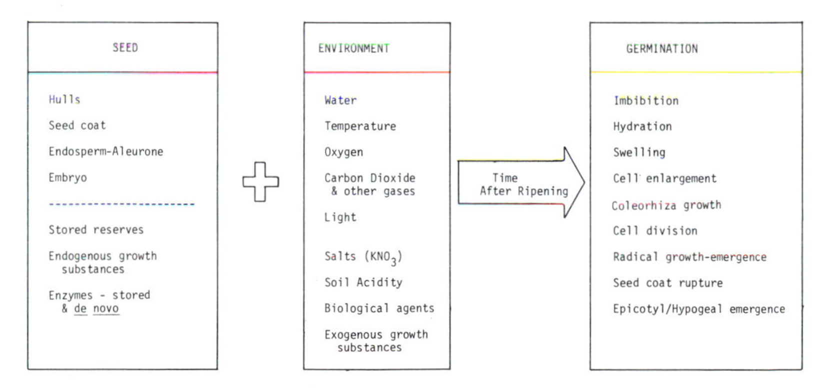

Fig. 9.16. Seed structures and environmental factors that in time lead to after-ripening. Seed and environmental factors interact to produce complex mechanisms that result in dormant or germinative seeds.

ary dormancy or *relative dormancy* (environmentally imposed in mature seed but germinable only at a narrow temperature range, 15–20°C), or *true dormancy,* as defined by Borriss (1949) (no germination even at the optimum temperature), are all internally controlled but may be imposed by the environment.

Emergence and Seedling Growth

Seedlings emerge either by (1) *epigeal emergence* (elongation of the hypocotyl, the upper part of the radicle) or (2) *hypogeal emergence* (elongation of the epicotyl, the first internode[s]) (Fig. 9.17). The structures that elongate in epigeous and hypogeous emergence are immediately below and above the cotyledonary node, respectively. In epigeal emergence the cotyledons push above the medium surface; in hypogeal emergence the cotyledons remain below the surface. Peanut is intermediate, since cotyledons emerge if planting depth is shallow but remain below the surface if relatively deep.

Fig. 9.17. Hypogeal emergence of maize and pea, epigeal emergence of bean. Seedling structures: first true leaf (*a*), apical bud (*b*), cotyledon (hidden in maize seed) (*c*), radicle (*d*), epicotyl (*e*), hypocotyl (*f*), mesocotyl (*g*).

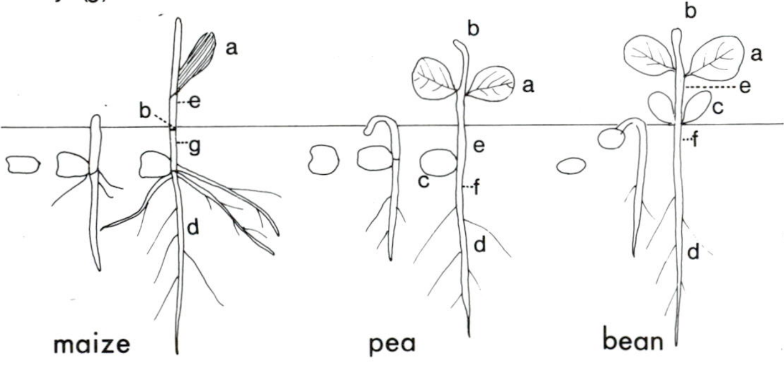

In grass seeds the cotyledons remain below the surface (hypogeal emergence) and absorb the hydrolyzed food reserves from the endosperm. Emergence of garden pea and pigeon pea are hypogeal, unlike most legumes, but food nutrients are drawn from the cotyledons as usual. However, cotyledons above the surface not only provide food reserves but, being densely packed with chloroplast, are a photosynthetic organ. Cotyledons normally begin to senesce soon after emergence and abscise in about 3 wk.

SEED SIZE AND DENSITY

Seed size is generally highly correlated to seedling weight. The heaviest seeds within a smooth bromegrass lot produced the most vigorous seedlings

(Kalton et al. 1959). When large- and small-seeded snap bean lots (Clark and Peck 1968) were planted with equal numbers of seeds per row foot, plants from large seeds outyielded plants from the small-seeded lots. When the same weight of seed was planted per row, small-seeded beans generally outyielded large-seeded. Seeds large enough to develop numerous transverse cracks in the seed coat had lower germination and poor stands. In these instances, large seed size gave some of the lowest yields of any size class used.

Soybean from different sources have shown a positive correlation between yield and larger seed size (Fehr and Probst 1971). In studies with soybean (Smith and Camper 1975), large seeds yielded more than an equal number of small seeds, while other agronomic characteristics did not differ with seed size. In another report soybean from large seeds yielded significantly more than those from small seeds, and seed size was also positively associated with emergence, leaf area, and plant height (Burriss et al. 1973). However, in the laboratory photosynthetic rate (CER) per unit of leaf area was greater in 7-day-old seedlings from small seeds. Black (1956) found a positive effect of seed size on cotyledon size. He observed that increasing planting depth decreased seedling weight of subterranean clover but did not decrease cotyledon area (Table 9.9). The large seed produced twice the cotyledon area and higher photosynthetic potential than the small seed.

TABLE 9.9. Effect of depth of sowing and seed size on seedling weight and cotyledon area of subterranean clover

Seed Size and Depth	Weight (mg)	Area (mm^2)
Large seed		
½ in.	3.3	16.4
1¼ in.	2.9	16.3
2 in.	2.5	16.3
Seed planted at ½-in. depth		
Small	1.2	7.8
Medium	2.1	11.8
Large	3.3	16.8

Source: Black 1956.

Placement of small-seeded forage crops is often too deep for good emergence. Seeding at much higher rates, many times more than the expected final population, is a common compensatory practice. In contrast, a small mortality rate is anticipated with cereals like maize, and seeding rates near the expected population are used.

With sorghum early performance was influenced more by seed density than by seed size (Maranville and Clegg 1977). Final stands were improved by dense seeds, but final seed yields were independent of seed size or density (Table 9.10). Overall, the evidence indicates that large and dense seeds are

TABLE 9.10. Effect of seed size and density on grain sorghum germination and yield

Seed Size or Density	Germination	Seed Wt	Final Stand	Grain Yield
Control	76 bc	24 c	61 b	4315 a
Large	81 a	31 a	64 b	4320 a
Small	74 c	23 d	58 b	4245 a
Dense	80 ab	25 b	71 a	4425 a
Light	52 d	19 e	63 b	4280 a

Source: From Maranville and Clegg 1977.
Note: a, b, c, etc., indicate significance expressed by Duncan's multiple range test.

superior in germination and early vegetative performance. Generally, however, this early advantage seems to disappear, and final yield outcome is about the same.

Several explanations seem plausible for the conflicting results reported on the advantage of large over small seeds. (1) In legumes, large seeds have a larger embryo and tend to be advantageous because larger cotyledons have greater initial photosynthesis (Black 1956, 1959). The cotyledons have a high specific leaf weight (SLW) and are highly active photosynthetically, but this initial advantage is easily compounded with other factors, such as a greater soil resistance to emergence of larger cotyledons. (2) In grasses, large seeds have more food reserves, but food reserves in excess of those needed for the plant to become independent are of little or no advantage. (3) Seed density appears to be as important as size, since it reflects embryo size and/or amount of stored nutrients. Most investigations, however, have considered size rather than density, so evidence is meager on the denser seed advantages. Within the limits of most acceptable cultivars of most crops, there appears to be little to be gained by emphasizing large or small seeds.

Summary

Biologically a seed is a ripe ovule, the wall of which forms the testa (seed coat). The seed (caryopsis) of grasses is a single-ovule dry fruit that ripens with fused ovary and ovule walls as the pericarp, or seed coat. Seeds of carrot and cocklebur are two-seeded, dry, nondehiscent fruits (schizocarps). The sunflower seed is a single-seeded, dry, nondehiscent fruit with the pericarp and testa not fused (achene). Sugar beet, except monogerm cultivars, has aggregate fruits that dry to produce a seed ball, or "seed."

Grasses store starches and protein in the endosperm and oil in the germ or embryo, whereas legumes store food reserves in the cotyledons, which absorb the endosperm. The principal food reserves of about one-half the legume species, such as cowpeas and other pulse crops, are starch and protein. The principal food reserves of the other legume species, such as soybean, are oil and protein. Seed proteins have an amino acid imbalance due to deficiencies of lysine, tryptophan, or methionine, and are therefore lower in biological value for monogastric animals (including humans) than is the vegetable matter produced from the seeds. Starches in grass seeds are usually amylopectin (branched) and amylose (straight chain). Most maize cultivars store both, but amylopectin is dominant. Other carbohydrates in seeds of certain species include hemicellulose (mannans and xylans), mucilages, pectin, and sugars. Stored lipids are generally unsaturated triglycerides or oils.

The protein of legume seeds is primarily the globulin type (water soluble) and that of cereal grains generally the prolamin type (alcohol soluble). Prolamin is deficient in lysine and tryptophan. Seeds may also contain alkaloids, phenols, and lactones, which generally act as germination inhibitors (dormancy mechanisms).

Germination (rupture of seed coat and emergence of the radicle) involves imbibition, rapid O_2 uptake, hydrolysis of stored reserves, and synthesis of new tissues. Auxins, GAs, and cytokinins are required in germination, as is ethylene. Gibberellins trigger the release of hydrolytic enzymes. Due to dormancy mechanisms, viable seeds may not germinate despite favorable conditions, that is, conditions favoring seedling growth. Dormancy can be caused by seed coats impervious to water (legumes) or to O_2 (certain grasses); by growth inhibitors; and by endosperm and embryo factors, which generally involve growth hormones. Seeds lose germinability and seedling vigor with time, especially under high temperature and/or humidity. Loss of vigor is even rapider than loss of germinability and results in weak seedlings predisposed to soil-borne diseases.

Seeds require water, O_2, and a moderately warm temperature for germination. Certain species require special conditions to break dormancy, that is, to cause after-ripening (readiness to germinate), such as any one or a combination of the following: cold-moist treatment for a sufficient period; light-moist treatment; exogenous chemicals, including such salts as thiourea or hydrogen peroxide; and hormones, including GAs and ethylene.

Visible light in the red (660-nm) range of the spectrum is most effective in promoting germination. Red light, GAs, and cold are interchangeable in breaking dormancy in photoblastic seeds such as lettuce, which can acquire an environmentally induced or secondary dormancy. Some seeds require scarification for water or O_2 absorption; some require leaching to remove seed coat inhibitors. Seed longevity may be as short as a few weeks or as long as centuries, depending on the intensity and type of dormancy and the storage environment. The interaction of seed structures, growth promoters, and inhibitors with environmental factors results in numerous complex dormancy mechanisms.

Seedling emergence is either hypogeal (expansion below cotyledons) or epigeal (expansion above cotyledons). In general large, dense seeds favor seedling emergence and growth, but they can have a negative effect in certain situations.

References

Abdul-Baki, A. A., and J. D. Anderson. 1970. Crop Sci. 10:31–35.
Abu-Shakra, S. S., and T. M. Ching. 1967. Crop Sci. 7:115–17.
Albrecht, K. A., E. A. Oelke, and M. L. Brenner. 1979. Crop Sci. 19:671–76.
Anslow, R. C. 1962. J. Br. Grassl. Soc. 17:260–63.
Amen, R. 1963. Am. Sci. 51:408–24.
______. 1968. Bot. Rev. 34:1–31.
Becquerel, M. P. 1934. C. R. Acad. Sci. [Paris] 199:1662–64.
Bewley, J. D., and M. Black. 1978. Physiology and Biochemistry of Seeds in Relation to Germination, vol. 1. New York: Springer-Verlag.
Black, J. N. 1956. Aust. J. Agric. Res. 7:98–109.
______. 1959. Aust. J. Agric. Res. 8:1–14.
Black, M., and H. M. Naylor. 1959. Nature 184:468–69.
Bloor, W. R. 1928. Chem. Rev. 2:243–300.
Bonner, J., and J. E. Varner, eds. 1965. Plant Biochemistry. New York: Academic Press.
Borriss, H. 1949. Jahrb. Wiss. Bot. 89:254–339.
______. 1967. In Physiologie, Okologie, und Biochemie der Keimung, 1, ed. H. Borriss. Greifswald: Ernst-Moritz-Arndt Universitat.
Borthwick. H. A., S. B. Hendricks, M. W. Parker, E. H. Toole, and V. K. Toole. 1952. Proc. Natl. Acad. Sci. 38:662–66.
Borthwick, H. A., S. B. Hendricks, E. H. Toole, and V. K. Toole. 1954. Bot. Gaz. 115:205–25.
Burriss, J. S., O. T. Edge, and A. H. Wahab. 1973. Crop Sci. 13:207–10.
Ching, T. M., and W. H. Foote. 1961. Agron. J. 53:183–86.
Ching, T. M., M. C. Parker, and D. D. Hill. 1959. Agron. J. 51:680–84.
Clark, B. E., and N. H. Peck. 1968. N.Y. Agric. Exp. Stn. Bull. 819.
Clark, L. E., J. W. Collier, and R. Langston. 1968. Crop Sci. 8:155–58.
Copeland, L. O. 1967. Principles of Seed Science and Technology. Minneapolis: Burgess.
Crocker, W. 1906. Bot. Gaz. 42:265–91.
Danielson, H. R., and V. K. Toole. 1976. Crop Sci. 16:317–20.
Daubert, B. F. 1950. In Soybeans and Soybean Products, ed. K. S. Markley. New York: Interscience.
Dexter, S. T. 1955. Agron. J. 47:357–61.
Dudley, J. W., and R. J. Lambert. 1969. Crop Sci. 9:179–81.
Early, E. B., and E. E. DeTurk. 1948. Proc. Am. Seed Trade Assoc. Chic., pp. 84–95.
Evenari, M. 1949. Bot. Rev. 15:153–94.
Fehr W. R., and A. H. Probst. 1971. Crop Sci. 11:865–67.
Flint, L. H., and E. D. McAlister. 1937. Smithson. Misc. Collect. 96:1–8.
Gadd, I. 1955. Proc. Int. Seed Test. Assoc. 23:41.
Grabe, D. F. 1956. Agron. J. 48:253–56.
Hay, J. R. 1967. In Physiologie, Okologie, und Biochemie der Keimung, 1, ed. H. Borriss. Greifswald, Ernst-Moritz-Arndt Universitat.
Hendricks, S. B., V. K. Toole, and H. A. Borthwick. Plant Physiol. 43:2023–28.
Jennings, A. C., and R. K. Morton. 1963. J. Biol. Sci. 16:318–31.
Khan, A. A., and N. E. Tolbert. 1965. Physiol. Plant. 18:41–43.
Kalton, R. R., R. A. Delong, and D. S. McLeod. 1959. Iowa State J. Sci. 34:47–80.
Kinzel, W. 1926. Frost und Licht, Neve Tabellen. Stuttgart: Eugen Ulmer.
Klingman, G. C., and F. M. Ashton. 1975. Weed Science: Principles and Practices. New York: Wiley.
Leopold, A. C., and P. E. Kriedemann. 1975. Plant Growth and Development. New York: McGraw-Hill.
McDonald, M. B., Jr., and A. A. Khan. 1977. Agron. J. 69:558–63.
Maranville, J. W., and M. D. Clegg. 1977. Agron. J. 69:329–30.
Martin, G. L., and M. E. Heath. 1973. In Forages, ed. M. E. Heath et al. Ames: Iowa State University Press.

Mayer, A. M., and A. Poljakoff-Mayber. 1963. The Germination of Seeds. New York: Macmillan.
______. 1967. In Physiologie, Okologie, und Biochemie der Keimung, 1, ed. H. Borriss. Greifswald: Ernst-Moritz-Arndt Universitat.
Meyer, B. S., and D. B. Anderson. 1949. Plant Physiology. New York: Van Nostrand.
Meyers, O., G. Gaffney, and D. Hall. 1979. Abstracts Ill. State Acad. Sci.
Morinaga, T. 1926. Am. J. Bot. 13:159–66.
Nutile, G. E. 1945. Plant Physiol. 20:433–42.
Ohga, I. 1926. Am. J. Bot. 13:754–59.
Osler, R. D., and J. L. Cartter. 1954. Agron. J. 46:267–70.
Olvera, E., S. H. West, and W. G. Blue. 1982. Submitted for publication.
Orthoefer, F. T. 1978. In Soybean Physiology, Agronomy, and Utilization, ed. A. G. Norman. New York: Academic Press.
Osborne, T. B. 1924. Monographs on Biochemistry: The Vegetable Proteins. 2d ed. London: Longmans, Green.
Poljakoff-Mayber, A., A. M. Mayer, and S. Zacks. 1958. Ann. Bot. n.s. 22:75–81.
Porsild, A. E., and Harrington, C. R. 1967. Science 158:113–14.
Rinker, C. M. 1954. Agron. J. 46:247–50.
Ryan, C. J. 1973. Annu. Rev. Plant Physiol. 24:173–96.
Simpson, G. M. 1978. In Dormancy and Development Arrest, ed. M. E. Cutter. New York: Academic Press.
Smith, A. K., and S. J. Circle. 1972. In Soybean Chemistry and Technology, ed. A. K. Smith and S. J. Circle. Westport, Conn.: AVI.
Smith, T. J., and E. M. Camper, Jr. 1975. Agron. J. 67:681–84.
Stone, J. F., and B. B. Tucker. 1969. Agron. J. 61:76–78.
Tilden, R. 1984. Ph.D. diss., University of Florida, Gainesville.
Toole, E. H., and S. Hendricks. 1956. Annu. Rev. Plant Physiol. 7:229–324.
Toole, V. K., and E. J. Koch. 1977. Crop Sci. 17:806–11.
Tukey, H. B., and R. F. Carelson. 1945. Plant Physiol. 20:505–16.
Van Overbeek, J. 1968. Sci. Am. 219:75–81.
Vegis, A. 1963. In Environmental Control of Plant Growth, ed. L. T. Evans. New York: Academic Press.
Wiesner, L. E., and R. C. Kinch. 1964. Agron. J. 56:371–73.

10 Root Growth

ROOTS as major vegetative organs supply water, minerals, and substances essential for plant growth and development. Despite these vital contributions they are often, probably too often, taken for granted because they are not visible—unfortunately, "out of sight, out of mind."

Research on roots is relatively limited, compared with that of other plant organs, due to a great extent to the difficulties involved in their study. However, there is more opportunity to promote plant growth by changing the root environment than by changing the shoot environment. Air, water, and mineral phases of the *rhizosphere* (root environment) are relatively easy to alter by agronomic practices; soil temperature can be influenced by tillage and mulching, moisture by irrigation, and nutrient status by fertilization. The shoot atmosphere of crop plants, on the other hand, is difficult, or practically impossible, to change. Root studies probably should be given more emphasis than is done at present.

Root Functions

Vigorous root growth is normally required for general vigor and growth of tops. If roots are damaged by biological, physical, or mechanical disturbances and become less functional, top growth will likely be less functional.

Roots serve the plant in the following important ways (Weaver 1926): (1) absorption; (2) anchorage; (3) storage; (4) transport; and (5) propagation. They are also a primary source of certain plant growth regulators.

The absorption of water and minerals occurs primarily through the root tips and hairs, although older and heavier parts of a root absorb some. Older roots perform the necessary functions of substance transport and storage, analogous to substance transport to and from leaves by stems and branches.

Anchorage is more than just holding the plant in place. Roots themselves need to be anchored against the force exerted by terminal parts that penetrate dense soil zones.

Roots often serve as the primary organ for storage of food reserves, especially in dicots. Dicot roots are well endowed with cortex, pith, or similar parenchymous tissue (e.g., sugar beet, alfalfa, and other plants with fleshy roots). Grass roots are fine and, by comparison, have little storage capacity.

Roots of numerous species can be used for propagation because of their capacity to develop adventitious shoots and to store food reserves that support new growth. Many obnoxious weeds, notorious for this type of propagation, can resist eradication by tillage.

Roots are believed to be the primary source of the growth regulators gibberellins and cytokinins, which influence overall plant growth and development (see Chap. 7).

Root Study Techniques

The difficulty of studying roots led to the development of a number of techniques to improve effectiveness. Basically, two methods are employed: determinations *in situ* (in position) and indirect methods, such as the use of radioactive isotopes or dye tracings (MacKey 1980).

TRENCH PROFILE METHOD

The trench profile method, which was introduced in the classic studies by Weaver (1926), is, with modification, still in common use (Kutschera 1960). It involves excavating a trench perpendicular to the row or beside an individual plant and mapping or photographing the visible roots.

FRAMED MONOLITH AND PINBOARD METHOD

The framed monolith method is the trench profile modified to quantify root extension and distribution. Pins inserted into holes equidistantly spaced in a board placed against the face of the trench in effect trap the roots in the soil profile in cubes. A soil monolith is cut and lifted from the trench. After the prisms are soaked in water, the roots are carefully washed free of soil, described, and quantified as to length, weight, or other parameters (Bohm et al. 1977) (Fig. 10.1). An application of Congo red stain stained differentially and permitted separation of living from dead roots (Ward et al. 1978). The framed monolith method can provide quantitative information, but it is laborious, time-consuming, and costly.

SOIL MOISTURE DEPLETION METHOD

Soil moisture depletion can be measured either *gravimetrically* (weighing core samples) or by a neutron probe. It indicates depth of moisture depletion. The depletion front usually exceeded rooting depth by about 15 cm (Stone et al. 1976) due to moisture migration in the soil. This method is rapid but is not reliable for determining root density or other root parameters because water extraction rates also depend on evaporative demand, soil water potential, and soil hydraulic characteristics.

CORE SAMPLING METHOD

Core sampling removes undistributed soil cores and roots contained

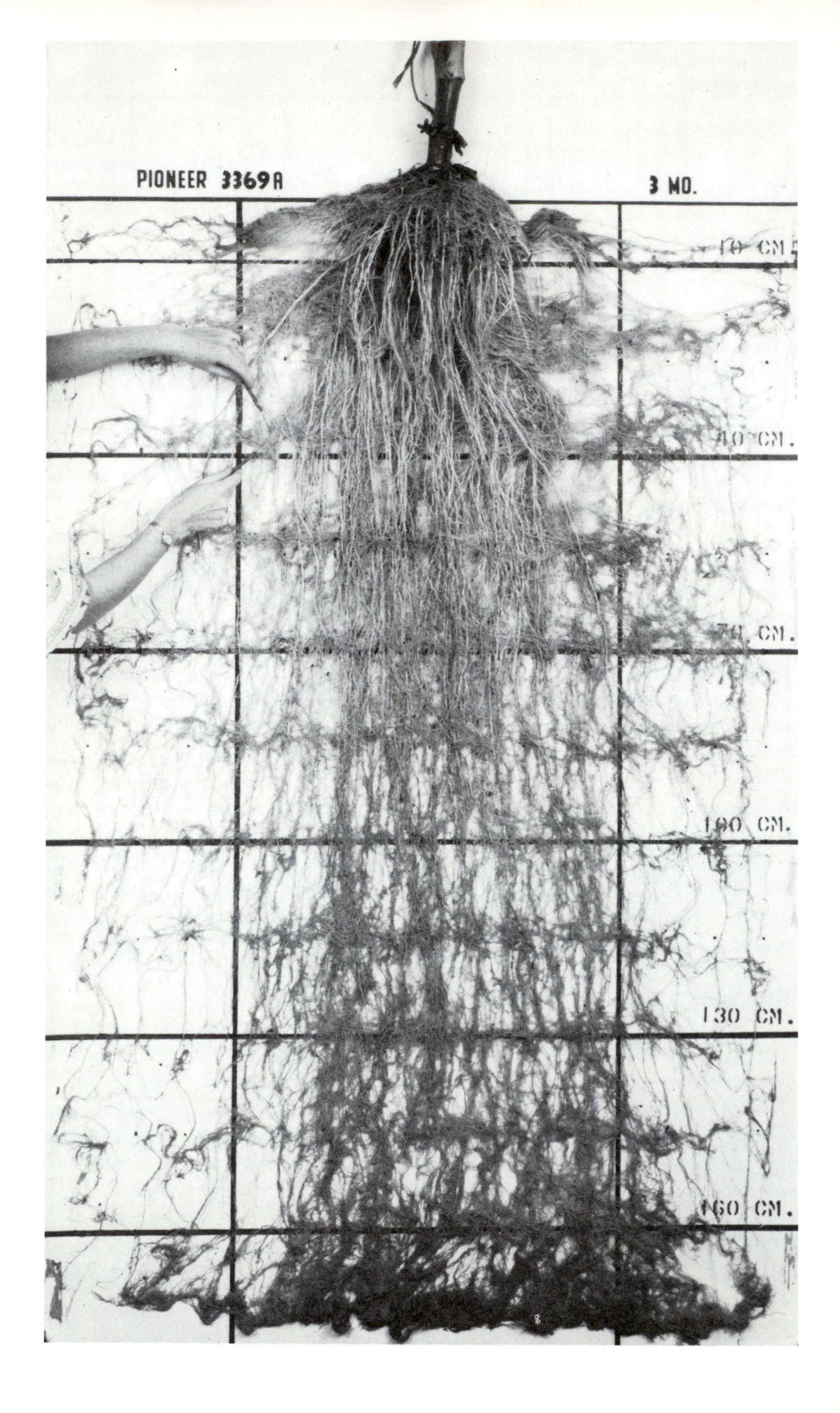
PIONEER 3369A
3 MO.
100 CM.
130 CM.
160 CM.

Fig. 10.1. Root system of a maize plant (framed monolith and pinboard method). Roots observed are the adventitious, or nodal, roots; the seminal roots are not visible. (Courtesy of M. Huck, USDA, National Tillage Machinery Laboratory, Auburn, Ala.)

therein from the root zone at predetermined locations relative to the plant and row (Newman 1966). Tube or bucket samplers are commonly used. Several cores per point of reference are required. This method can be mechanized but is only semiquantitative (Bohm et al. 1977).

MINIRHIZOTRON METHOD

A glass tube positioned in the soil profile can be used, with the aid of a light and mirror or TV camera, to observe rooting as roots grow against the side. The findings are only qualitative. A planting box with one tilted glass side or translucent tubes tilted at about 20° to 30° from upright and kept dark can indicate growth rate and depth penetration, as roots contact the glass and follow it downward.

RADIOACTIVE ISOTOPE METHOD

Isotope ^{32}P, placed at variable depths, can indicate the rooting depth from the counted and recorded isotope uptake. This method is not effective in determining root density. A similar method, consisting of placing perforated plastic bags of nitrogen (N) fertilizer at variable depths and observing the foliar N changes, yields comparable results, especially with grass plants.

ALLOMETRY METHOD

This method assumes proportionality of the logarithms of weights of plant parts, in this case the shoot-root (S-R) ratio. Large top growth indicates proportionally large root growth. Measurements of plant height reliably indicated the depth of rooting of soybean on irrigated or rain-fed, barrier-free soils (Mayaki et al. 1976). Rooting depth was twice plant height at the V_3 stage of growth (vegetative with three expanded leaves), and this relationship continued until pod development, when it became 1.4 times plant height. Irrigated plants had approximately 15% more roots by weight in the 0- to 15-cm zone. The rooting of 22 strains of perennial grasses was allometric (Troughton 1956). This method can be misleading in that barriers in the soil profile do not significantly change the S-R ratios but drastically alter the locations of the root mass (Taylor 1963).

To characterize rooting most root studies have reported either fresh or dry weight, but root weight may be poorly correlated with water and nutrient uptake, the primary interest in root studies. The fine and young roots, principally the root hair zone, function most effectively in nutrient uptake. Root hairs are confined to a section of a few millimeters, at most a few centimeters, of root near the tip. Root hairs form just after elongation of epidermal cells (Cormack 1962). The number and rate of root hair formation was greater at 26°C than at 15°C, but the life span was 40 hr and 55 hr for the two temperatures, respectively (McElgunn and Harrison 1969). This suggests a somewhat constant rate of occurrence for a root segment of a given species. The length of the root hair zone depends to a degree on genotype and environment. Effectiveness of root uptake is better estimated by determination of *root density* (root length per soil volume) (Barley 1970) or by root surface area (Barber 1978) than by root length.

Root Initiation and Growth

Root length results from elongation of cells behind the apical meristem; breadth greater than that of enlargement of apical cells results from lateral meristem or cambium formation, which initiates secondary growth from the cambium meristem. Length and girth growth is generally analogous to that of shoots. However, lateral branching is not analogous, since it arises from the pericycle deep within older or differentiated tissues, a morphogenesis markedly different from the surface origin of branches from the apex in the shoot (Fig. 10.2). A detailed comparison of the morphogenesis of root and shoot is given in Table 10.1.

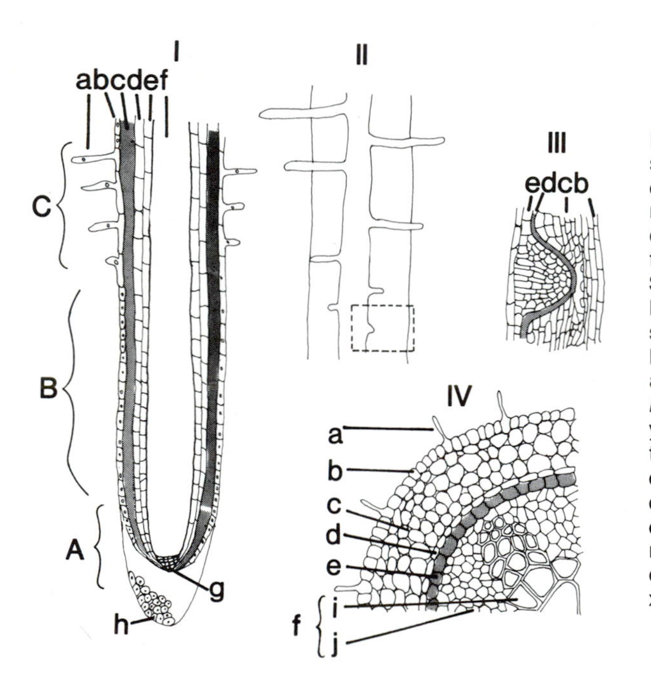

Fig. 10.2. Longitudinal section of herbaceous dicot root. *I.* Root tip with regions of cell division (*A*), elongation (*B*), and maturation (differentiation) (*C*). *II.* Section of mature root with lateral roots in varying stages of development. *III.* Meristem of a lateral root arising from the pericycle. *IV.* Cross section of a young root. Differentiated tissues: root hair (*a*), epidermis (*b*), cortex (*c*), endodermis (*d*), pericycle (*e*), central cylinder or stele (*f*), meristem with quiescent center (*g*), root cap (*h*), xylem (*i*), phloem (*j*).

TABLE 10.1. Comparison of stem and root development

Stem	Root
Meristem apical	Meristem subapical
Lateral organs arise near apex of meristem	Lateral organs arise some distance from apex
Lateral organs arise from surface layers	Lateral organs arise from internal tissue layers, normally the pericycle
Nodes and internodes present due to regularity of lateral organ formation	Nodes and internodes not present
Primary xylem and primary phloem are on same radius	Primary xylem and primary phloem are on alternate radii; star-shaped
Vascular cambium arises from parenchyma cells between primary xylem and phloem	Vascular cambium arises both from parenchyma cells between primary xylem and phloem and from pericycle
Epidermis with stomata	Epidermis without stomata
Pericycle generally absent in seed plants	Pericycle generally present in seed plants
Endodermis generally absent	Endodermis almost universally present
Quiescent center not present in apical meristem	Quiescent center present

Based on ATPase enzyme activity indicative of a high metabolic rate characteristic of meristems, a subapical meristem has been located a few millimeters from the root tip (Fig. 10.2). ATPase activity in soybean roots was observed to begin near the tip and continue for 27.5 mm, with a maximum at 3.5 mm (Travis et al. 1979). Root elongation was greatest in a zone between 5.0 and 15.0 mm. The zone of differentiation, including root hairs, xylem, phloem, pericyle, and other specialized cells, began at about 15 to 25 mm (Fig. 10.2). The faster a root grows, the greater the length of the zone of differentiation.

New cells from the root apical meristem may be partitioned to root extension or to renewal of the root cap. The root cap plays an important role in protecting the root meristem from physical damage during soil penetration and possibly in guiding its direction. The sloughed root cap cells also provide lubrication for the growing tip, substrate for microbes, and additional soil organic matter. The root cap also produces abscisic acid, a plant growth substance.

The root apical meristem differs from the shoot apical meristem in its relatively low DNA, RNA, and mitotic activity (Milthorpe and Moorby 1974). In case of damage or decapitation, the quiescent center (Fig. 10.2) regenerates a new meristem and restores the geotropic character within 36 hr at a favorable temperature (Clowes 1969). Root extension and cap renewal can then continue as before.

EXTENSION

Root meristems are capable of continous, indeterminate growth that results in root extension for potentially indefinite periods. Growth may occur for the entire growing season or longer, amounting to penetration of up to 2 m per season. Excised roots were found to grow for 40 to 50 wk, but only if the sucrose content of the medium was relatively low and the culture solutions changed frequently (Street 1959). High sucrose levels promoted ageing and shortened root extension. Grasses (*Agropyron*) adapted to dry areas elongated at a rate of as much as 15 cm per wk. At 49 days there were large variations in total root length among species as follows (Kittock and Patterson 1959): *A. desertorum,* 73.8 cm; *A. intermedium,* 72.5 cm; *A. cristatum,* 48.9 cm; and tall fescue (*Festuca arundinacea*), only 12.2 cm. These values reflect genetically controlled morphological variations that impart differences in drought tolerance. Rate of root growth is generally believed to decrease with maturity. With soybean, total root lengths per unit of leaf area were 630 m • m^{-2} leaf area at the V_6 stage, 1190 m • m^{-2} at $V_{12}R_2$, and 345 m • m^{-2} at $V_{15}R_5$ (Sivakumar et al. 1977). In another study soybean root length increased for 70 to 80 days, remained constant to 100 days, and decreased thereafter (Barber 1978). Even though rooting density may decrease with maturity, rooting depth continues to increase until stage R_7 in soybeans (Kaspar et al. 1978).

The decline of root density during pod fill in these studies appears to be of particular importance physiologically. Reduced mineral uptake at the time of maximum demand is indicated. Senescence of the vegetative parts and redistribution of minerals and assimilate to fruits may be a consequence or may itself be the cause of reduced root growth. Also, the loss of roots means loss of new tips and meristematic activity in the root and probably a decline in cytokinin export from roots to shoots. The cytokinin decline may be the mediating mechanism in senescence.

LATERAL ROOTS

As discussed previously, lateral roots have their origin in meristems that form in the pericycle several centimeters from the root tip (Fig. 10.2). The lateral or new root breaks through the endodermis and cortex as cell division and elongation push the new root tip toward the root surface (Clowes 1969). In dicots lateral root formation is opposite the points of the *xylem star* (the pattern of xylem formation in a cross section of the root). The xylem star of sugar beet root has two points and hence two rows of lateral roots. A four-point xylem star in soybean root gives rise to four rows of lateral roots. Cotton roots have four-, five- or six-point xylem stars (McMichell 1983).

Formation of lateral roots is under genetic control but is also highly influenced by environment. Genetic control can result from three factors: (1) production of β-inhibitor in the root tip, which is related to apical dominace (Street 1959; Clowes 1978); (2) production of growth-promoting substances in the shoot, which are transported to the roots (e.g., auxin, thiamine, nicotinic acid, and adenine; and (3) a balance or interaction between growth-promoting and growth-inhibiting substances. Injury to or removal of the root tip removes apical dominance and promotes lateral root formation. In cultured, excised root sections, auxin promoted formation of lateral roots on root sections of field bindweed (Westmore and Steeves 1971). Carbon dioxide and gibberellic acid promoted lateral root formation, as did the so-called "stoppered bottle" effect, which is believed to be due to ethylene production (Street 1959) but could be due to carbon dioxide.

DIFFERENTIATION

Specialized cells or tissues are first evident in the undifferentiated tips in the formation of *root hairs,* lateral extensions of epidermal cells (Fig. 10.2). Root hairs may achieve a length of several millimeters and number 200 • mm^{-2}. Their life span is about 50 hr at moderate temperatures and less at high temperatures (McElgunn and Harrison 1969). A new root hair zone a few centimeters in length is formed as new growth increments are produced. Root hairs produce mucigel, which invites microbial activity. Most important, theoretically root hairs provide an extremely large surface area to interface with large volumes of soil fractions for mineral uptake.

A few millimeters from the root tip the amorphous cells begin to differ in size, shape, and structure, becoming specialized or differentiated. The central or vascular cylinder, consisting of xylem and phloem tissues, is ringed by a specialized, one-cell-thick layer of cells, termed the *pericycle.* The thin-walled parenchyma cells of the cortex are bounded on the inside by the endodermis and on the exterior by the epidermis (Fig. 10.2). Dicot roots usually have the capacity to grow in diameter from the vascular cambium (Table 10.1). A balance of auxins and cytokinins admitted at the basal end of root sections of radish was found to be essential for secondary thickening from the vascular cambium (Torrey and Loomis 1967). Either natural or synthetic auxin (including 2,4-D), combined with cytokinin, was equally effective in stimulating vascular cambium activity and secondary thickening. In addition to losing root hairs, older, differentiated root sections may lose absorptive capacity by becoming *suberized* (impregnated with phenolic compounds).

Root Systems

In a homogeneous and barrier-free rooting medium, which is rare or nonexistent in nature, root growth produces geometric configurations: a hemisphere, cylinder, cone, or inverted cone, depending on the genotype. This configuration and its components at any point in the life cycle are referred to as the *root system.* Several factors contribute to characteristic differences in the architecture of root systems, such as fineness, branching habit, and geotropism. Soil factors also strongly influence root growth and the architecture of the system.

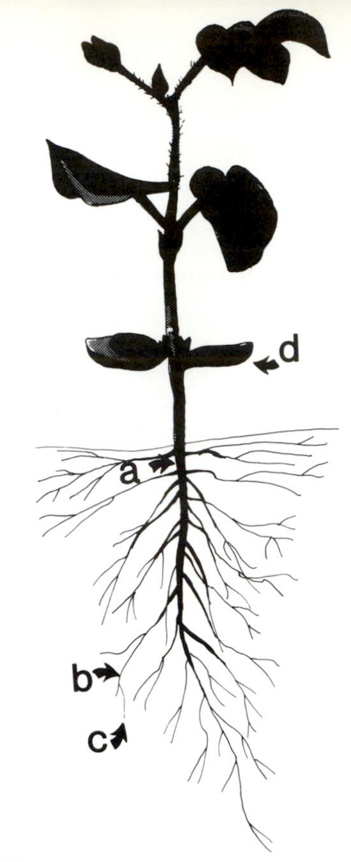

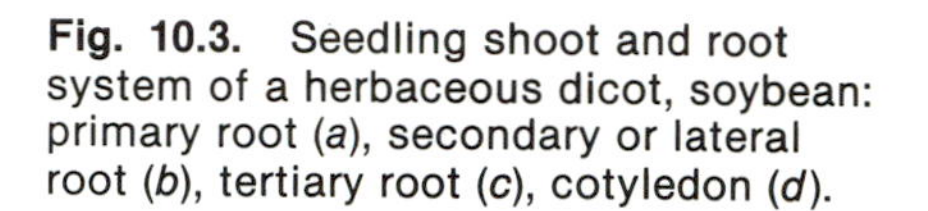
Fig. 10.3. Seedling shoot and root system of a herbaceous dicot, soybean: primary root (*a*), secondary or lateral root (*b*), tertiary root (*c*), cotyledon (*d*).

DICOTS

The root system of dicot species generally consists of a large, positively geotropic, primary root with fine branching laterals (Fig. 10.3). Fineness increases with branching order, that is, tertiary branches are finer than secondary branches. Often the primary root (taproot) has so much secondary thickening that it obscures the fine secondary or lateral branches (e.g., carrot). Between the extremes of the typical taproot and the typical fibrous system (e.g., grass plant) are a number of intermediate types (Fig. 10.4). Species such as radish and turnip have an unusually large swelling or secondary thickening in the hypocotyl area of the primary root. The swelling of the taproot of sugar beet and carrot is more generally distributed along the length of the root. The inverted cone of these roots has a thick cortex adapted to carbohydrate storage.

The perennial forage legume alfalfa typifies a taproot system, whereas birdsfoot trefoil has a more or less branching taproot (Fig. 10.5). Branching taproot systems are common in all legumes since they can be induced by soil barriers or injury to the primary root apex.

Early in the season lateral roots of soybean were found to be much less positively geotropic than the primary root (Mitchell and Russell 1971). The angle formed by laterals with the subtending root was generally obtuse. Later in the growth cycle these and any newly formed roots became strongly posi-

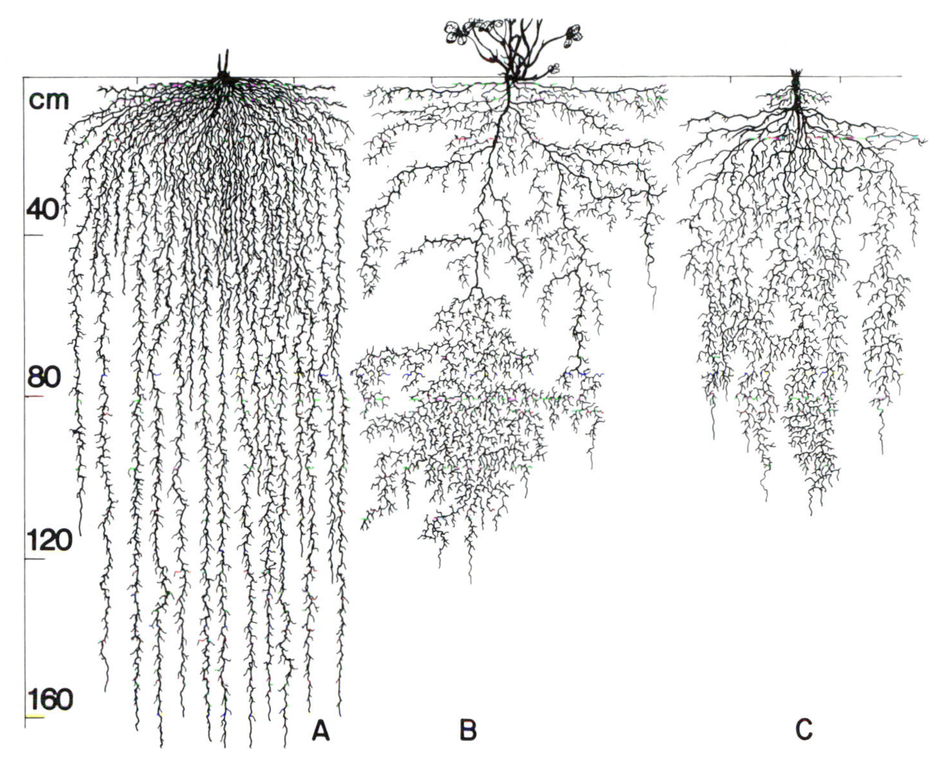

Fig. 10.4. Comparison of the root systems of three species: *A.* Wheat plant. *B.* White clover. *C.* Carrot. (From Kutschera 1960)

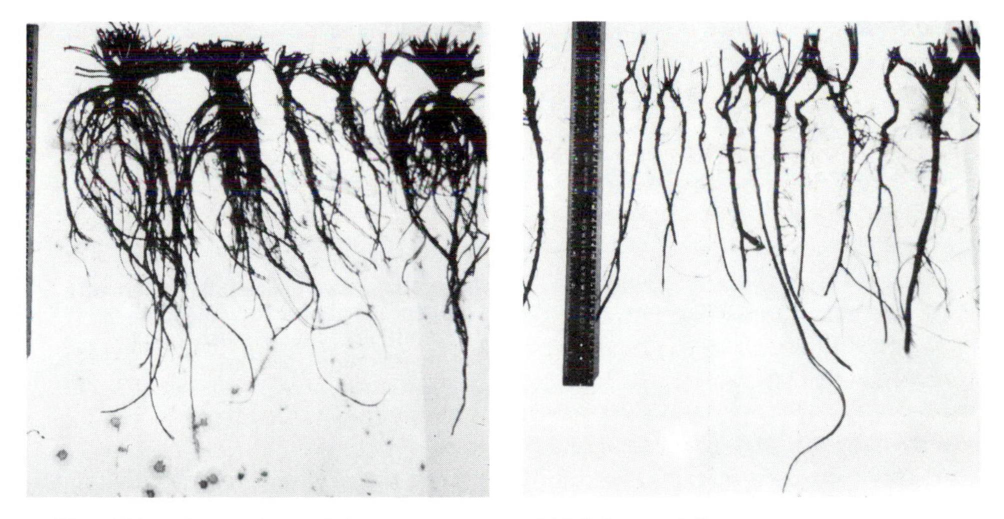

Fig. 10.5. Comparison of the root systems of birdsfoot trefoil (*left*) and alfalfa (*right*). The trefoil represents a branched taproot, the alfalfa an unbranched taproot. (Photo courtesy of Louis Greub)

tively geotropic, growing vertically from an acute angle with the mother root. Mitchell and Russell (1971) described the ontogeny and rooting pattern of soybean in Iowa (group II) as having three definable phases:

1. Vegetative growth, from 0 to 31 days; a positively geotropic primary root to a depth of 5 to 60 cm and horizontal laterals, primarily in the top 10 cm of soil
2. Pod filling, from 67 to 80 days; new and positively geotropic laterals, secondary and tertiary branching; nodules present on the primary root and coarser laterals; about 85% of the total root weight in the top 15 cm of soil
3. Rapid pod fill, from 80 to 100 days; primary root growth rate decline, lateral root growth on a strong, positively geotropic course to a depth of 120 to 180 cm; root weight increase in top 8 cm and below 120 cm

MONOCOTS

Roots of *monocots* (grasses) are fine and lack a cambium for secondary thickening. Collectively these are referred to as a fibrous root system. The monocot root system has two phases (Fig. 10.6):

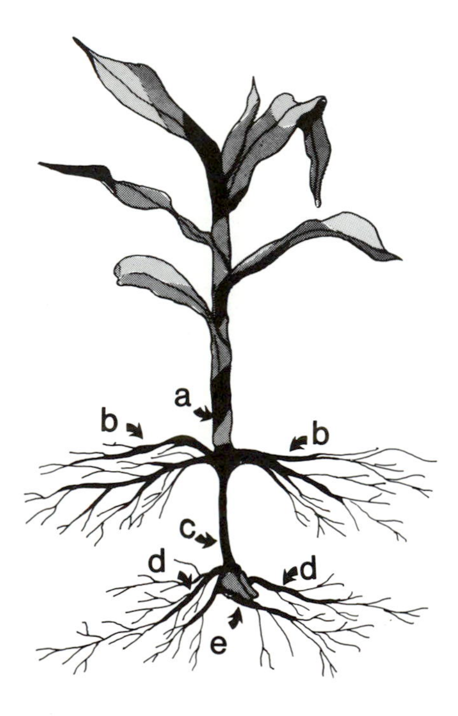

Fig. 10.6. Two-phase root system of a monocot (maize) plant: coleoptile (*a*), adventitious root (*b*), mesocotyl (*c*), seminal root (*d*), seed (*e*).

1. *Seminal roots* (also referred to as *seed roots*) emerge, along with the radicle, from the scutellar (first) node of the seed embryo axis. The scutellum (cotyledon) and subtending node remain embedded in the seed, hence the term *seed roots*. In wheat the coleoptilar, or second, node and the scutellar node remain in the seed, since emergence is from elongation of the second internode (see Chap. 11). Therefore this node also contributes to the seminal root number. Seminal roots of wheat comprise the radicle and one to seven roots from these two seed nodes. Number varies widely with genotype. Differences in seminal root number appear to contribute to adaptation and competitive advantage, particularly in certain environments (Pavlychenko 1937).

2. Adventitious roots, also referred to as *nodal, coronal,* or *crown roots,* emerge from the basal nodes of the grass shoot just below the soil surface. In grasses three to six nodes without internode growth form the crown, which gives rise to successive whorls of adventitious roots referred to as the crown roots. Since grass emergence is by elongation of the first internode or mesocotyl (Fig. 10.7) (in the case of wheat, the second internode), the crown is positioned near the soil surface regardless of seeding depth. Mesocotyl elongation stops just below the soil surface due to the phytochrome–red light control mechanism in the emerging coleoptile. In maize aerial adventitious roots also emerge from four or more nodes well above the soil surface. These are usually referred to as the *prop,* or *brace,* roots.

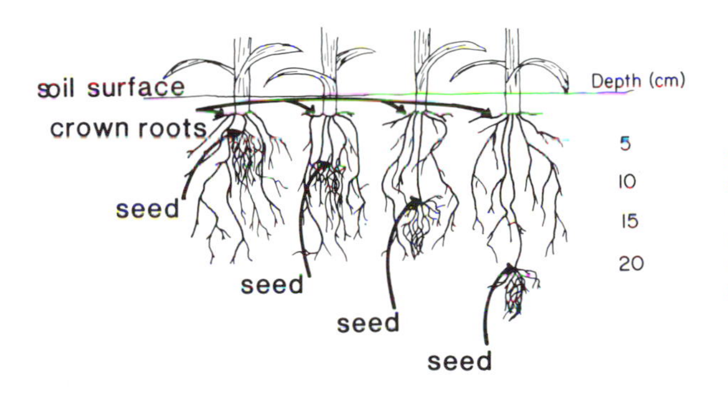

Fig. 10.7. Maize planted 5, 10, 15, and 20 cm deep. The crown was formed at nearly the same depth regardless of planting depth.

The first nodal roots to appear on grasses were found to be fine and only slightly positively geotropic (Brouwer 1966), growing downward at an angle of about 25° from horizontal and densely branched. Nodal roots that arise later are thicker and more positively geotropic, generally growing at an angle of about 45° from horizontal. The last-formed nodal roots are coarse and grow vertically. These coarse, brace roots may produce fine branches upon entering the soil and become functional in absorption as well as anchorage. Other observations (Mosher and Miller 1972) revealed that in maize the direction of growth was highly related to soil temperature surrounding the seed. The angle of radicle growth was 30° (from horizontal) at 18°C and 61° at 36°C.

The fineness of grass roots depends on the order of the internode of origin and, in the case of lateral roots, on the branching order. Very fine seminal roots arise from the scutellar node. Whorls arising from successive nodes are progressively coarser and brace roots are probably 10 times the circumference of seminal roots. On the other hand, successive orders of lateral branches from these primary roots reverse the order of size and become progressively finer. Laterals will be long if the primary root is short and vice versa.

CONTRIBUTION OF SEMINAL AND CROWN ROOT SYSTEMS

The question of the longevity and contribution of seminal roots to the total system seems unresolved. Although it is generally believed that seminal roots are short-lived and make a minor contribution, some studies have shown that they are long-lived and make a major contribution. Both views are correct, depending on the species and environment. Pavlychenko (1937) observed that in a number of cool-season cereals under western Canadian conditions the seminal roots not only were important but were the sole root system because crown roots did not develop in drought years. Competitive ability early in the season was related to the development of seminal roots. For example, 'Hannachen' barley at 80 days had 6.6 seminal roots per plant, compared with 3.0 for wild oat and 4.6 for wheat, and was the most competitive of the three. Even in heavily wild oat–infested soils, a reasonably good crop could be expected from barley, which Pavlychenko attributed to an early competitive advantage due to more seminal roots. Later in the season, at 80 days, barley and wheat had 11 to 13 crown roots, respectively, compared with 17 for wild oat, a shift in number and probably in competitive ability. Spacing of plants did not appear to affect the number of seminal roots, but close spacing drastically reduced the number of primary and secondary branches on crown roots. Total root length of plants in drilled rows was greatly reduced, compared with that of individually spaced plants.

Boatwright and Ferguson (1967) observed that the early tillering, phosphorus uptake, and grain yields of wheat were all significantly greater if plants had both seminal and crown root phases, since removal of either phase decreased their values. However, grain yields were greater from plants with crown roots alone than from plants with seminal roots alone. Seminal roots of the perennial grass timothy were of little functional significance, since plants with adventitious roots alone performed as well as plants with both root phases (Williams 1962) (Table 10.2). The absorptive capacity of seminal roots of timothy were estimated to be 50–fold greater than that of adventitious roots. Therefore the seminal roots might be expected to be important in the early growth phases.

The importance of seminal roots to wheat and certain other cool-season cereals is more apparent, since more seminals are produced because of the

TABLE 10.2. Relative dry matter production and nutrient uptake efficiency per tiller of timothy

	Root System, %		
	Both	Seminal	Adventitious
Dry Matter	100	100	100
Nitrogen	100	50	100
Phosphorus	100	100	100
Potassium	100	100	100
Calcium	100	100	100

Source: From Williams 1962.

emergence from elongation of the second internode, as mentioned earlier. However, the potential for crown root numbers is correspondingly decreased by one whorl as a result of one less node in the crown. The early advantage from more seminal roots seems to outweigh this loss.

There is general agreement that in maize under field conditions the seminal roots are short-lived and make a relatively small contribution to the total because (1) the mesocotyl disintegrates after a few weeks, separating the seminals from the plant, and (2) the magnitude of the weight, volume, and length of adventitious roots is enormous, compared with seminal roots. Nevertheless seminal roots are important to maize, especially for early support. And the fineness and frequency of branching of the seminals results in high uptake efficiency, important in the early stages.

Root Efficiency

While older roots are vital to the plant, absorption is severely reduced because (1) root hairs are no longer present, (2) old roots often have deposits of phenolic substances, and (3) older roots occupy exploited soil spaces. The last is true for minerals but may not be true for water, which is periodically recharged. New roots, primary or lateral, push into unexploited soil spaces and develop root hairs in large numbers, resulting in a large surface area. Successive whorls of crown roots showed optimal activity on successive dates, which was believed to be due to root impermeability with age and/or environment exhaustion (Brouwer 1966).

In view of the large number, length, density (cm root length • cm^{-3} soil), and surface area of root hairs, they would seem to be the most effective component of the system for mineral uptake (Jungk and Barber 1975). However, root hairs are thought to be scarce under natural conditions in which roots are infested by mycorrhiza. The mycelia of mycorrhiza greatly increase active root surface and mineral uptake so the loss of root hairs to mycorrhiza may not be important.

Deep penetrating roots may grow into unexploited moist soil layers, which generally have a low content of certain minerals. On the other hand, new roots and branches near the surface find a higher mineral content but in areas frequently low in moisture. Since mineral nutrients, especially N and phosphorus (P), are usually concentrated in the plow layer, a plant that is frequently watered should not need deep rooting. In fact it might be better to invest the assimilate in fruits or harvest products, and under irrigation this is usually the natural course of development.

Factors Affecting Root Growth and Distribution

Differences in rooting habit, although inherent, are also highly influenced by soil environment, both directly and indirectly. Above-ground factors affecting shoot growth, especially the transport of carbohydrates to the roots, can have a major effect on root growth, as can rhizosphere factors (e.g., moisture, temperature, nutrient levels, toxic substances, soil strength, and biological agents).

GENOTYPE

Large differences in rooting exist among genotypes, as does the opportunity for breeding and selection (MacKey 1980). Apparently most root characteristics are quantitatively inherited, that is, controlled by a number of genes. These inherent differences then interact with the soil environment. Large and highly heritable differences were observed in the ratio of branch roots to main roots of maize inbred lines (Fig. 10.8). Sorghum roots were more geotropic, with more secondary branching (Weaver 1926) (Fig. 10.9). 'Harosoy 63' cultivar of soybean had a more extensive root system and twice the root surface of 'Aoda' (Raper and Barber 1970). Depth distribution of roots varies widely among forage species. Compared with bluegrass (*Poa travialis*) and

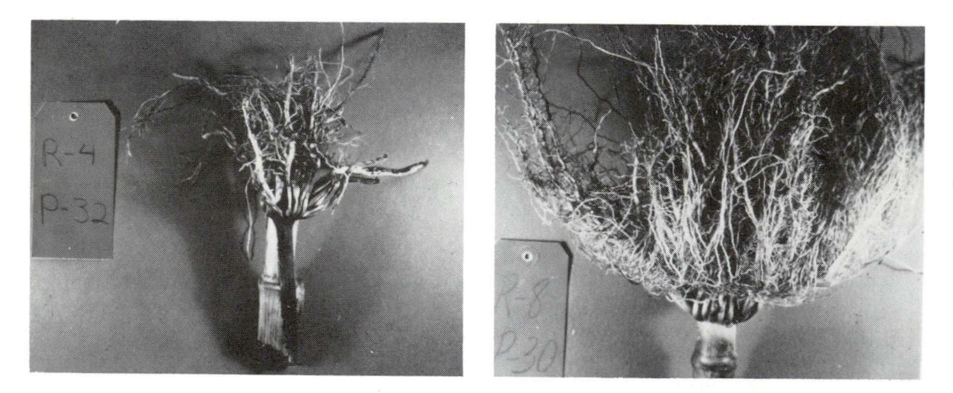

Fig. 10.8. Root systems of two inbred lines of maize demonstrate the wide inherent variance in root systems within a species. (Courtesy of F. F. Dicke)

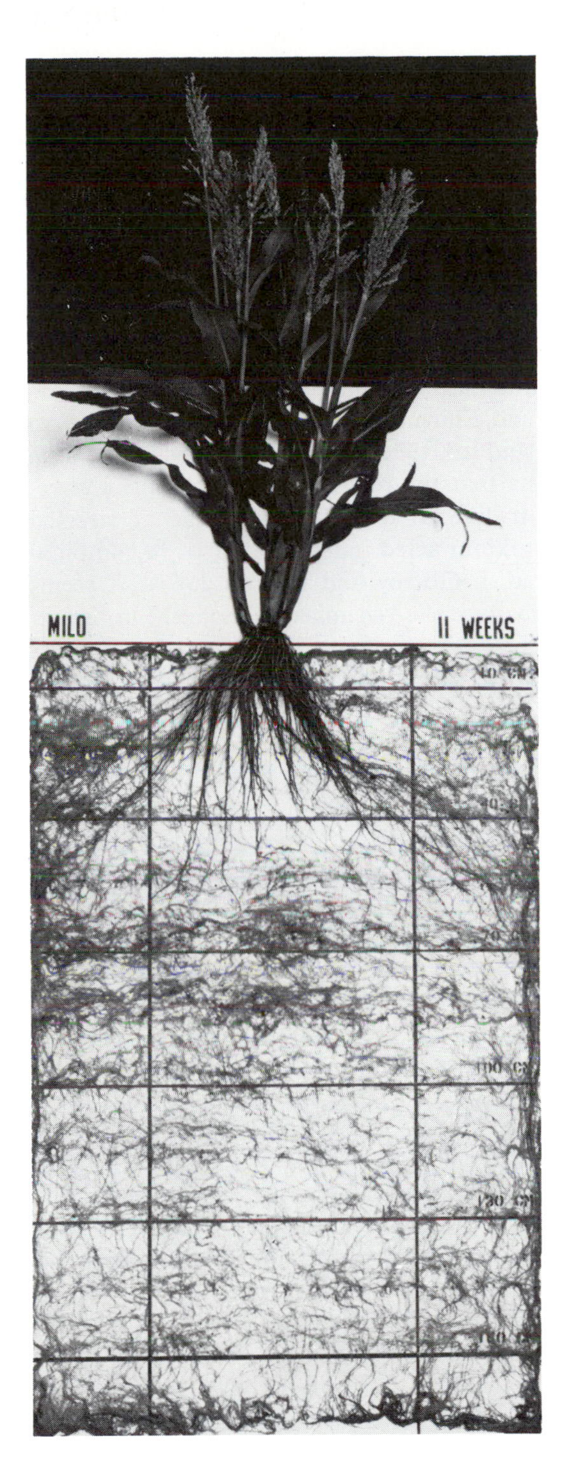

Fig. 10.9. Root habit of grain sorghum. (Courtesy of M. Huch, USDA, National Tillage Machinery Laboratory, Auburn, Ala.)

perennial ryegrass, orchardgrass had 20 to 30% less root mass in the top 10 cm and a greater percentage of roots in deeper strata (Troughton 1956).

The mechanisms of genetic control of rooting are complex, but as in the shoot, the action of growth hormones is indicated. Auxins (IAA) promote root growth but only in low concentrations (Vaadia and Itia 1969). The requirement for auxin is evidenced by the leaf factor that is necessary for the rooting of cuttings. Many species require some leaf tissue or an active bud, apparently to produce the diffusable growth-promoting subtance(s). Girdling the stem piece nullified the leaf factor effect (Hess 169). A rooting cofactor, isolated and identified as catechol and pyrogallol, acted synergistically with IAA to promote rooting. Ethylene, produced during germination of certain species, retarded root growth (Nakayama and Shirmura 1973); cytokinins tended to inhibit rooting (Hess 1969; Vaadia and Itia 1969). Roots of stressed plants were lower in cytokinins, which suggests that the reduction in cytokinins and reduced supply to leaves may contribute to drought senescence of stressed leaves. Auxin, gibberellin, and cytokinin acted independently or in combination to control root growth in radish (Torrey and Loomis 1967). It seems evident that growth hormones constitute the "chemical messenger" in the expression of genotype.

PLANT COMPETITION

The competitive advantage of barley over wild oats is due, at least in part, to the greater number or density of barley seminal roots, as discussed above. In drilled rows of 'Marquis' wheat, for example, the number of crown roots was greatly reduced, from 73 to 12, compared with spaced plants; the total root length was reduced from 70,000 to 900 m per plant (Pavlychenko 1937). Competition from close spacing had only a minor effect on the seminal roots.

Increasing the plant population of maize from about 12,000 to 62,000 plants per ha resulted in a 72% decrease in root dry weight per plant (Norden 1964). However, the total root weight per ha increased up to a population of about 50,000.

The competitive ability of many species appears to be due to the secretion of toxic or inhibitory substances by roots, a phenomenon known as *allelopathy* (Rice 1974). A thick, vigorous tall fescue sod is practically immune to weed encroachment, which has been attributed to root density and resulting root competition. Root competition probably is an operative factor, but it has recently been demonstrated that tall fescue roots excrete allelopathic chemicals (Wheeler and Young 1978). Quite likely tall fescue is allelopathic to competing species but not to itself, allowing dense sod formation from its own roots while suppressing growth of competitors.

DEFOLIATION

The old saying "To prune the shoots is to prune the roots" is a valid assessment, since roots are dependent on shoots for assimilate. Pruning the top of sudangrass periodically at 10 cm reduced root weight by 85%. Pruning the roots also reduced root and top growth. Clipping at frequent intervals significantly reduced root weight of blue panicgrass (Wright 1962). The effect of clipping varies with species and is related to the amount of photosynthetic area remaining after defoliation, which may still maintain a critical leaf area index (95% of light absorption). For example, close and continuous clipping can be practiced on a low-growing species such as creeping bentgrass.

Most perennial species exhibit cyclic rooting, that is, annual dieback and regeneration of a portion of the root system. As defoliation occurs in autumn from frost the root zone is still warm and supportive of respiration, which exhausts food reserves and results in root dieback. The annual cyclic dieback is a plausible explanation for the high humus content under tall grasses of prairie sod, since the top growth of species in this ecosystem is killed by frost and the fine roots have little food reserve storage. A cool-season species like orchardgrass exhibits less dieback and regeneration (Sprague 1933) and appears not to build humus content so rapidly.

SOIL ATMOSPHERE

The atmosphere of roots is usually quite unlike the atmosphere of shoots. Both oxygen (O_2) and carbon dixoide (CO_2) levels in the rhizosphere may differ greatly from ambient atmosphere and both can have a direct effect on root growth (Fig. 10.10). Generally, the effect of either one is modified by the presence of the other (Geisler 1967). Nitrogen is an inert gas and has no negative effect.

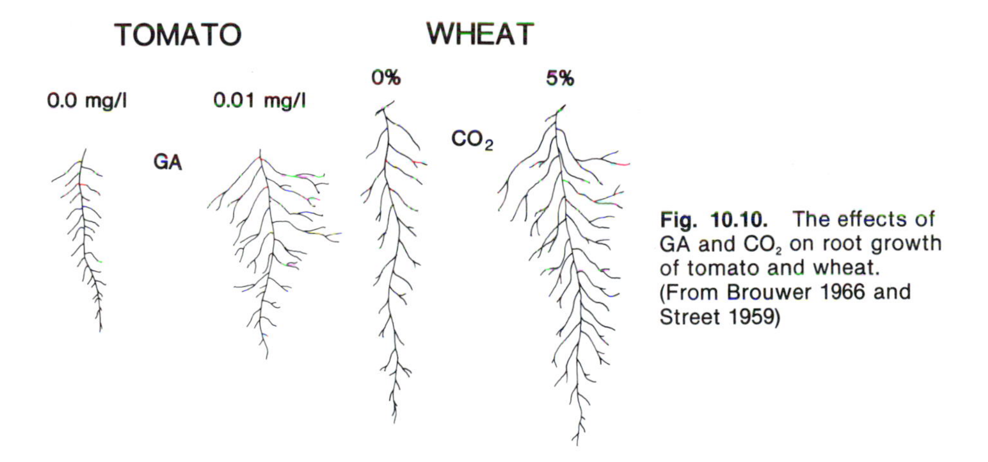

Fig. 10.10. The effects of GA and CO_2 on root growth of tomato and wheat. (From Brouwer 1966 and Street 1959)

Oxygen is essential for metablic processes, including active absorption and transport. In soybean, the O_2 requirement and nutrient uptake were greater during vegetative stages than during reproductive stages (Jones et al. 1978). Soil water removal by barley roots was increased by increasing O_2

(Letey et al. 1965), which suggests that water uptake is active or perhaps additional roots are stimulated. Some species, such as rice, can absorb adequate O_2 through the leaves and transport it to the roots via the *aerenchyma* (air cells); therefore O_2 is not always required in the rhizosphere. Maize was observed to have this capability (Jensen et al. 1964) but evidently not sufficiently for normal growth under prolonged flooding. Oxygen in the rhizosphere has indirect effects, such as stimulation of microorganism activity, which in turn influences nutrient availability to roots. Some CO_2 is evidently beneficial for optimum root growth (Fig. 10.10).

Concentrations of CO_2 up to 2%, or nearly 10-fold greater than ambient atmosphere, stimulated root growth of barley and peas, but 8% retarded root length (Geisler 1967). The effect of CO_2 depended on the partial pressure of O_2 in the root atmosphere. In general an O_2 level one-third that of normal air (21%) was adequate for growth, unless the CO_2 concentration was too high.

SOIL pH

Soil pH outside the range of 5.0–8.0 has a potentially direct effect in limiting root growth; within this range, as occurs under most field conditions, the effect is usually indirect. Soil pH of less than 6.0 increases the solubility of aluminum, manganese, and iron, which can be toxic and limit root growth. Plant breeders have been successful in selecting aluminum-tolerant lines in a number of crops (Fig. 10.11). Tolerant lines raise the pH in the immediate

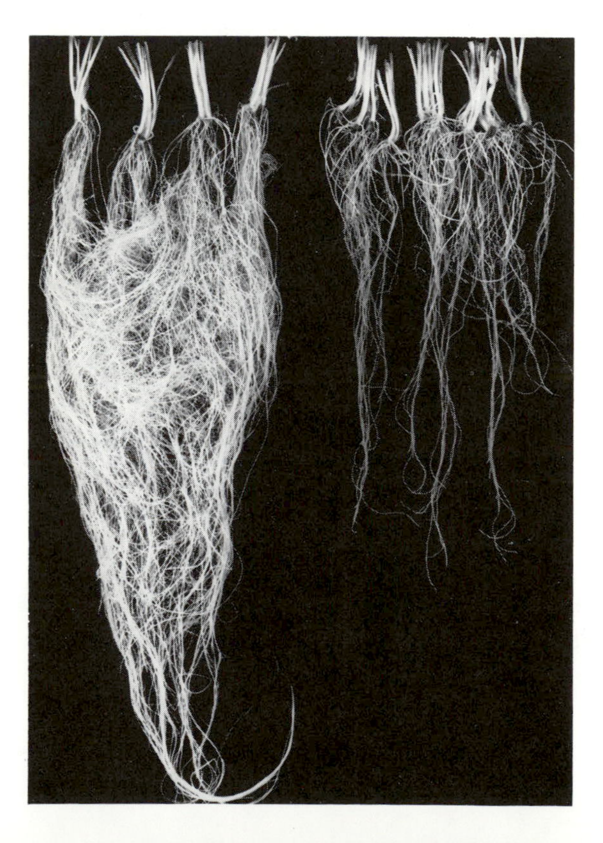

Fig. 10.11. Variation in aluminum tolerance in two lines of wheat. (Courtesy of C. D. Foy)

vicinity of the root. Species and cultivars vary in capacity to alter the pH of the immediate root environment (Olsen et al. 1981).

SOIL TEMPERATURE

Optimum temperatures were generally lower for roots than for shoots (Brouwer 1966), which is consistent with natural growth; during spring root temperatures under a sod or vegetation are lower than the above-ground temperatures. Temperatures optimal for species growth vary widely, of course. Increasing the root temperature by hot water pipes favored warm-season grasses, such as sudangrass, more than cool-season grasses, such as tall fescue (Rykbost et al. 1975). Temperature affected the growth of roots more than the growth of shoots (Aldous and Kaufman 1979). The direction of growth is highly related to temperature, as indicated previously.

SOIL FERTILITY

Roots require adequate mineral nutrients for growth and development, as do other plant parts. Because they are closer to the source than shoots, roots have the first opportunity for minerals and water, although they have the last opportunity for assimilate formed in the shoots. For this reason, a water or mineral deficiency generally affects roots less than tops (decreasing the S-R ratio) unless it interferes directly with photosynthesis (e.g., an iron deficiency, which reduces chlorophyll). A light deficiency also interferes directly with photosynthesis, resulting in shoot priority (increasing the S-R ratio).

Generally fertilization favors expression of the inherent tendencies of roots (Fig. 10.12). Maize roots tend to proliferate in zones containing organic matter and fertilizer (e.g., a fertilizer band) (Duncan and Ohlrogge 1958), particularly if the band contains N and P. Brouwer (1966) suggested that the determining factor is not the presence of fertilizer elements in the root environment per se but rather the overall nutrient status of the plant. Fertilized maize in a Muscatine soil rooted to a depth of 1.7 m, compared with 1.4 m for unfertilized maize (Fehrenbacher et al. 1969) (Fig. 10.12).

Roots that contact a fertilizer band may become injured, exhibit deformities, and be shorter than untreated roots (Isensee et al. 1966). Apparently, seminal roots and first order branches are deformed or killed in a fertilizer band, or by other chemicals in sufficient concentration to be toxic, but higher orders of roots may proliferate more as the fertilizer concentration in the band declines with time.

Increasing the N level favors top growth in relation to root growth, that is, increases the S-R ratio (see Chap. 8). Thus high N may allow top growth to usurp available carbohydrates; the increased top growth causes more shading of lower leaves, which further aggravates the situation. In addition, a greater N supply tends to increase the auxin levels (Wilkinson and Ohlrogge 1962), which may inhibit root growth. However, N fertilizer increases the total dry

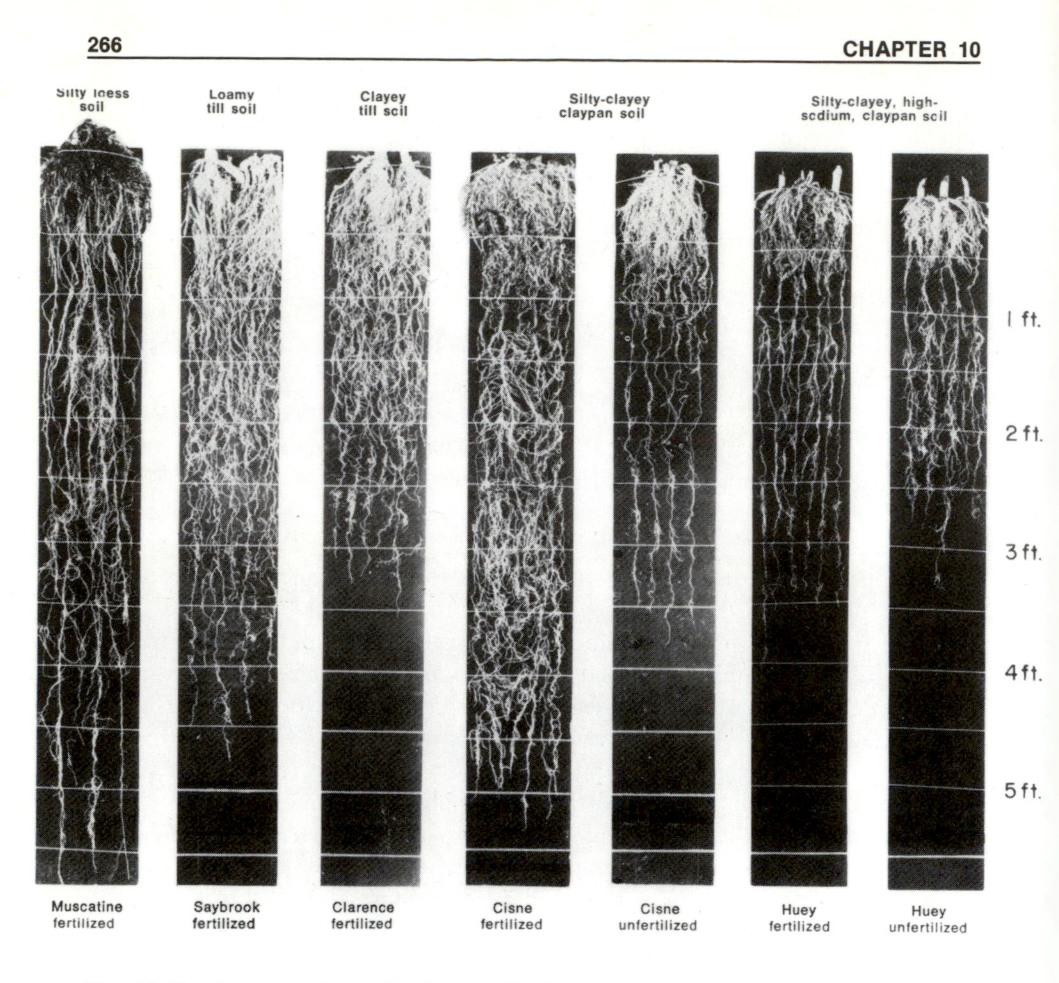

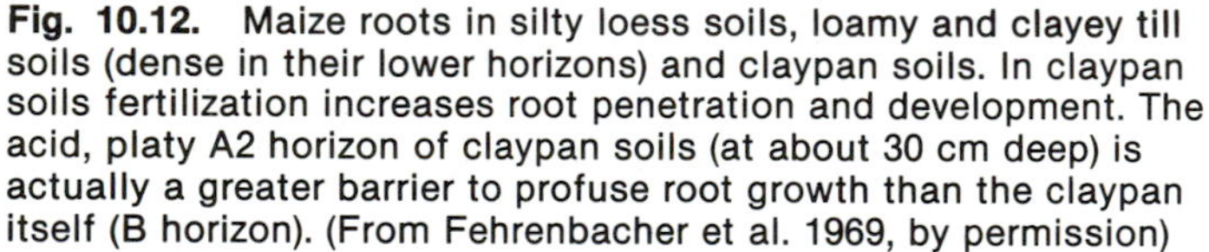

Fig. 10.12. Maize roots in silty loess soils, loamy and clayey till soils (dense in their lower horizons) and claypan soils. In claypan soils fertilization increases root penetration and development. The acid, platy A2 horizon of claypan soils (at about 30 cm deep) is actually a greater barrier to profuse root growth than the claypan itself (B horizon). (From Fehrenbacher et al. 1969, by permission)

weight of roots. Cereal crops that have a high early level of N and a diminishing concentration as the season progresses generally produce a greater leaf area early and more photosynthate later for the roots. Nitrogen-fertilized maize plants have been observed to have a greater root development and to use considerably more water in drought conditions. Nitrogen fertilization seems to promote deeper and more profuse rooting early in the season, probably due to increased leaf area and more assimilate for root growth.

Phosphorus-fertilized plants develop more roots than do unfertilized plants, but this is probably not a direct effect; P availability first increases photosynthesis, which in turn increases root growth. Also, extracts from P-fertilized roots had less auxin activity and theoretically less inhibition than extracts from N-fertilized roots (Wilkinson and Ohlrogge 1962). However, P caused a direct increase in root hair proliferation. Phosphorus did not have to be at the site of growth to provide normal development (Pearson 1966), which suggests that P in the subsoil may have no advantage over P in the surface layer in promoting deep rooting.

Considerable work has been done to evaluate the best N-P ratio in fertilizer mixtures, especially for applications made at planting. A 1:5 ratio of N to P in a fertilizer band appeared to be optimum for maize root development (Duncan and Ohlrogge 1958).

Potassium (K) seems to have no direct effect on rooting in either elongation or branching. It is, however, important to certain physiological functions of the root; inadequate K may cause a weak translocation system, poor cell organization, and loss of cell permeability. In general the effects of K and other fertilizers are primarily indirect, increasing root growth only after increasing top growth.

Weaver (1926) was interested in the potential benefit of deep fertilization to stimulate root growth. Experimental work since then has produced both positive and negative results. However, deep fertilization experiments have most commonly been done in conjunction with deep tillage, and the effects of each have not been clearly separated.

WATER

Water is essential for root growth, evidenced by the fact that roots do not grow through dry soil layers. However, roots have what might be regarded as a water-stress adjustment mechanism whereby solutes accumulate in the tip and elevate the turgor pressure, which can sustain growth for a limited time (Sharp and Davis 1979). Soil moisture stress significantly reduced root weight of blue panicgrass (Wright 1962), and root length of soybean was significantly reduced by water potentials less than -2 bars, or 16% (Sivakumar et al. 1977).

A moisture-deficient soil also modified rooting patterns: a smaller percentage of the total roots was found in the surface layer (0–15 cm) and a higher proportion in deeper strata (Mayaki et al. 1976). Irrigation reversed this trend.

MECHANICAL AND PHYSICAL FORCES

Roots encounter mechanical resistance to growth from a variety of causes, such as particle size, lack of aggregation, soil strength, and compaction. Decreasing porosity or increasing bulk density decreased root growth (Fig. 10.12). Root penetration and proliferation were greater in undisturbed soil cores of a loess than in cores of three soils higher in clay content and bulk density (Davis and Runge 1969). Root growth was not confined to interpedal spaces, however, suggesting that soil strength affects root entry into soil space. High soil bulk density, as in compacted sandy loam soil, greatly reduced the

TABLE 10.3. Estimates of rooting density of 'Havanna' tobacco in a Merrimac sandy loam in relation to compaction

Bulk Density Soil	Number of Cores Observed with		
($g \cdot cm^{-3}$)	Many roots	Some roots	Few roots or none
<1.40	30	2	0
1.41–1.44	3	4	3
1.45–1.48	2	6	3
1.49–1.52	4	7	1
1.53–1.56	1	3	1
1.57–1.60	0	0	5
>1.60	0	0	11

Source: From De Roo 1969.
Note: Made on 89 soil core samples collected over 3 yr at harvest time.

root growth of tobacco (Table 10.3). Maize roots were observed to penetrate about 2 m in a coarse-textured glacial till but seldom more than 1 m in fine-textured soils (Pearson 1966). Increasing the bulk density of the soil medium from 1.65 to 1.96 $g \cdot cm^{-2}$ not only reduced soybean root growth but altered the root anatomy by increasing the thickness of cell walls and casparian strip and by distorting the shape of the central cylinder (Baligar et al. 1975). Such anatomical aberrations indicate impairment of the absorption function. Root density of maize and cotton was related to soil strength, and the relationship of rooting density to absorption of soil moisture was linear (Grimes et al. 1975).

Machinery rolling between plant rows caused compaction and reduced water availability (Nelson et al. 1975); the first pass with machinery accounted for the greatest percentage of compaction. Whether compaction reduced root growth and water uptake by mechanical impedance, by reduced root growth due to reduced O_2 supply, or by reduced active water absorption due to reduced O_2 was not clarified. Some researchers (Phillips and Kirkham 1962) attributed restricted root growth in compacted soils to mechanical impedance. Others have pointed not only to mechanical impedance but to lack of O_2 (Rickman et al. 1966). While both factors have direct and indirect effects, a limiting O_2 supply appeared to be more important over a wider range of conditions (Bertrand and Kohnke 1957). The fact remains that low porosity or high bulk density due to soil compaction causes restricted root growth and function. Much research effort has been given to methods of shattering compacted or hardpan areas, but results have been variable and benefits usually short-lived.

Summary

Vigorous root growth is normally essential for high-yielding crop plants. Roots function in water and mineral absorption, anchorage, transport, storage, propagation, and as a source of growth hormones. Rooting habit, controlled genetically, varies widely among species and is under environmental control, as well. Excavation to observe rooting is often destructive and laborious, so rooting research is more limited than that on aerial plant parts. Many convenient methods to characterize root parameters have been developed, but each has limitations.

The dicot root system is derived from ontogenetic orders of branching, beginning at the primary, or main, root. The monocot root system is two-phase: (1) the seed or seminal roots, which may be temporary, and (2) the adventitious, nodal, coronal (crown) roots, which arise from nodes in the crown and form the principal component of the system. Roots of both types branch by one to several orders. Crown roots are produced from stem nodes (crown), which are located just below or near the soil surface. Monocot roots are generally fine and function less as storage organs.

Lateral roots of dicots and monocots are initiated from meristems that form in the pericycle. The geotropic habit of growth varies with species, order of branching, and plant age. Root branches tend to be more horizontal, or less positively geotropic, than the primary root, but the geotropic response can be altered somewhat by soil temperature. Absorption, especially of minerals, is primarily restricted to young roots, principally the root-hair zone. Older roots lose their root hairs and become impregnated with phenol-type compounds, or suberized. Soil factors, such as bulk density, water, O_2, minerals, pH, temperature, and toxins highly influence root growth; top growth and available photosynthate are also essential. Compaction and the resultant increased bulk density increase impedance to roots and reduce O_2 or the O_2-CO_2 ratio, which adversely affects root growth. Competition for space (light) and defoliation reduce the assimilate available for root growth.

References

Aldous, D. E., and J. E. Kaufmann. 1979. Agron. J. 71:545–47.
Baligar, V. E., V. E. Nash, M. L. Hare, and J. A. Price, Jr. 1975. Agron. J. 67:842–44.
Barber, S. A. 1978. Agron. J. 70:457–61.
Barley, K. P. 1970. In Advances in Agronomy, vol. 22, ed. N. C. Brady. New York and London: Academic Press.
Bertrand, A. R., and H. Kohnke. 1957. Soil Sci. Soc. Am. Proc. 21:135–40.
Boatwright, G. O., and H. Ferguson. 1967. Agron. J. 59:299–302.
Bohm, W., H. Maduakor, and H. M. Taylor. 1977. Agron. J. 69:415–19.
Brouwer, R. 1966. In The Growth of Cereals and Grasses, ed. J. D. Ivins and F. L. Milthorpe. London: Butterworth.
Clowes, F. A. L. 1969. In Root Growth, ed. W. J. Whittington. London: Butterworth.
______. 1978. Ann. Bot. n.s. 42:801–6.
Cormack, R. G. H. 1962. Bot. Rev. 28:446–64.
Davis, R. B., and E. C. A. Runge. 1969. Agron. J. 61:518–21.
De Roo, H. C. 1969. In Root Growth, ed. W. J. Whittington. London: Butterworth.
Duncan, W. G., and A. J. Ohlrogge. 1958. Agron. J. 50:605–8.
Fehrenbacher, J. B., B. W. Ray, and J. D. Alexander. 1969. Crops Soils 21:14–18.
Geisler, G. 1967. Plant Physiol. 42:305–7.
Grimes, D. W., R. J. Miller, and P. L. Wiley. 1975. Agron. J. 67:519–23.
Hess, C. E. 1969. In Root Growth, ed. W. J. Whittington. London: Butterworth.
Isensee, A. R., K. C. Berger, and B. E. Struckmeyer. 1966. Agron. J. 58:94–97.

Jensen, C. R., J. Letey, and L. H. Stolzy. 1964. Science 144:550–52.
Jones, C. A., A. Reeves III, J. D. Scott, and D. A. Brown. 1978. Agron. J. 70:751–55.
Jungk, A., and S. A. Barber. 1975. Plant Soil 42:227–39.
Kaspar, T. C., C. D. Stanley, and H. M. Taylor. 1978. Agron. J. 70:1105–7.
Kittock, D. L., and J. K. Patterson. 1959. Agron. J. 51:512.
Kutschera, L. 1960. Wurzelatlas Mitteleuropäischer und Ackerunkrauter und Kulturpflanzen. Frankfurt: DLG-Verlags-GmbH.
Letey, J., W. F. Richardson, and N. Valoras. 1965. Agron. J. 57:629–31.
McElgunn, J. D., and C. M. Harrison. 1969. Agron. J. 61:79–81.
MacKey, J. 1980. In Plant Roots: A Compilation of Ten Seminars. Iowa State University, Ames, unpublished.
Mayaki, W. C., I. D. Teare, and L. R. Stone. 1976. Crop Sci. 16:92–94.
McMichell, B. L. 1983. Personal communication.
Milthorpe, F. L., and J. Moorby. 1974. An Introduction to Crop Physiology. London: Cambridge University Press.
Mitchell, R. L., and W. J. Russell. 1971. Agron. J. 63:313–16.
Mosher, P. N., and M. H. Miller. 1972. Agron. J. 64:459–62.
Nakayama, M. K., and Y. O. Shirmura. 1973. Proc. Crop Sci. Soc. Jpn. 42:493.
Nelson, W. E., G. S. Rahi, and L. Z. Reeves. 1975. Agron. J. 67:769–72.
Newman, E. I. 1966. J. Appl. Ecol. 3:139–45.
Norden, A. J. 1964. Agron. J. 56:269–73.
Olsen, R. A., R. R. Clark, and J. J. Bennett. 1981. Am. Sci. 69:378–84.
Pavlychenko, T. K. 1937. Ecology 18:62–79.
Pearson, R. W. 1966. In Plant Environment and Efficient Water Use, ed. W. H. Pierre et al. Madison, Wis.: American Society of Agronomy and Soil Science Society of America.
Phillips, R. E., and D. Kirkham. 1962. Soil Sci. Soc. Am. Proc. 26:319–22.
Raper, C. D., Jr., and S. A. Barber. 1970. Agron. J. 62:581–84.
Rice, E. L. 1974. Allelopathy. New York: Academic Press.
Rickman, R. W., J. Letey, and L. H. Stolzy. 1966. Soil Sci. Soc. Am. Proc. 30:304–7.
Rykbost, K. A., L. Boersma, H. J. Mack, and W. E. Schmisseur. 1975. Agron. J. 67:733–38.
Sharp, R. E., and W. J. Davis. 1979. Planta 147:43–49.
Sivakumar, M. V. K., H. M. Taylor, and R. H. Shaw. 1977. Agron. J. 69:470–73.
Sprague, H. B. 1933. Soil Sci. 36:189–209.
Stone, L. R., I. D. Teare, C. D. Nickell, and W. C. Mayaki. 1976. Agron. J. 68:677–80.
Street, H. E. 1959. In Root Growth, ed. W. J. Whittington. London: Butterworth.
Taylor, H. M., and H. R. Gardner. 1963. Soil Sci. 96:153–56.
Torrey, J. G., and R. S. Loomis. 1967. Am. J. Bot. 54:1098–1106.
Travis, R. L., S. Geng, and R. L. Berkowitz. 1979. Plant Physiol. 63:1187–90.
Troughton, A. 1956. J. Br. Grassl. Soc. 11:56–65.
Vaadia, Y., and C. Itia. 1969. In Root Growth, ed. W. J. Whittington. London: Butterworth.
Ward, K.J., B. Klepper, R. W. Rickman, and R. R. Allmaras. 1978. Agron. J. 70:675–77.
Weaver, J. E. 1926. Root Development of Field Crops. New York: McGraw-Hill.
Westmore, R. H., and T. A. Steeves. 1971. In Plant Physiology: A Treatise, ed. F. C. Steward. New York and London: Academic Press.
Wheeler, G. L., and J. F. Young. 1978. Ark. Farm Res., p. 6.
Wilkinson, S. R., and A. J. Ohlrogge. 1962. Agron. J. 54:288–91.
Williams, D. 1962. Ann. Bot. n.s. 26:129–36.
Wright, N. 1962. Agron. J. 54:200–202.

11 Vegetative Growth

LEAVES supported by stems and branches are the carbohydrate factories of crop plants. They are necessary for the interception and transformation, via photosynthesis, of light energy to growth and yield. Leaves are also sources of nitrogen (N) for fruiting, mobilizing it and redistributing it to fruits.

Vegetative organs (including buds, leaves, and stems) have their origin in the apical and lateral buds of stems, starting with the embryo axis in the seed. Lateral or axillary buds are located in leaf axils. New growth can also arise from adventitious buds that may form in the stem cambium or the root pericycle.

Regardless of species or origin, buds are morphologically and functionally homologous (Fig. 11.1). They can conveniently be viewed as a series

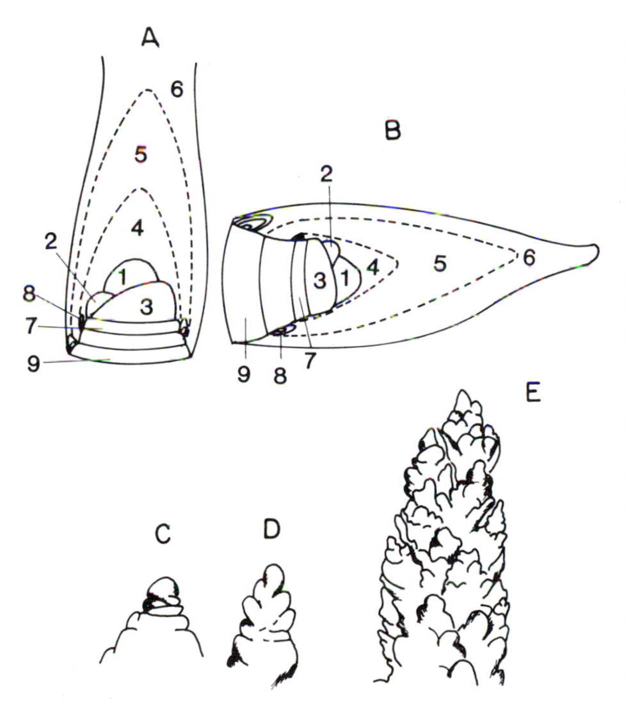

Fig. 11.1. Typical vegetative and floral buds of a grass plant. *A.* Bud of aerial stem or apical tiller. *B.* Apical bud of rhizome or rhizome branch. *C.* Vegetative shoot apex of oat. *D.* Inflorescence initial of oat at the 4-leaf stage. *E.* Inflorescence primordium of oat at the 6-leaf stage. Homomorphic structures in *A* and *B:* growing point (*1*), primordial leaves (*2* and *3*), expanding young leaves of an aerial shoot or expanding cataphylls of a rhizome (*4* and *5*), mature leaf sheath or rhizome cataphyll (*6*), unexpanded node and internode (*7*), axillary bud (*8*), and expanding internode of a rhizome (*9*) (aerial shoot internodes of temperate grasses do not elongate until after initiation of an inflorescence).

of structural subunits varying in degree of development from base to tip (*acropetally*). Each subunit of structure, the *phytomer,* has three components: (1) stem node and internode, (2) leaf, and (3) axillary bud. In some species more mature phytomers also have root initials. Phytomers develop acropetally and indeterminately. However, floral initiation terminates vegetative phytomers, as described above, and the components are modified as parts of the inflorescence. In temperate grasses, floral initiation also signals the beginning of stem (internode) elongation, which separates the leaves and elevates and exposes the inflorescence to a solar energy–rich environment at some height above the ground.

Leaves

INITIATION AND EMERGENCE

Leaf initials (*primordia*) begin with certain cells in the apical dome, which divide (become meristematic) and produce swellings or protuberances on the stem apex. The protuberances spread and encircle the apex, particularly the sheath primordium of a grass leaf (Fig. 11.1). After the leaf collar is formed, cells in the subhypodermis become meristematic and produce an axillary bud. Subsequent growth of leaf blades (*lamina*) and sheaths or petioles and stem internodes is from *intercalary* meristems (those between differentiated tissues).

In a grass leaf the intercalary meristem is divided into two parts by formation of the ligule. The upper part was found to contribute to lamina growth, the lower to the sheath (Jewiss 1966). Growth of a grass leaf occurs while the leaf is enclosed in the *pseudostem* (the rolls of older leaves). In dicots leaves emerge from short bud scales; therefore growth by expansion is principally in the open.

In a constant environment leaf initials appear on the stem apex at a constant rate for a given genotype (Mitchell 1953). The interval of time between the appearance of successive leaf primorida is termed the *plastochron.* The interval of time between the appearance of the tip of successive leaves is termed the *phyllochron* (Bunting and Drennan 1966) and may differ from the plastochron. Time intervals in which phyllochrons are longer than plastochrons result in longer shoot apexes (Langer 1972). In wheat the emergence of a given leaf tip is at plastochron 5; that is, the fifth leaf is initiated as the first leaf emerges. Since emergence in dicots is from bud scales, the distinction between phyllochron and plastochron is not as useful as in grasses in which leaf growth occurs in the pseudostem.

Research on leaf initiation rate and appearance in crop plants is limited. Temperature, light, and other factors have been shown to influence plastochron development (Table 11.1). With ryegrass it was observed that high temperatures (18–25°C) and light intensity increased plastochron and phyllochron rates. This is not surprising, since rate of plant development is temperature driven. Raising the temperature from 15 to 20°C increased leaf appearance

TABLE 11.1. Rates of leaf appearance in perennial ryegrass

Temperature	Other Conditions	Rate (days • leaf^{-1})
25°C	...	5.8
18°C	21,530 lx	6.4
12°C	...	9.4
10°C	21,530 lx	10.0
Winter	Unheated greenhouse	15.5
Winter	Heated greenhouse	9.5

Source: R. H. M. Langer, *How Grasses Grow, Studies in Biology 34,* © 1979, by permission of Edward Arnold, Ltd.

rate in wheat by over 50% and shortened plastochron rate by 50%, from 5 or 6 days to 2 or 3 days (Langer 1972). Rate of leaf appearance in barley was linear as light increased from 7.8 to 32.5 W • m^{-2} (Aspinall and Paleg 1963), but these effects could have been influenced by temperature changes.

LEAF NUMBER

The number of leaves produced on a shoot or tiller is determined by inflorescence initiation. Formation of leaf initials on the apex gives way to floral initial formation (Fig. 11.1D) (Sharman 1945; Bunting and Drennan 1966), which fixes leaf number. Secondary and higher order tillers or branches generally have one to two fewer leaves than the primary shoot, since they emerge later and receive the same environmental cues to flower. Thus floral initiation is at a lower leaf number.

Characteristic leaf numbers are 7 to 9 for wheat, oats, and barley; 7 to 14 for sorghum; 14 to 21 for most U.S. maize hybrids; and 10 to 16 for upper-latitude U.S. soybean cultivars. Maize cultivars adapted from 50° latitude to the equator vary from 7 to 48 leaves. Height and maturity of maize are highly correlated to leaf number (Cross and Zuber 1973). The number of primordial leaves present in a mature seed embryo is characteristic of the species. Most cereal grains, such as wheat, have three leaves in the mature seed, while five leaves can be recognized in the embryos of maize seeds (Sass 1951). Plastochron 6 is during emergence or early seedling growth.

FACTORS AFFECTING LEAF GROWTH

Leaf number and size are affected by genotype and environment (Humphries and Wheeler 1963). The position of the leaf on the plant (plastochron number), which is principally controlled by genotype, also has a pronounced effect on leaf growth rate, final dimensions (Bunting and Drennan 1966), and capacity to respond to improved environmental conditions, such as available water (Ralph 1982).

Leaf length, width, and area generally increase progressively with ontogeny up to a point; then in certain species these parameters decrease progressively with ontogeny so that the largest leaves are near the center of the plant,

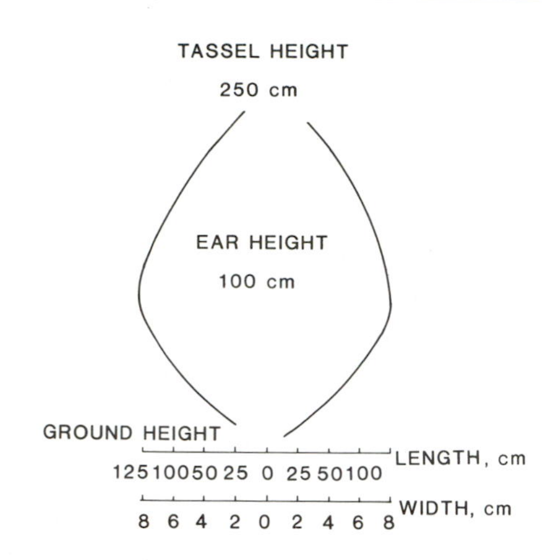

Fig. 11.2. Profile of leaf sizes of a maize plant.

such as on a maize plant (Fig. 11.2). The *flag* (uppermost) leaf of maize is shorter and narrower and has less area than the ear leaf. This type of profile is characteristic of many species. In other grasses, such as barley, the length of lamina decreased with flower initiation but the width increased, resulting in a broad flag leaf (Goodin 1972). The cause of the diminution of upper leaves is not known but appears to be competition with the inflorescence for nutrients. The relative growth rate of leaves decreases with leaf number (Milthorpe and Moorby 1974).

At stage 5 of soybean, 70% of the total plant dry weight was leaves (Hanway and Weber 1971). Leaf growth peaked at stage 6 and remained constant until stage 10, while total plant dry weight increased rapidly due to stem and fruit growth. Size and weight of new leaves decreased after stage 6; after stage 10 total leaf weight decreased due to senescence of bottom leaves. The maximum weight and area of the leaves of a plant is reached early in the life cycle, after which the gain in leaves just equals the loss, a status referred to as the critical leaf area.

Even though the lower leaves on plants are smaller and frequently lost due to environmental stresses and senescence, they are important to the vegetative growth. For example, ^{14}C fed to leaf 3 of a grass plant was active in leaf 4, 5, and 6 (Bunting and Drennan 1966). The lower pods of soybean were primarily supplied by the subtending leaves (Johnson et al. 1960). In barley the sheath plus stem contributed 50 to 70% as much of the apparent photosynthesis for grain production as did the blade (Thorne 1959). In dicots leaves with a long petiole and large petiole base make a significant photosynthetic contribution.

Nitrogen (N) fertilization had a pronounced effect on leaf expansion, especially on leaf width and area (Humphries and Wheeler 1963). With low N, leaf 4 of wheat was the largest in size; with high N, leaf 5 was largest. The shift in maximum size to an upper leaf was thought to have resulted from a reduction in the competition for N between upper leaves and emerging stem and inflorescence (Bunting and Drennan 1966). A N deficiency also causes a reduction in leaf area due to senescence of older leaves.

Other minerals appear to have less effect than N on leaf growth and senescence, although the competition for most nutrients between new and old leaves and between fruits and leaves is evident.

For reasons not completely understood, elongation of wheat leaves was significantly less during the night than during the day (Christ 1978) and dropped almost to zero if the period of darkness was extended. This decline was also related to lower irradiance during the previous light period. A night (far-red) effect is indicated, which probably interacted with organic nutrition.

Irrigation in a humid climate (Columbia, Mo.) promoted rapid leaf elongation of tall fescue (*Festuca arundinaceae*) during summer months (Nelson et al. 1978). However, leaf growth in nonirrigated plots was greater than that in irrigated plots in the fall and the following spring, when moisture was available naturally.

Ralph (1982) showed that late-maturing cultivars of sunflower, unlike early cultivars, were benefited in leaf expansion by water stress during the vegetative phase. Leaves of late cultivars were less determinate with less competition from inflorescences, and expanded more once irrigation was renewed. The stressed plants of late cultivars, such as 'Stenchurian,' produced 60% more leaf area than plants under full irrigation. The leaf area of nonirrigated early cultivars was less than that of irrigated plants.

With 'Marquis' wheat, high temperatures (25°C), long days, and low irradiance (about 14–42 W • m^{-2}) resulted in long, slender, thin leaves (Friend 1966) (Fig. 11.3). On the other hand, low temperatures (15°C), high irradiance, and short days resulted in wider, shorter, thicker leaves. The greatest

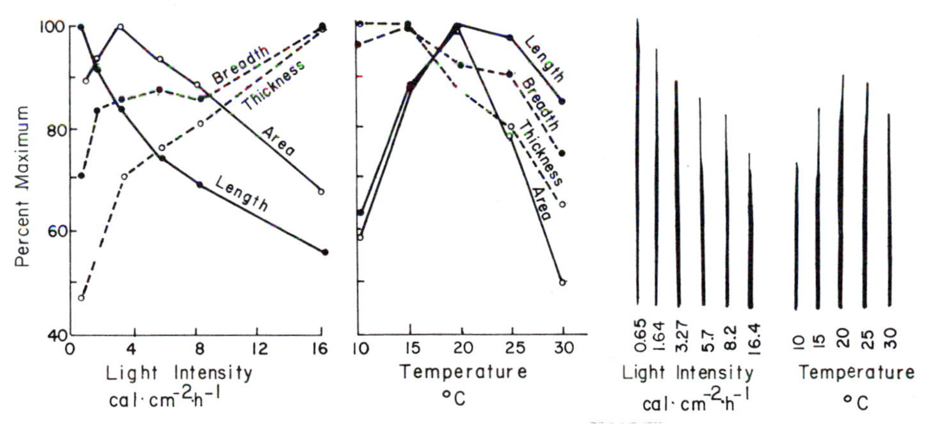

Fig. 11.3. Effect of light and temperature on growth of wheat leaves. (From Friend 1966, by permission)

leaf area was obtained at intermediate levels of these factors. Studies with timothy (Langer 1954) supported these conclusions; moderately warm (warmer than outdoors) greenhouse temperatures increased leaf length 2.5-fold. Phyllochrons were 9.3 days and 13.5 days for indoors and outdoors, respectively. Increasing day length increased leaf growth rate. *Vernalized* (exposed to a cold period) wheat plants produced leaves with shorter lamina than nonvernalized (Westmore and Steeves 1971).

The effect of photoperiod on leaf appearance rate is harder to evaluate, since longer photoperiods are often associated with greater heat input, which is the major driving force in plant development. Thus evidence on photoperiod effect is often conflicting and inconclusive.

LEAF SENESCENCE

The leaf number and leaf area index (LAI) peak and then remain quite constant until general senescence begins. This equilibrium in LAI results from loss of lower leaves at a rate equaling the production of new (upper) leaves. Therefore LAI tends to plateau at a maximum of about 4 to 7 for crop canopies, regardless of plant population above a moderate level. Forage grasses with narrow, upright leaves usually have LAIs higher than 7.

Langer (1972) reported the mean numbers of living leaves as 4.5 to 5.8 per shoot for several different forage grasses growing in a heated greenhouse in England, compared with means of 3.1 to 3.7 in an unheated greenhouse. Additional N raised the number slightly at the higher temperature. Grass plants low in N tend to *fire* (senesce) the lower leaves.

Senescence of an individual grass leaf begins at the oldest part of the leaf (the tip) and progresses downward. Senescence of an individual plant begins at the basal (older) leaves and progresses upward. By the time maize has produced 10 to 12 leaves, 4 to 5 leaves have been lost to senescence (the loss in LAI is relatively small since the lost leaves are small). Usually leaf 5 is the first healthy leaf on a maize plant at tasseling.

The cause of senescence is generally thought to be mobilization and redistribution of mineral and organic nutrients to more competitive sinks, such as young leaves, fruits, tillers, and roots. Contribution from leaves to these organs declines progressively with senescence. There is no evidence of reverse flow (*parasitism*), by aging leaves, as was once commonly thought to occur.

Rapid production and expansion of leaves are highly important in crop production in order to maximize light interception and assimilation. A full canopy also reduces weed competition and sheet erosion. Seeding rates of peanut are inordinately high, partly to reduce intrarow weed competition. Interestingly, assimilation rates are usually maximized at a LAI of 3 to 5 for most cultivated crop plants.

Stems

The stem consists of internodes spaced between the nodes, with attached leaves. The number of nodes and internodes is equal to the leaf number, all three having a common origin in the phytomer. Shoots of temperate grasses have compacted or untelescoped nodes (without internode elongation), which, until elongation after floral initiation, are positioned below the soil surface. At flowering, four to five of the upper internodes elongate and vertically space the upper leaves. A similar number of internodes remain short and confined at or below the soil surface (referred to as the *crown*). Many dicot species are stemless until flowering. On the other hand, tropical species may produce vegetative stems, that is, internode growth without flowering.

Westmore and Steeves (1971) classified plants based on internode length as follows: (1) short stem (those without conspicuous internodes, such as plantains and the first-year growth of biennials) and (2) long stem (those with conspicuous internodes, such as maize and the second-year growth of biennials).

INTERNODE ELONGATION

Growth in height of stem occurs in the intercalary meristems of the internodes (see Chap. 8). Internodes lengthen both by increases in cell number and (primarily) by cell expansion, the latter resulting in an increase of up to 25 cm or more. Growth by cell division is at the internode base (i.e., intercalary) rather than in apical meristems. However, intercalary meristematic activity is distributed throughout the length of the leaf lamina, sheath, and internode at the primordial stage (Fig. 11.4). With maturation, the meristem activity moves to the basal regions and eventually terminates (Sharman 1942).

The *peduncle* (internode supporting the inflorescence in grasses) and the flowering stalk (of dicots) grow from intercalary meristems. Generally internode growth is determinate for reasons not fully understood but apparently due to a limitation of the potential number of active cells. An exception is found in the *mesocotyl* (first internode in grasses) (Vanderhoef et al. 1979),

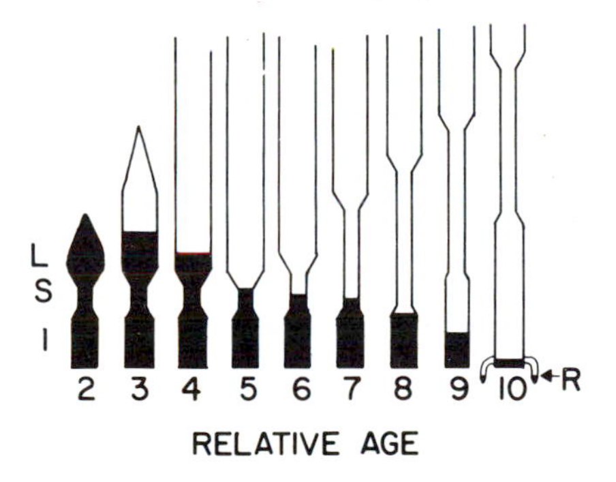

Fig. 11.4. Flux of intercalary meristematic activity (shaded) in relation to advancement of age of leaf stem tissue of a grass plant. Meristematic zone in leaf lamina (*L*), sheath (*S*), and internode (*I*) is reduced with time to a small area at the internode base and the tips of root initials (*R*). (From Sharman 1945. Reprinted by permission. © University of Chicago Press. All rights reserved.)

which, within food reserve limits, continued to elongate indefinitely in darkness or in infrared light. Growth of the mesocotyl is inhibited immediately by exposure to red light; that is, growth is phytochrome (pigment) controlled but modifiable by organic nutrition. In addition to growth limitation because of the number of active cells, the amount of growth hormones in the intercalary meristems may be limited since they are not self-generating as in apical meristems. Consequently plant growth regulators (PGRs) must be supplied from plant parts outside the meristem. Dwarf plants can respond to an exogenous (external) source, generally to gibberellin (GA) applications.

Maize remains stemless until reaching a height of about 40 cm and developing eight fully expanded leaves, which arise from the *pseudostem* (a vegetative shoot). At this stage there is no perceptible internode growth (Fig. 11.5). Due to compacted nodes and internodes, biennials produce a stemless rosette during the first year. Until flower initiation, temperate grasses produce pseudostems. At floral initiation the internodes of grass and biennial shoots that bear an inflorescence elongate. In early season a grass plant usually has both vegetative and reproductive tillers (*culms*).

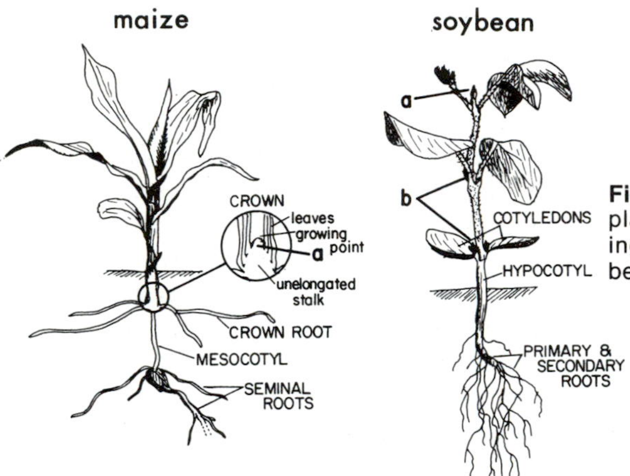

Fig. 11.5. Young maize and soybean plants showing position of apical growing points (*a*) and axillary buds of soybean (*b*). (From Crookston et al. 1976)

In dicots with no stems (e.g., *Plantago*) the last internode below the inflorescence greatly elongates to produce a flowering stalk (Sachs 1965). Long flowering stalks are evident in certain other species, such as white clover. The *gynophore* (peg) of peanut can be considered a nodeless fruiting stalk although its morphogenesis by originating from the flower differs somewhat from the typical flowering stalk. In monocots and dicots with long stems internode length generally increases acropetally, but patterns alternating long and short internodes are characteristic of some species. Basal internodes of many species may be short enough to escape observation, whereas the uppermost internodes, especially the peduncle of a grass culm, may exceed 25 cm. Normally two or more internodes elongate simultaneously, but in sunflower a new internode does not initiate elongation until the preceding internode has completed it (Sachs 1965). Meristematic activity to cause elongation of internodes, except in the gynophore of peanut (Jacobs 1947), is concentrated at the basal end, as indicated by the presence of mitotic activity in stained cells.

CROWN DEVELOPMENT

The lower, closely spaced nodes of a plant form the crown, which is located at or just below the soil surface. In grasses these densely spaced nodes give rise to the successive whorls of adventitious roots called the nodal, crown, or coronal root system (Fig. 11.5). The lower nodes of perennial legume plants such as alfalfa form a crown but without adventitious root development.

The location of growing points in the grass crown below the soil surface and resultant exposure of new leaves from sheaths of older leaves (pseudostem) have crop management implications. Since maize maintains this condition for 4 or more wk (until approximately eight leaves are fully exposed) (Fig. 11.5), an early frost or clipping usually injures only the above-ground foliage, the oldest and smallest leaves. Little permanent injury is caused by such early defoliation, since a new canopy of leaves soon emerges from the unspent intercalary meristems protected by the leaf roll and from newly formed leaves. The common practice of pasturing wheat during winter and early spring in the southern United States does not seriously damage grain production as long as the growing points of shoots remain vegetative, that is, below the soil surface. After floral initiation and concomitant stem elongation with the onset of longer spring days, grazing can remove the inflorescence and destroy the grain production potential. Unlike temperate grasses, dicots and many tropical grasses grow from exposed buds of aerial stems (Fig. 11.5, soybean). Hence, freezing or destruction of above-ground shoots can destroy axillary buds and regrowth potential. If a killing frost should occur on soybean, for example, growth potential is destroyed because no buds are present below the cotyledon axils, which are above ground; reseeding of the crop is necessary.

FACTORS AFFECTING STEM GROWTH

Growth Regulators. The effect of plant growth regulators, especially GAs, on stem growth is well documented. They can overcome dwarfism in genetic dwarfs, such as dwarf maize and pea, promoting increased internode growth and normal height presumably by correcting an endogenous GA deficiency (see Chap. 7). However, the dwarf habit in dwarf 'RS 610' sorghum was not corrected by GA sprays; only the below-ground nodes (mesocotyl and second internode) and the coleoptile responded (Gardner and Kasperbauer 1961). This comparative lack of response in sorghum was probably due to the fact that sorghum dwarfing is controlled by several genes, and maize and pea dwarfing by a single gene (Windscheffel et al. 1973). Evidently GA is more effective in correcting dwarfing that is inherited simply.

Leopold (1949) showed that auxin has a pronounced effect on *tillering* (growth of shoots from crown buds) in barley (Table 11.2). When the shoot apex and source of auxin was destroyed, 'Wintex' barley tillered profusely unless given an application of the auxin naphthaleneacetic acid (NAA). The NAA-treated plants with destroyed apexes tillered about the same as normal plants, that is, those with undisturbed apexes.

Light. Light has a pronounced effect on stem growth. In the dark, *etiolation* (elongation of internodes) is extreme and similar to that of the mesocotyl internode. The internodes of shaded plants, such as in dense stands, are more etiolated. The shade effect is believed to be due to auxin enhancement, probably acting synergistically with GA. Theoretically photodestruction of auxin is less in shaded stands, since high irradiance decreases auxin and plant height.

Day length affects stem growth usually less conspicuously than it affects flowering. Consequently photoperiodic responses of stem development are not often reported. Long days cause increases in internode length and plant height, especially on short-day plants. Soybean cultivars adapted to northern latitudes had fewer and shorter internodes and flowered earlier when grown at lower latitudes (Shibles et al. 1975). Planting the same cultivars at higher latitudes than those to which they are adapted has the reverse effect and would also probably result in immature seeds at harvest. Much earlier seeding dates in an area of adaptation tend to result in the same response as to shorter days characteristic of lower latitudes. For example, early planting produces shorter internodes in maize, which results in sturdier plants.

As with leaf growth, internode growth of grasses is influenced by light quality. Both grow from intercalary meristems of common origin and are shielded from light in the roles of older leaf sheaths, producing a dark or far-red effect (Fig. 11.5). Far-red light (maximum effectiveness at 730 nm) promotes, and red light (maximum effectiveness at 660 nm) inhibits, mesocotyl elongation (Vanderhoef et al. 1979), the mechanism controlling emergence from variable planting depths. In wheat, emergence is primarily from elongation of the second rather than the first internode or mesocotyl; distinct effects of monochromatic light on upper internodes has not been demonstrated. However, in grasses these internodes and young leaves are trapped in the darkness of the sheaths of older leaves for the greatest part of their total growth. Consequently the phytochrome–far-red response appears to be operative. Growth is inhibited when exposed to light.

Dicot internodes are not so enclosed in leaf rolls, probably indicating little or no phytochrome–monochromatic light response, but this relationship has not been well established. However, a bush-type garden bean under long night–far-red radiation (see Chap. 12) in effect assumed a climbing habit (internode elongation) due to development of long internodes (Kretchmer et al. 1977), a single-gene response. Flowering, which was independent of the climbing habit, was promoted by short nights (long photoperiods). It was concluded that flowering response and climbing habit are both phytochrome controlled but are discrete, independently inherited responses.

Mineral nutrient and water availability affect internode growth, especially by cell enlargement, as in any vegetative or fruiting organ. Nitrogen and water, particularly, increase plant height, but the effect is complex since a larger leaf size results in more shading. Shading tends to increase auxin levels, which could affect internode length.

Branching

Whether buds in leaf axils resume growth to produce side branches (e.g., in dicots) or tillers in grasses depends on genotype and environment. The potential for axillary branches is always present, since there is a bud in each leaf axil. Modern maize hybrids do not tiller, except for the development of ear shoots, despite favorable environments because of strong internal (genetic) control (Duncan 1975). Ear shoots can be forced to develop at several lower nodes if apical dominance is broken. For example, removal of a young ear shoot on a maize plant stimulates development of the next lower ear. The potential number of ears is equal to leaf number, since axillary buds and leaves are phytomer components. The number of shoots that actually develop on grass plants is always less than the potential, due to genetic and environmental controls.

Three types of tillers of grasses have been described by Arber (1934):

1. Upward, or *apogeotropic.* These tillers are similar in appearance to the primary shoot but have one or two fewer leaves and often remain vegetative even though the primary shoot and even sister tillers are reproductive. These *intravaginal* tillers emerge from live leaf sheaths (Fig. 11.6).

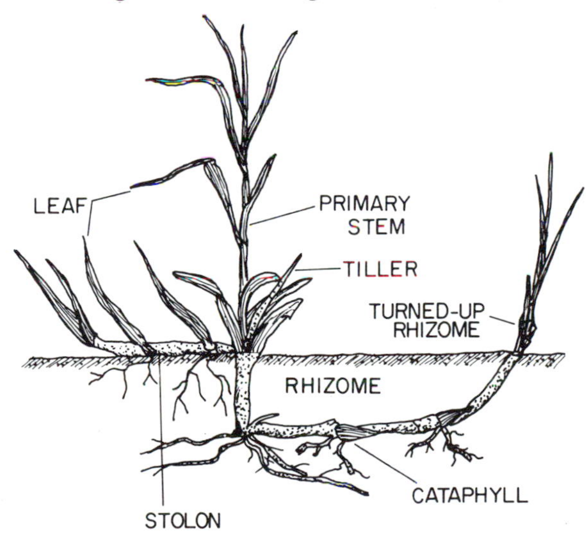

Fig. 11.6. Branching habit of a perennial grass plant, showing the main shoot in relation to the axillary branches: stolon, aerial tiller (intravaginal), and rhizome (with cataphylls) turned up to produce a new primary shoot.

2. Horizontal, or *diageotropic.* Stolons and rhizomes belong to this type. They differ from upright shoots in certain morphological details. Generally, rhizome and stolons emerge from dead sheaths of the most basal nodes at or below the soil surface. Stolons grow horizontally above ground and produce normal stems and leaves (Fig. 11.6). Rhizomes grow below ground and produce modified leaves without laminas (*cataphylls*), on stems that have normal nodes and internodes (Fig. 11.6).

3. Downward, or *geotropic.* These stem types are not common in occurrence.

TILLERING

Upward, or intravaginal, axillary shoots in grasses are commonly referred to as *tillers.* Shoots from axillary buds on stems of dicots are usually referred to as side branches. Their origin and development morphogenetically are the same: both emerge from leaf axils, generally from the lower nodes, if the shoot is not apically dominant. Tillers emerge acropetally, beginning at the lowest nodes. In wheat the first tiller emerges from the coleoptile axil, in rice from the leaf-3 axil (Murata and Matsushima 1975). Regardless of species, the lower leaf axils of the main stem give rise to the primary tillers. These in turn give rise to the secondary tillers, which give rise to the tertiary tillers, and so on. In general all primary tillers emerge before any secondary and tertiary tillers. Perennial grasses tiller throughout the season. This tillering habit, along with accumulation of food reserves, is the primary basis for *perennation* (living from season to season). Rice and sorghum, temperate annuals, have these characteristics and perennate in tropical climates. This perennial habit of rice and sorghum is often exploited in tropical areas to produce *ratoon crops* (regrowth from stubble).

Tiller production in cool-season perennial grasses is subject to wide seasonal fluxes. In 'S 170' tall fescue tillering increased exponentially during spring, was static during summer, and increased again during autumn, accumulating a total of 300 tillers per plant (Robinson 1968). In March of the following year the number declined to 250, in June to 100. The tillering habit in timothy is much weaker than in tall fescue, as is the capacity to perennate and persist. Temperate grasses produce large numbers of reproductive tillers in spring to early summer. They also produce an equal or greater number of vegetative tillers then and produce only vegetative tillers later in the season. The vegetative tillers produce leaf growth during the growing season, overwinter, and become reproductive tillers the next season, having been vernalized by the cool temperatures and short days of the previous fall (Gardner and Loomis 1953). Tropical grasses generally produce vegetative stems with distinct nodes and internodes as tillers. Their vegetative and reproductive tillers are similar in appearance prior to panicle emergence. Reed canarygrass, although temperate, often produces vegetative aerial stems. Dallas and bahiagrass (*Paspa-*

lum spp.), which are semitropical species, tend to be intermediate between typical temperate and tropical species in terms of the production of vegetative and reproductive stems.

In alfalfa and many other perennial legumes tillers emerge from the basal buds (crown). Emergence of the new crop of tillers is signaled by loss of apical dominance in the mother shoots, generally by flowering and senescence or mechanically by grazing or haying.

Annual legumes, such as soybean, tiller (produce side branches) from the lower six nodes, depending on cultivar and environment (particularly high light or wide spacing) (Fig. 11.7). Axillary buds that remain quiescent, whether on branches or on the main stem, are activated by photoperiod later in the season to produce flowering *racemes* (inflorescences). The racemes emerge from the bud prophyll and scales (modified leaves) at leaf axils more or less simultaneously in determinate varieties and by an acropetal hierarchy in indeterminate types.

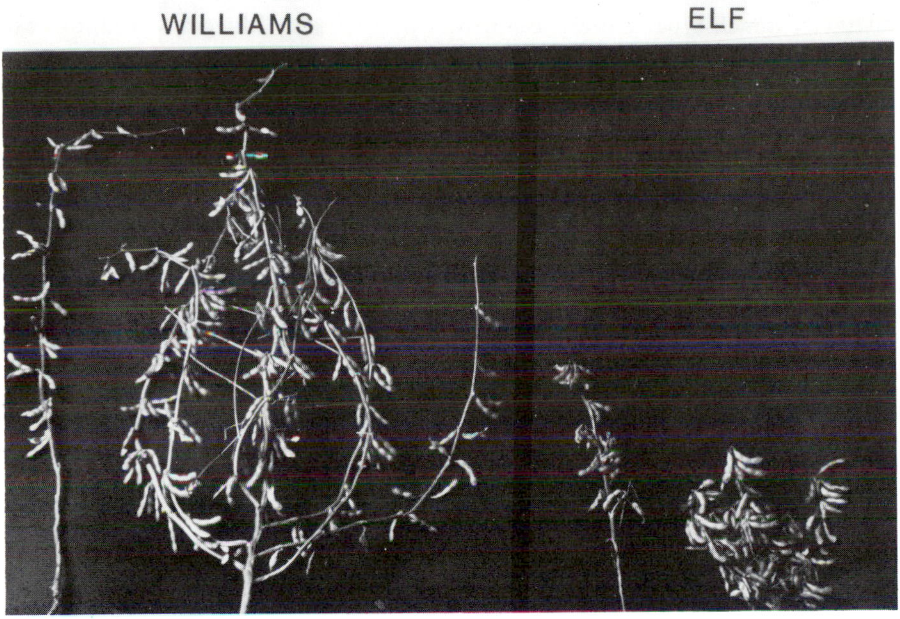

Fig. 11.7. Branching from axillary buds of soybean, in relation to cultivars ('Williams' and 'Elf') and plant density (seeding rates: plant on left, 555,000/ha; plant on right, 185,000/ha). Williams is an indeterminate, maturity group III; Elf is a semi-dwarf determinate, maturity group III.

RHIZOMES AND STOLONS

The pattern of rhizome and stolon development in perennial grasses differs little from that of upright tillers, since they also arise from basal axillary buds (Etter 1951) (Fig. 11.6). Both upright tillers and rhizomes and stolons emerge from a modified protective leaf, termed the *prophyll,* which is analogous to the coleoptile in embryo emergence. On a rhizome, cataphylls (Fig. 11.1) assume a normal leaf morphology after the rhizome turns up from the soil (Fig. 11.6). Leaves on stolons are morphologically normal.

The subterranean duration of rhizomes depends on the genotype. Rhizome branches tend to turn up earlier than the primary rhizome, which generally grows for a considerable distance before turning up (Sharman 1945; Etter 1951). In the process of surfacing, one or more cataphylls become transitional leaves with a short lamina (about 1 mm long). Leaves produced above ground are normal.

Rhizomes of tall fescue turn up in a short distance; the effect is a bunch-type habit. In other grasses the turning up of rhizomes is delayed, resulting in a sod habit. Since tillers of orchardgrass and timothy are intravaginal, growth is distinctly bunch type. Different cultivars of Kentucky bluegrass seedlings exhibited sod, bunch, or intermediate types (Nittler and Kenny 1976). Cultivars of alfalfa differed in this aspect (Cowett and Sprague 1962); some shoots grew horizontally prior to turning up, forming a creeping habit.

FACTORS AFFECTING BRANCHING

Research on branching has been concentrated on upright tillers, so information on rhizome and stolon production is comparatively limited. This is understandable, since upright tillers may contribute directly to grain or forage yield and are relatively easy to observe. Branching is a function of the genotype interacting with a host of physical and biological environmental factors.

Genotype. Although the potential number of branches is directly related to leaf number, all leaf axils do not produce branches. Some species produce more than one branch from a single axil, thus appearing to have more than a single axillary bud. Tobacco is a classic example; this habit is a major problem in tobacco production inasmuch as the axillary shoots must be removed by hand, or chemically with a growth inhibitor. In tobacco new shoots are stimulated by the practice of *topping* (removal of inflorescence), allowing retention of N and carbon assimilate in the leaves, which are the harvest product (Beatty 1982). However, after the tops are removed two or three shoots emerge from each leaf axil rather than the single shoot found in most plants. The emergence of more than a single shoot from an axil does not present a divergence in bud development homomorphy; rather, additional axillary buds are formed and emerge from the leaf axils of the new branch.

Genetic control of axillary branching was demonstrated in a study of oat cultivars by Frey and Wiggans (1957a). Low-tillering types of spring oat in 12-

in. spacings developed a mean of 5.7 to 7.1 tillers per plant, compared with a 9.7 to 13.0 mean from high-tillering cultivars. Winter-type cultivars produced 9.7 to 14.2 tillers per plant. With wheat, tiller numbers varied with genotype but the range appeared to be narrower than with oat (Bunting and Drennan 1966). Pasture species differ greatly in both tiller number and growth duration.

For reasons not completely understood, position of the axil on the stem has a strong influence on expression of tillers or branches. Despite the axillary shoot potential and a favorable environment, buds did not develop into tillers at the top three leaves of timothy (Langer 1956). All axillary buds of timothy were dormant until a minimum of five leaves were fully expanded (i.e., a juvenile factor). The basal and older axils of a shoot produce tillers first. Since maize tillers do not usually make a worthwhile contribution to grain yield, breeding and selection pressures have been for nontillering genotypes, unlike the breeding objectives for most cereal crops. For example, the Green Revolution wheat and rice cultivars were selected for high tillering, which in favorable environments is often positively correlated with yield. Tiller number in rice increased for approximately 1 mo after planting, reaching a maximum number about 2 wk prior to heading (depending considerably on the nutrient status of the mother plant) (Murata and Matsushima 1975).

While axillary buds on maize stems do not produce *suckers* (vegetative shoots) on modern hybrids, usually the bud at leaf 11 and frequently the bud at leaf 10 produce a reproductive ear shoot. Ear shoots are modified, compared with normal tillers, in that all the nodes are compacted due to the short internodes (depending somewhat on genotype) and the leaves (husks) are much shorter than normal (Fig. 11.8). The ear shoot is terminated by a spike inflorescence instead of the *tassel* (panicle) of the main shoot. Under optimum conditions, more than two ear shoots can develop on some genotypes, or if the

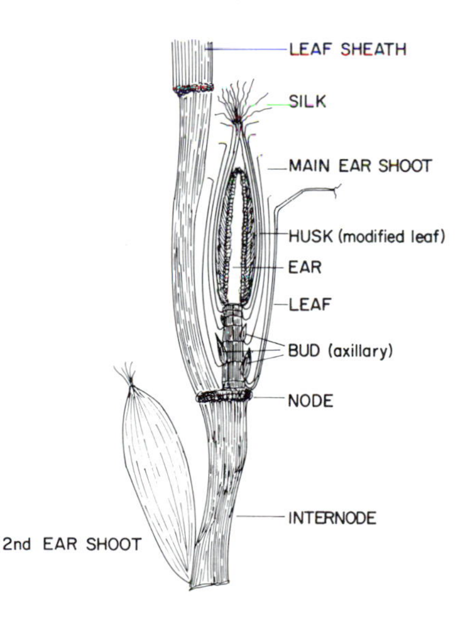

Fig. 11.8. Diagram of longitudinal stem section and ear branch with component structures. (From Sharman 1942)

top ears are removed at an early stage, ear shoots emerge from the lower leaf axils. It is interesting that buds at axils above the first ear are completely quiescent, showing no signs of development regardless of environment and plant management.

Growth Hormones. Growth hormones, particularly the auxin NAA, exercised a strong control over axillary branching (Leopold 1949; Cowett and Sprague 1962; and Laude 1975), as shown by dissecting 'Wintex' barley and 'Chalco' teosinte shoot apexes (Tables 11.2, 11.3).

TABLE 11.2. Effect of auxin and apex destruction on tillering in barley

	Number of Plants	
Treatment	Tillering	Not tillering
Untreated	3	7
Apex destroyed	9	1
Apex destroyed plus auxin	3	7

Source: From Leopold 1949, by permission.

TABLE 11.3. Tiller development on alfalfa treated with growth regulators

Treatment	Buds • $Plant^{-1}$	Stems • $Plant^{-1}$
TIBA	8.3	5.3
NAA	2.3	1.3
Control	3.7	2.3

Source: From Cowett and Sprague 1962, by permission.

Long days decreased tillering in barley and alfalfa, as did auxin applications (Leopold 1949). Barley and alfalfa tillered vigorously in the short autumn days, which suggests an auxin factor because auxins have been shown to increase with long days. Removal of apical buds and even young leaves of wheat by clipping apparently removed rich auxin sources, resulting in increased tillering (Laude 1975).

Rhizomes were observed to branch more in a dry soil; this might be attributable to auxin inhibition by ethylene, which may be generated by soil resistance (see Chap. 7).

Light and Plant Density. Plant density (and the resultant availability of light to the lower canopy) is known to be an overriding factor in axillary shoot development (Fig. 11.6; Table 11.4). The necessity for high irradiance at the base of plants has been hypothesized. However, Mitchell and Coles (1955) concluded from their studies with ryegrass that lighting of the whole plant, or the integral of light from tip to base, was the operative factor. Langer (1972) showed that the relationship between tillering and light intensity was essentially linear for two grasses. Increasing light intensity from 5380 to 18,840 lx (about 38–132 W

TABLE 11.4. Effect of nitrogen fertilizer and seeding rate on tillers per oat plant

	Seeding Rate (bu • acre^{-1})		
Nitrogen (lb • acre^{-1})	1	3	5
0	1.02	1.00	1.02
20	1.18	1.12	1.03
40	1.40	1.35	1.16
80	1.68	1.35	1.19

Source: Frey and Wiggans 1957.

• m^{-2}) tripled tillering per plant for biweekly periods at 21°C and 16°C (day and night) temperatures. If conditions are favorable, tillering of small-grain cereals increases until a maximum number of culms per unit of land is normalized regardless of seeding rate.

Increasing plant population reduces maize axillary branches that form ear shoots. All ears may be eliminated if the population is too high. This barrenness has been attributed to competition for assimilate, since photosynthesis is reduced in thick stands. Apical dominance due to increased auxin in shaded plants is also a plausible explanation. A lowered activity of the nitrate reductase enzyme also has been suggested (Zieserl et al. 1963), since nitrate reduction is dependent on photosynthesis. Barrenness has also been shown to result from a failure of silk growth (Sass and Loeffel 1959), which could reflect either inadequate photosynthate or a failure in protein synthesis due to limitations on nitrate reduction, or both. Failure of ear shoots to initiate growth from leaf axils is quite likely related to apical dominance and controlled by auxin. Once growth is initiated the failure to develop into normal ears, as often occurs, may be due to competition for assimilate.

Photoperiod and Temperature. Photoperiod and temperature interact to influence tillering. Generally cool-season grasses, including wheat, tiller in response to the short days and cool temperatures of fall. Tillering of the subtropical grass *Paspalum dilatatum* was favored by warm temperatures, and there was no adverse effect from temperatures as high as 35°C (Youngner 1972).

Long photoperiods decreased tillering significantly in *Oryzopsis miliaca,* a tropical plant (Koller and Kigel 1972). In alfalfa, a long-day plant, short days promoted tillering in stands of established plants but long days promoted tillering in seedlings (Cowett and Sprague 1962). However, in first-year biennial sweet clover seedlings (*Melilotus,* a long-day plant), tiller bud formation and fleshy taproot development were strongly favored by the short days of autumn (Fig. 11.9). Temperature did not influence bud formation significantly (Kasperbauer et al. 1962). In general increasing temperature, especially under long days, decreases branching in a wide range of temperate species. High temperatures tend to favor tillering in tropical species. Branching of rhizomes in Kentucky bluegrass is evidently promoted by short autumn days, as is upright tillering (Table 11.5). Rhizomes did not branch in spring (Etter 1951),

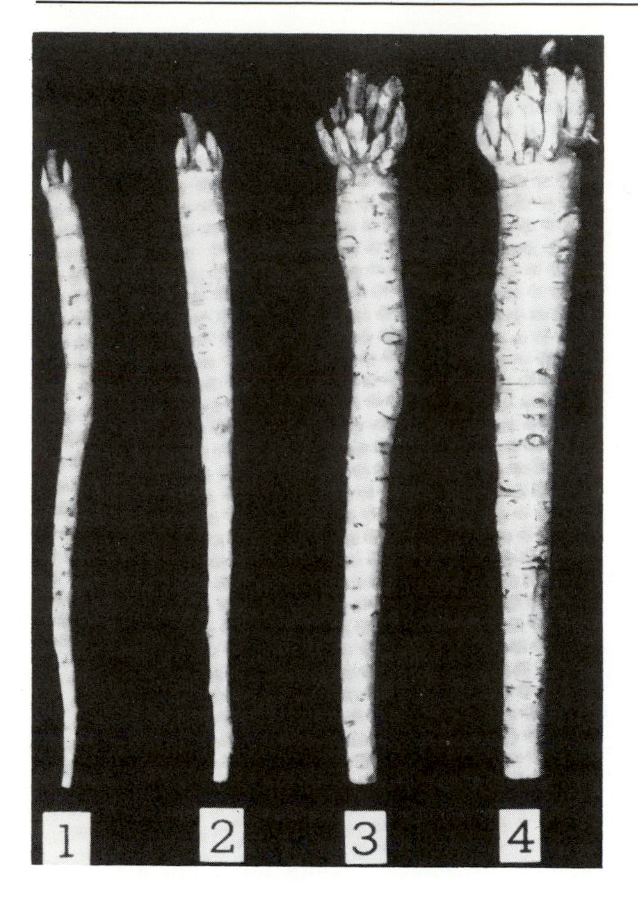

Fig. 11.9. Crown buds on sweet clover roots dug in (*1*) August, (*2*) September, (*3*) October, and (*4*) November (Kasperbauer et al. 1962, by permission).

TABLE 11.5. Effect of day length and temperature on the number of tillers per 100 Kentucky bluegrass plants

	Number of Tillers		
Temperature	11 hours of light	15 hours of light	19 hours of light
Cool	160	114	116
Warm	142	116	106

Source: Peterson and Loomis 1949.

probably due to the effect of long days. The flush of new rhizome branches in autumn and their early turnup enhance sod thickening.

Waters and Minerals. Tillering is highly dependent on factors that favor rapid vegetative growth, particularly adequate moisture and N, if light (spacing) and other inputs are adequate (Table 11.4). Nitrogen has a dominant effect on tiller production in grasses. Langer (1972) reported that the number of tillers on timothy in 3 wk with 150 ppm N was twice as great as with 6 ppm N; 4 wk were required for phosphorus (P) to show this much increase. Increasing potassium (K) rates never causes tiller numbers to double. There was no response to P or K at low-N levels. Tillering in grasses is normally arrested prior to flowering, but Aspinall (1961) found that tillering in barley continued until heading (after anthesis) with adequate nutrient levels. Phosphorus and zinc increased tillering in wheat, but K had no effect (Fuehring 1969). It seems reasonable to expect tillering responses to N and water, as they are required in generous amounts to support rapid vegetative growth. Replacing other deficient mineral elements can often stimulate tillering.

Clipping or Grazing. Any treatment, mechanical or otherwise, that removes the stem apex may also destroy apical dominance and stimulate tillering or branching unless the level of cutting is below the axillary buds, as in the case of *epigeous* (cotyledons forced above the surface during germination) dicots such as soybean. Removal of apical buds by clipping of soybean increased branching but had no positive effect on seed yields (Bauer et al. 1976). Similar results were obtained with grain sorghum (Singh and Colville 1962). Removal of only the young leaves of wheat stimulated tiller production (Laude 1975). Both buds and young leaves are sources of auxin, which promotes apical dominance. Clipping to eliminate apical dominance and promote tillering usually produces negative results in terms of increased grain yields, probably due to loss of leaf area and N.

Grazing of foliage and N fertilization resulted in shorter rhizomes and earlier turnup in Kentucky bluegrass (Etter 1951), but rhizome growth and development in brunswickgrass (*Paspalum micorae*) was unaffected by defoliation and N fertilization (Beaty et al. 1970).

Vegetative Regrowth

Forage plants are subjected to partial or complete defoliation, and regrowth is necessary to sustain production. Animal grazing usually leads to continuous but partial defoliation. Clipping is not as selective as grazing and frequently results in complete defoliation, depending on species, growth habit, and past management. Plant responses, including regrowth, may differ significantly between the two defoliation systems.

REGROWTH OF GRASSES

Langer (1972) pointed out that regrowth from vegetative or stemless tillers of grass plants, especially of perennials, can and usually does occur from several origins (Fig. 11.10): (1) young, expanding leaves emerging from the leaf roll (fully expanded leaves do not regrow); (2) new plastochrons; (3) new tillers from leaf axils, which may have been stimulated by clipping; and (4) newly emerged turned-up rhizomes, which become normal tillers after surfac-

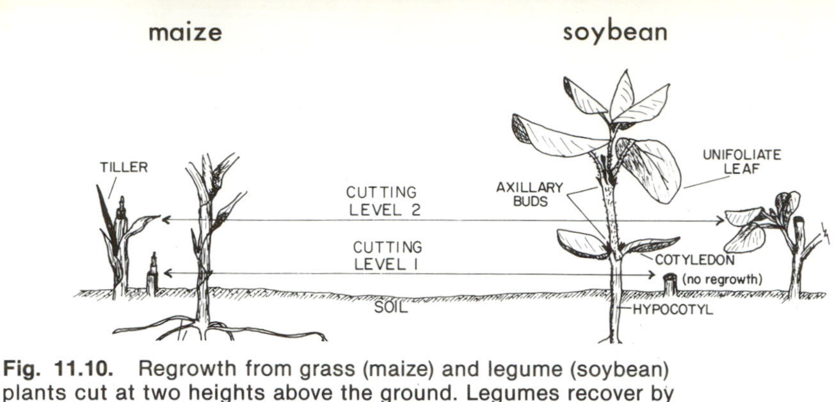

Fig. 11.10. Regrowth from grass (maize) and legume (soybean) plants cut at two heights above the ground. Legumes recover by new shoot growth from axillary buds unless these are destroyed. Grass regrowth is by continued extension of unexpanded leaves, regardless of clipping height, and from production of new tillers, which often tend to be stimulated by clipping.

ing. The multiple pathways of regeneration of a new canopy explain the persistence and usefulness of perennial grasses for pasture and turf.

REGROWTH OF LEGUMES

Regrowth of legumes is limited to new shoots developing from quiescent buds of basal nodes (crown) (Fig. 11.10). Defoliation can be so severe as to remove the lower axillary buds and regrowth potential, such as happens with a frost on a legume with epigeous emergence (e.g., soybean). With alfalfa, a perennial, clipping stimulates new growth from the crown buds located just below or near the soil surface. If alfalfa harvesting is delayed past the flowering stage, apical dominance is lost (as if clipping had occurred); new shoots emerge from crown buds, which may also be removed at harvest. Removal of the new shoots in the alfalfa crop gives cause for concern, but it is commonly believed that no damage is done to the next crop, evidently because of the rapid succession of new, unclipped tillers. Of course, repeated removal of new alfalfa tillers, as in continuous and close grazing, could deplete food reserves and weaken stands.

FOOD RESERVES

Organic food reserves are necessary for initiation of new growth. Dark growth studies have demonstrated which carbohydrates function as food reserves. Smith (1962) showed that root-reserve starch and sugars (i.e., nonstructural carbohydrates) were metabolized during dark growth of alfalfa, whereas hemicellulose and other dry matter fractions (structural carbohydrates) were not metabolized (Table 11.6).

A new annual plant is normally nourished from the seed, which has a generous supply of carbohydrates, fats, and proteins. Perennials commit few food reserves to seed production, so that regeneration of new plants is primarily from food reserves stored in various types of vegetative structures, such as rhizomes, stolons, stems (sugarcane), corms (timothy), leaf bases or stubble (orchardgrass), and roots (alfalfa) (Table 11.6), or a combination of these. Sprague and Sullivan (1950) concluded that regrowth from orchardgrass (*Dactylis glomerata*) was initially dependent on reserves stored in lower leaf bases and roots. Biennials do not produce seeds during the first season but instead store generous quantities of food reserves in fleshy taproots, in the case of celery, or in the leaf petioles.

TABLE 11.6. Composition of alfalfa roots before and after growth in the dark

Components	Before	After
Dry matter	34.2 g	26.4 g
Starch	10.8%	0%
Dextrins and soluble sugars	3.3%	1.8%
Total sugars	7.9%	1.4%
Hemicelluloses	10.1%	16.5%
Total nitrogen	2.6%	2.3%

Source: Smith 1962.

Carbohydrate reserves (total nonstructural carbohydrates, or TNC) accumulate during periods favorable for net photosynthesis but suboptimal for foliage growth, for example, during the warm, bright days and cool nights of autumn. Certain climatic and edaphic factors, particularly high soil N, negatively influence accumulation. Storage increases with plant age.

Storage of food reserves is an important management strategy for forage, pasture, and turf crops. Species vary greatly. Some have the capability of adapting to close, continuous defoliation, due to a prostrate growth habit and retention of a comparatively large LAI near the ground surface after defoliation (e.g., turf-grass species). Alfalfa has much less or no capacity to do this. Therefore frequency of defoliation should be based on a compromise between forage quality, yield, and maintenance of sufficient leaf area duration for food reserve accumulation adequate for persistence. Frequent harvests increase forage quality, but they decrease yield, accumulation of food reserves, and persistence.

Numerous studies with legumes and grasses have shown a marked decline in carbohydrate reserves following defoliation (Youngner 1972). In Wisconsin Graber (1927) showed that a harvest schedule that allowed for adequate accumulation of reserve carbohydrates was essential for alfalfa yield and persistence (Fig. 11.11). Carlson (1966) observed that the rate of leaflet unfolding was greater in clipped red clover plants than in unclipped and suggested that growth regulators may also be a factor in regrowth.

Legumes like alfalfa would appear to depend more on root reserves for regrowth because lower leaves tend to senesce, leaving little or no healthy leaf area after mowing as hay. On the other hand, birdsfoot trefoil maintains more basal leaf area under close defoliation than does alfalfa, and regrowth depends heavily on the remaining foliage. It stores less carbohydrate in the roots but adapts to a prostrate habit with a high basal leaf area even under continuous

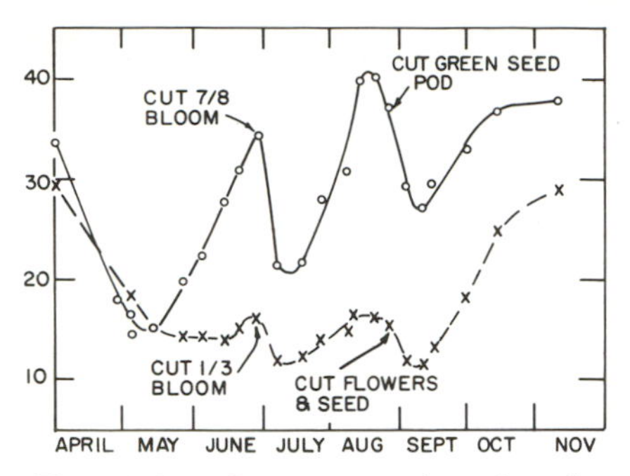

Fig. 11.11. Percentage of total available carbohydrates in the roots of alfalfa (*o-o*) and birdsfoot trefoil (*x-x*) cut twice during the growing season (Smith 1962).

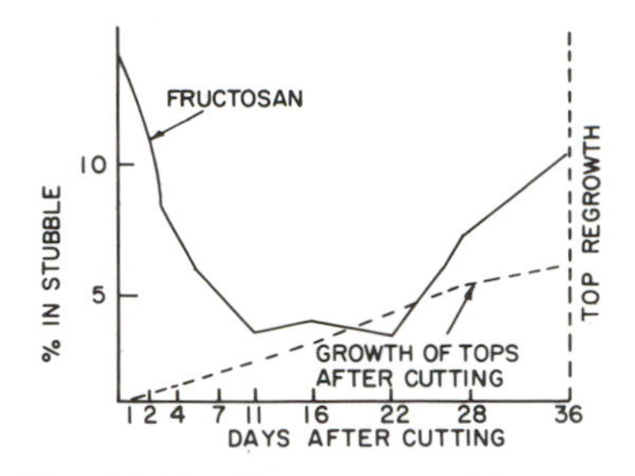

Fig. 11.12. Percentage of fructosan in ryegrass stubble at different intervals following cutting, and top growth at the same intervals (Hughes et al. 1962).

defoliation (Fig. 11.11). As a result, this species is much better suited than alfalfa to close and continuous defoliation in a pasture, and less suited as a hay crop (Smith 1962).

The necessity for carbohydrate reserves in legume regrowth has been widely recognized, but in grasses their role appears to be relatively minor (May 1960). The accumulation and depletion trends after clipping are similar to those of alfalfa (Fig. 11.12). The predominant TNC in temperate grasses (e.g., ryegrass) is fructosan rather than the glucosan in legumes. Davidson and Milthorpe (1966) have suggested that TNC storage in orchardgrass leaf bases is important for only the first 2 or 3 days following defoliation, that is, only to initiate new growth. In studies by Ward and Blaser (1961) regrowth in orchardgrass was dependent both on the stored carbohydrate in the stubble and on the leaf area remaining after clipping. Individual tillers with high carbohydrate reserves produced more dry matter during the first 25 days than the low-reserve tillers. Regardless of amount of reserves, plants with two leaf blades remaining produced more dry matter during a 35-day period than plants with all blades removed. In grasses the LAI after defoliation is highly important.

Following severe defoliation of orchardgrass, root respiration and extension and nutrient uptake dropped to nearly zero (Fig. 11.13). Labile carbohy-

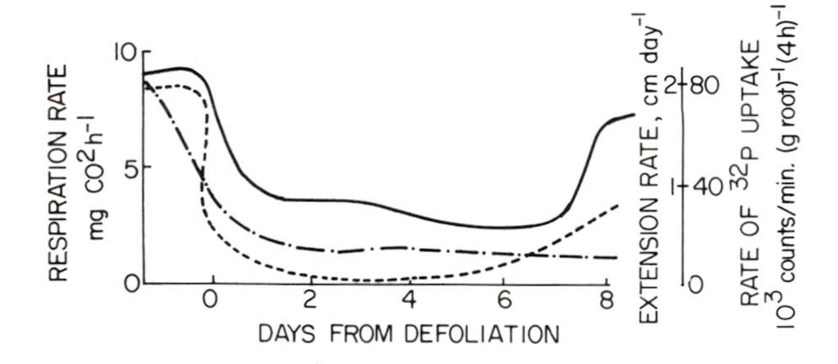

Fig. 11.13. Rates of root respiration (——), root extension (— — —), and ^{32}P uptake as radioactive counts (—•—•—••—) in plants of *Dactylis glomerata* before and after severe defoliation. (From Davidson and Milthorpe 1966)

drates may be necessary to maintain these metabolic activities during early phases of regrowth. Evidently the energy requirement for regrowth can be partially or completely supplied if sufficient lamina remains after clipping to supply newly emerging leaves, which soon become autotrophic.

Summary

Vegetative growth originates in the meristems of apical and lateral buds and in intercalary meristems of young leaves and internodes. Growth from intercalary meristems is usually limited by a fixed number of active cells and/or by the necessity for hormones to be supplied from other sources (buds and young leaves). Dwarf plants result when the intercalary meristems in the internodes are deficient in growth hormones, particularly in GAs, but revert to normal size if exogenous GAs are supplied. Intercalary meristems respond both to GAs and to infrared light. For example, the first internode (mesocotyl) of grasses elongates in darkness = far-red light (730 nm) or if treated with GAs and is inhibited by red light (660 nm). Other intercalary meristems indicate phytochrome control.

Leaves originate as lateral or encircling protuberances of the growing point, or bud apex, at a fixed time interval referred to as the plastochron. The plastochron number is indeterminate until floral initiation, at which time production of leaf initials gives rise to inflorescence initials.

A unit of the growing point, the phytomer, contains stem of node and internode, leaf sheath (or petiole) and lamina, and an axillary bud. The number of phytomers produced (i.e., the number of leaves formed), whether determined by photoperiod, as with soybean, or by temperature (vernalization), as with winter cereal grains, varies with species; it usually consists of 7 to 9 for most small grains, 10 to 16 for soybean and 14 to 21 for U.S. maize hybrids. From 3 to 5 leaves are differentiated in the embryo of a mature grass seed. Maize embryos generally contain 5 leaves, and wheat 4.

Final leaf size varies with vertical location on the plant. High temperatures and low irradiance favor long, thin leaves; lower temperatures and high irradiance favor short, thick leaves and a high specific leaf weight. The bottom leaves on plants generally have relatively less area and are usually thinner and senesce before plant maturity, particularly in crop stands. Leaf sheaths, petioles (to a lesser extent), and other nonlaminar parts contribute substantially to photosynthesis, depending on species.

Growth of stems is primarily from the intercalary meristem at the base of the internodes. Some species (stemless) exhibit no internode growth except for elongation of the last internode as a flowering stalk. Temperate grasses are stemless until floral initiation. Buds in leaf axils may assume active growth to produce vegetative tillers or side branches or to produce reproductive shoots, such as ear shoots of maize and racemes of soybean, depending on age of plant and photoperiod. New shoots or tillers from live leaf sheaths of grasses (intravaginal) grow upright and resemble the main stem. Others may emerge from dead leaf enclosures, usually at the lower axils on the stem, and exhibit diageotropic growth as stolons or rhizomes. Rhizomes turn up in time to form normal upright shoots, depending somewhat on genotype and environment. Tillering habit to a great extent determines perennation of the species.

Light quality has a pronounced effect on internode growth, particularly red (660-nm) and far-red (730-nm) light. The mesocotyl (first internode) on grass seedlings elongates to the limit of organic reserves in darkness (far-red light). In grasses intercalary meristematic growth by elongation in young stems is protected from light in the leaf roll (pseudostem). Accordingly the effect of darkness, or far-red radiation, is mimicked and growth is probably under phytochrome control, as is observed in the mesocotyl. Genotype, light availability (related to stand density), temperature, moisture, fertility, and PGRs highly influence apical dominance, axillary bud growth, and the production of tillers and branches. Clipping, grazing, or inflorescence initiation can alter the hormone balance and associated apical dominance, resulting in tillering from the crown or axillary buds. Regrowth in grasses is generated from intravaginal tillers, turned-up rhizomes, and the continuation of leaf extension from a basal intercalary meristem of emerging leaves. Food reserves in grasses are significant primarily in the initiation of new leaf surfaces, since the plant becomes autotrophic in a few days. Photosynthetic tissues at the base of the plant remaining after clipping, such as in turf species, may lessen or even eliminate the requirement for storage of organic food reserves. To the contrary, plants like alfalfa that have few healthy basal leaves after clipping need large amounts of food reserves to support regrowth and to persist in the stand.

References

Arber, A. 1934. The Gramineae: A Study of Cereals, Bamboo, and Grass. New York: Macmillan.
Aspinall, D. 1961. Aust. J. Biol. Sci. 14:493–505.
Aspinall, D., and L. G. Paleg. 1963. Bot. Gaz. 124:429–37.
Bauer, M. E., J. W. Pendleton, J. E. Beuerlein, and S. R. Ghorashy. 1976. Agron. J. 68:709–11.
Beatty, D. 1982. AGR 240-14. Murray State University.
Beaty, E. R., J. D. Powell, and R. M. Lawrence. 1970. Agron. J. 62:363–65.
Bunting, A. H., and D. S. H. Drennan. 1966. In The Growth of Cereals and Grasses, ed. J. D. Ivins and F. L. Milthorpe. London: Butterworth.
Carlson, G. E. 1966. Crop Sci. 6:419–22.
Christ, R. A. 1978. J. Exp. Bot. 29:603–10.
Cowett, E. R., and M. A. Sprague. 1962. Agron. J. 54:294–97.
Crookston, R. K., D. R. Hicks, and G. R. Miller. 1976. Crops Soils 28:7–11.
Cross, H. Z., and M. S. Zuber. 1973. Agron. J. 65:71–74.
Davidson, J. L., and F. L. Milthorpe. 1966. Ann. Bot. n.s. 30:185–98.
Duncan, W. D. 1975. In Crop Physiology, ed. L. T. Evans. London: Cambridge University Press.
Etter, A. G. 1951. Mo. Bot. Gard. Annu. 38:293–375.
Frey, K. J., and S. C. Wiggans. 1957a. Agron. J. 49:48–50.

______. 1957b. Proc. Iowa Acad. Sci. 64:160–67.
Friend, D. J. C. 1966. In The Growth of Cereals and Grasses, ed. J. D. Ivins and F. L. Milthorpe. London: Butterworth.
Fuehring, H. D. 1969. Agron. J. 61:591–94.
Gardner, F. P., and M. J. Kasperbauer. 1961. Iowa State J. Sci. 35:311–18.
Gardner, F. P., and W. E. Loomis. 1953. Plant Physiol. 28:201–17.
Goodin, J. R. 1972. In The Biology and Utilization of Grasses, ed. V. B. Youngner and C. M. McKell. New York: Academic Press.
Graber, L. F. 1927. Univ. Wis. Res. Bull. 80.
Hanway, J. J., and C. R. Weber. 1971. Agron. J. 63:227–30.
Humphries, E. C., and A. W. Wheeler. 1963. Annu. Rev. Plant Physiol. 14:385–410.
Jacobs, W. P. 1947. Am. J. Bot. 34:361–70.
Jewiss, D. R. 1966. In The Growth of Cereals and Grasses, ed. J. D. Ivins and F. L. Milthorpe. London: Butterworth.
Johnson, H. W., H. A. Borthwick, and R. C. Leffel. 1960. Bot. Gaz. 122:77–95.
Kasperbauer, M. J., F. P. Gardner, and W. E. Loomis. 1962. Plant Physiol. 37:165–70.
Koller, D., and J. Kigel. 1972. In The Biology and Utilization of Grasses, ed. V. B. Youngner and C. M. McKell. New York: Academic Press.
Kretchmer, P. M., J. L. Ozbun, S. L. Kaplan, D. R. Laing, and D. H. Wallace. 1977. Crop Sci. 17:797–99.
Langer, R. H. M. 1954. Br. J. Grassl. Soc. 9:275.
______. 1956. Ann. Appl. Biol. 44:167–87.
______. 1972. How Grasses Grow. London: Edward Arnold.
Laude, H. M. 1975. Crop Sci.15:621–24.
Leopold, A. C. 1949. Am. J. Bot. 36:437–40.
May, L. H. 1960. Herb. Abstr. 30:239–45.
Milthorpe, F. L., and J. Moorby. 1974. An Introduction to Crop Physiology. London: Cambridge University Press.
Mitchell, K. J. 1953. Physiol. Plant. 6:425–43.
Mitchell, K. J., and S. T. G. Coles. 1955. Herb. Abstr. 25:235.
Murata, Y., and S. Matsushima. 1975. In Crop Physiology, ed. L. T. Evans. London: Cambridge Unversity Press.
Nelson, C. J., K. J. Treharne, and J. P. Cooper. 1978. Crop Sci. 18:217–20.
Nittler, L. W., and T. J. Kenny. 1976. Agron. J. 68:395–97.
Peterson, M. L., and W. E. Loomis. 1949. Plant Physiol. 24:31–43.
Ralph, W. 1982. CSIRO Q. Rep., pp. 4–9.
Robinson, M. J. 1968. J. Appl. Ecol. 5:575–90.
Sachs, R. M. 1965. Annu. Rev. Plant Physiol. 16:73–96.
Sass, J. E., and F. A. Loeffel. 1959. Agron J. 51:984–86.
Sharman, B. C. 1942. Ann. Bot. n.s. 6:245–82.
______. 1945. Bot. Gaz. 106:269–89.
Shibles, R., I. C. Anderson, and A. H. Gibson. 1975. In Crop Physiology, ed. L. T. Evans. London: Cambridge University Press.
Singh, S. S., and W. L. Colville. 1962. Agron. J. 54:484–86.
Smith, D. 1962. Crop Sci. 2:75–78.
Sprague, V. G., and J. T. Sullivan. 1950. Plant Physiol. 25:92–102.
Thorne, G. H. 1959. Ann. Bot. n.s. 23:365–70.
Vanderhoef, L. H., P. H. Quail, and W. R. Briggs. 1979. Plant Physiol. 63:1062–67.
Ward, C. Y., and R. E. Blaser. 1961. Crop Sci. 1:366–70.
Westmore, R. J., and T. A. Steeves. 1971. In Plant Physiology: A Treatise, vol. 1A, ed. F. C. Steward. New York: Academic Press.
Windscheffel, J. A., R. L. Vanderlip, and A. J. Cassady. 1973. Crop Sci. 13:215–18.
Youngner, V. B. 1972. In The Biology and Utilization of Grasses, ed. V. B. Youngner and C. M. McKell. New York: Academic Press.
Zieserl, J. F., W. L. Rivenbark, and R. H. Hageman. 1963. Crop Sci. 3:27–32.

12 Flowering and Fruiting

THE PRODUCTION of seeds is often the primary objective in crop production. Seed production results from numerous physiological and morphological events that lead to flowering and fruiting in response to *photoperiod* (length of day) and temperature. Flowering and fruiting responses to these environmental factors have been the subject of intensive investigation for more than 50 years.

Studies in the 1920s (Garner and Allard 1920, 1923, 1925) led to the identification of photoperiod as a controlling environmental factor in flowering. Later studies demonstrated that the *nyctoperiod* (length of night), rather than length of day, was actually the operative factor in control of plant responses. If the dark period is interrupted by short exposures of light at low energy, the result is a long-day effect. Interruption of the light period by dark exposures, however, has no effect on flowering. Later studies by other USDA scientists identified *phytochrome* as the photoreceptor (pigment) in the control of developmental processes such as flowering, and demonstrated how phytochrome reacted in relation to light quality in the red part of the spectrum. The role of mature leaves as the production site of a flowering stimulus (hormone) and its transport to and activation of the meristems have been matters of keen interest in research since the pioneer work.

Latitude and time of year (sun declination) are the determinants of photoperiod and temperature, both of which vary widely from season to season and from the equator to the poles. Although nearly constant year-round at the equator, photoperiod can vary by 24 hr per day between June and December (summer and winter solstices) at the poles (Fig. 12.1).

The dominant role of photoperiod and temperature on flowering and fruiting and ultimately on seed production emphasizes the importance of cultivar selection. Photoperiodically sensitive cultivars of soybean are adapted to a narrow latitudinal range, often no more than 200 to 250 km. Water, nutrients, and other factors at most can only modify the response to photoperiod or temperature. On the other hand, certain crops, such as tomato, are insensitive to photoperiod and can be produced in any latitude within broad temperature limits.

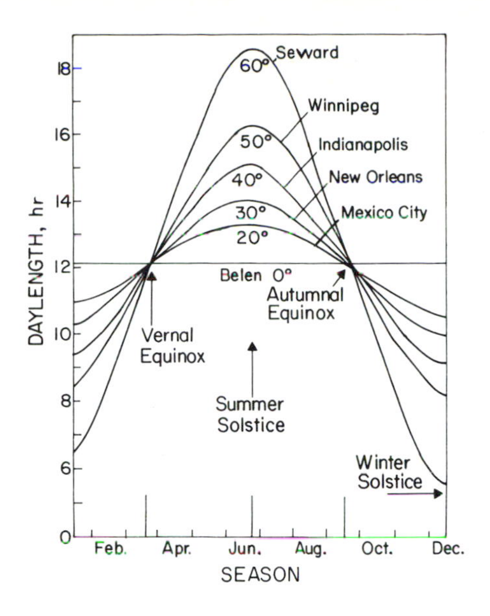

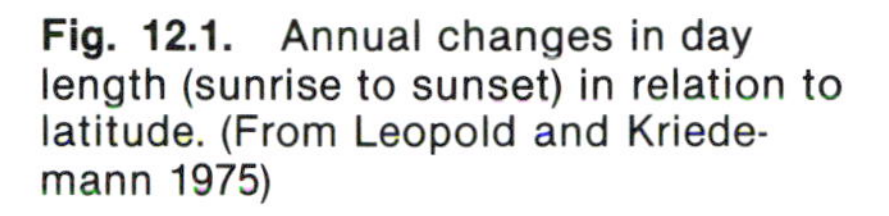

Fig. 12.1. Annual changes in day length (sunrise to sunset) in relation to latitude. (From Leopold and Kriedemann 1975)

Transition to Flowering

Shoot meristems produce either leaf or inflorescence primordia, depending on photoperiod and possible interactions with temperature. Indeterminate growth initially produces leaves. In a few species the same bud first produces leaves, then floral structures, and then leaves again. In *monocarpic* plants (annuals), the transformation of a vegetative (leaf producing) bud to flowering ends further leaf production. The initiation of flowering in these plants may be thought of as a final commitment of energy resources. After flowering and fruiting the plant dies. Such growth is determinate. On the other hand, only partial commitment is made to sexual reproduction in perennials; vegetative growth may continue indefinitely, separately or concurrently with flowering. Axillary buds that have adequate food reserves renew vegetative growth if leaf formation is terminated by flowering of older shoots, such as in alfalfa, or the old shoots may continue to grow, as in trees. Biennials characteristically produce a stemless rosette growth in the first year and stems, flowers, and fruits in the second year, a monocarpic habit much like an annual.

Photoperiodism

The effect of day length on flowering was first alluded to by Tournis in France and by Klebs in Germany just after the turn of the century (Evans 1969). Although these researchers came close to the recognition of photoperiodism, the actual discovery is credited to two USDA scientists, W. W. Garner and H. A. Allard (1920, 1923, 1925), working near Washington, D.C. They coined the term *photoperiodism* to define plant response to day length. Their observations were on two short-day crop cultivars, 'Maryland Mammoth' tobacco and 'Biloxi' soybean (maturity group VIII). The tobacco plants did not produce flowers during the growing season in the field at the Washington, D.C., latitude but flowered later in the fall and winter after the plants were moved into the greenhouse. Axillary shoots from stumps of the greenhouse plants flowered in the short days of winter but remained vegetative if emergence occurred in spring, when days were long again.

Garner and Allard observed field-grown 'Biloxi' soybean plants. Field plantings made from early spring to midsummer matured at about the same time in autumn. They concluded that both 'Maryland Mammoth' tobacco and 'Biloxi' soybean flowered in response to day lengths below a certain critical length; that is, they were short-day plants. Their observations on a range of species clearly illustrated that a combination of short days and long nights in a 24-hr cycle promoted flowering in many species, whereas the opposite favored flowering in others. Some species were indifferent or insensitive to day length. However, even photoperiodically sensitive plants did not require a specific day length for flowering but flowered optimally over a wide range of day lengths, generally becoming less sensitive with increasing plant age (Fig. 12.2). Day lengths longer than optimum delayed flowering of short-day plants until a critical length was reached, above which the plant remained vegetative. Similarly, day length below a critical length caused long-day plants to remain vegetative. Both types of crop species became sensitive to photoinductive conditions after a required basic vegetative phase (BVP) (Vergara and Chang 1976; Major 1980).

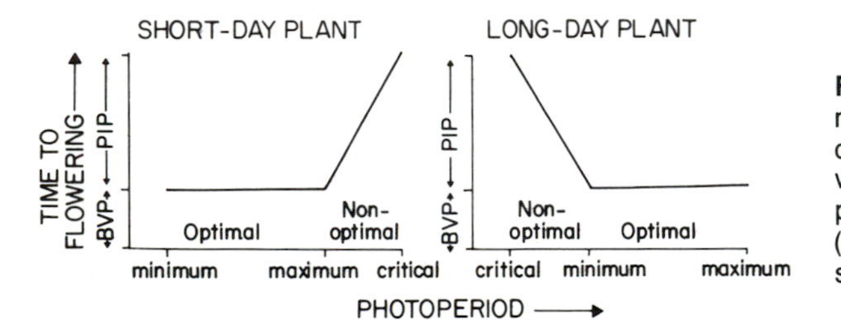

Fig. 12.2. Generalized model of plant response to day length. BVP is basic vegetative phase; PIP is photoperiod-induced phase (Major 1980, by permission).

In spite of the complexity of responses to a range of day lengths and the various interactions with other environmental factors, the following classification scheme outlined by Hillman (1962) is useful in understanding the day-length responses of crop and wild species.

1. Short-day plants (SDPs). Flowering is promoted by day lengths shorter than a critical maximum (which varies among species and varieties) and usually is influenced by other environmental factors, such as temperature (Kasperbauer 1973). 'Maryland Mammoth' tobacco, 'Biloxi' soybean, and cocklebur are classic SDPs. Since they are highly sensitive to photoperiod, they have been used in numerous photoperiod studies.

2. Long-day plants (LDPs). Flowering is promoted by day lengths longer than a critical minimum (which is influenced by genotype and other environmental factors). 'Wintex' barley and black henbane (*Hyoscyamus niger*), a biennial, are classic LDPs and have been used extensively in photoperiod research.

3. Short-long-day plants (SLDPs). Flowering is promoted by exposure to a sequence of short days prior to an exposure to long days. Many temperate-zone, perennial grasses (e.g., orchardgrass) fit this category, although their responses are more complex than even this indicates because a cold period (vernalization) is also required between exposures to short and long days (Gardner and Loomis 1953).

4. Long-short-day plants (LSDPs). Flowering is promoted by exposure to a sequence of long days prior to exposure to short days. Night jasmine (*Cestrum nocturnum*) is regarded as a LSDP.

5. Day-neutral plants (DNPs). Flowering is insensitive to photoperiod but is associated with an age factor. Generally flowers commence after a minimum age or size is attained. Dandelion, tomato, and buckwheat are DNPs. These species are adapted to any latitude within broad temperature limits. 'Big Boy' tomato cultivar, for example, can be produced from Mississippi to Canada. Many plants of tropical origin are DNPs, but others are SDPs (e.g., tropical soybean cultivars).

As mentioned before, classification of plant response is made on the basis of the relationship of flowering to the length of day (light period in a 24-hr cycle), but the actual controlling factor is the length of uninterrupted darkness, or *nyctoperiod,* rather than the photoperiod. Some plants do not fit any of these categories. For example, sunflower is a LDP in juvenile stages but becomes insensitive with plant age.

Generally it can be assumed that crop and wild species that flower and fruit in midsummer are LDPs, and those that flower and fruit in autumn SDPs. Winter annuals (e.g., wheat), biennials (e.g., sugar beet), and numerous perennials (e.g., orchardgrass), are obligatory LDPs. These plants, however, flower only after vernalization, or a cold period. Maize, sorghum, and soybean are SDPs. There is considerable overlap between the ranges for short- and long-day plants.

Categorizing the flowering response of soybean cultivars adapted to northern latitudes, which have long summer days, presents an interesting problem. These cultivars (maturity groups 0 and 00) can flower under 16- to 18-hr days, whereas the critical day length for 'Biloxi' cultivar is about 12 hr. However, a classification other than SDP for such cultivars would not be correct, since flowering occurs even earlier or in fewer days on plants with fewer nodes as photoperiods shorten (e.g., from 16 to 8 hr).

Thermoperiodism (Vernalization)

Although temperature generally altered or modified photoperiodic response of species and varieties (Thomas and Raper 1982), many species were found to require a 2- to 6-wk period of cold or near freezing temperature (10°C or less) in order to flower under long photoperiods of spring (Lang 1951; Schwabe 1957). These cold treatments, called *vernalization,* are usually effective between 2 and 10°C (Fig. 12.3). Response to the cold exposure is qualitative (absolute); either flowering occurs or it does not. The cold period must be several days to several weeks in duration, depending on species. Winter annuals, biennials, and many temperate perennial species require such vernalization for flowering. Seeds, bulbs, and buds of many temperate species require *stratification* (several weeks of cold, wet storage) to break dormancy and induce growth (see Chap. 9).

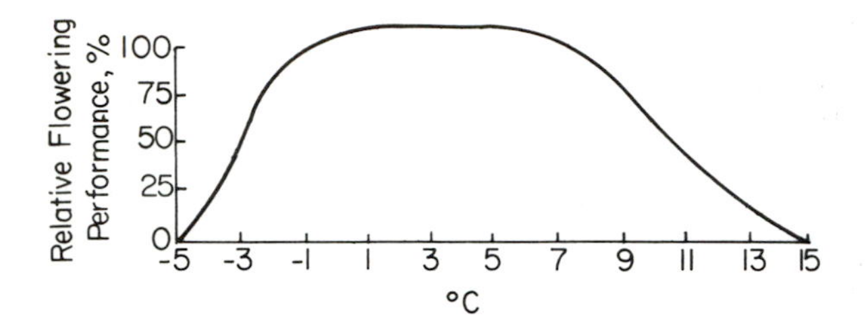

Fig. 12.3. Generalized model of plant response to temperature during vernalization.

Vernalization literally means to make springlike, that is, to promote flowering in response to the long days during spring. Winter cereal grains, such as wheat and rye, respond as spring types after vernalization; both flower under long photoperiods after producing a minimum of seven leaves (Purvis and Gregory 1937). Floral development in the shoot apex of rye in relation to vernalization was as follows (Fig. 12.4): (1) shoot apex primordia 1 to 7 were obligate leaf plastochrons; (2) primordia 8 to 21 were facultative, that is, capable of producing either leaves or floral structures, depending on intensity of vernalization; (3) primordia 21 and above were obligate flowering.

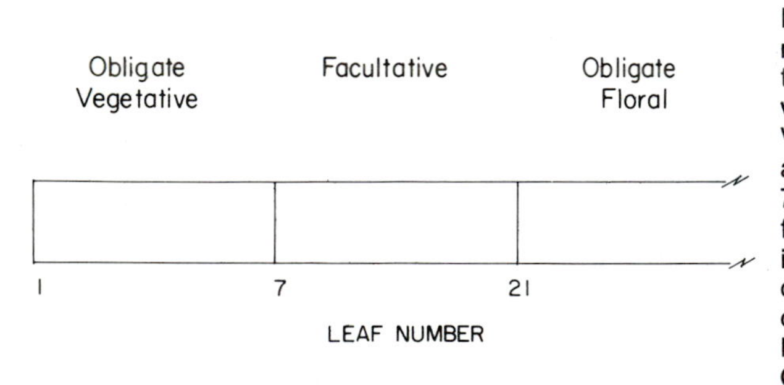

Fig. 12.4. Floral development of winter rye in relation to the amount of vernalization stimulus. Vernalized and spring types are obligate vegetative for 7 leaves, responsive to flowering after 7 leaves if induced (vernalized) in the case of winter types, and obligate floral after 21 leaves. (From Purvis and Gregory 1937)

Modern cultivars of such biennial crops as sugar beet and celery have been selected for a high vernalization requirement because stems bearing inflorescences the first year (*bolting*) are undesirable in a commercial crop. Annual types (which flower without vernalization) have been selected from biennial sugar beet, black henbane, and sweet clover.

LOCUS OF VERNALIZATION

Evidence that the cold stimulus is produced in the meristems or buds rather than in the leaves is derived from four phenomena: (1) imbibed seeds are readily vernalized; (2) cold exposure of only leaves, roots, or stems was not effective (Salisbury 1963); (3) developing seeds on a mother plant can be, and sometimes are, vernalized if cold persists before the seeds become dry; and (4) plants regenerated from adventitious buds from a vernalized leaf were induced to flower (Wellensiek 1962).

LOSS OF VERNALIZATION

Vernalization in seeds can be nullified by exposure to adverse conditions, such as desiccation or high temperatures (30–35°C) for a period of days (Purvis and Gregory 1937; Lang and Melchers 1947). It is difficult to reconcile these findings with the agricultural practice advocated by Lysenko in the USSR of vernalizing winter cereal grains and holding them for spring seeding. It would seem that holding the seeds in a dry state would devernalize them. At any rate, the Lysenko practice has not persisted anywhere, probably because adapted spring-type cultivars became available.

Vernalization in certain perennial grasses is more complex; in addition to cold, some short photoperiods are necessary (Peterson and Loomis 1949; Cooper 1950). In orchardgrass, floral induction occurred naturally by November 15 at Ames, Iowa (42°N) (Gardner and Loomis 1953). A requirement for short days in conjunction with vernalization has not been observed in biennials and winter annuals; that is, only cold is required for floral induction in these species.

Flowering

FLORAL INDUCTION

Gardner and Loomis (1953) recognized three discrete phases in orchardgrass flowering, each with distinctive photoperiodic and thermal requirements (Fig. 12.5):

1. Floral induction: production of the flowering stimulus (a chemical change in the shoot apex) in response to the cold (nongrowth) temperatures and short days of autumn
2. Floral initiation: transformation of induced but morphologically vege-

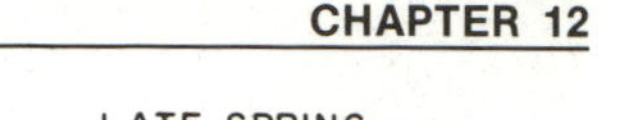

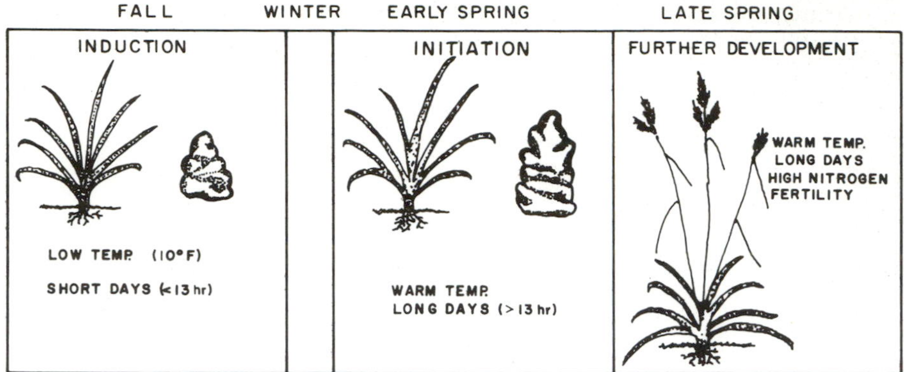

Fig. 12.5. Flowering of orchardgrass in relation to seasonal temperature and photoperiod (Gardner and Loomis 1953).

tative growing points to floral primordia in response to the long days and moderately warm temperatures of spring

3. Further floral development: growth and development of floral initials into mature flowers and inflorescences in response to the long days and moderately warm temperatures of spring (also favored by high nitrogen fertility)

In orchardgrass the flowering stimulus produced in the shoots exposed to induction conditions was not transferred to tillers that did not receive both cold and short days, although an organic connection is presumed to exist (Fig. 12.6). The fact that only the tillers exposed to light flowered, while the many hidden in leaf sheaths did not, supports the concept that leaves are the photoperiodic receptors.

Such a three-phase response usually is not delineated for flowering in most other studies, although it is probably universal among species. Generally emphasis has been on *floral induction* (production of the flowering stimulus) and *floral expression* (development). Much of the classic work on microdissected buds of soybean and cocklebur described early morphological changes, defined as floral initiation. The requirements for floral induction and floral initiation are the same in a SDP, such as soybean; consequently these two phases are not separable as they are in orchardgrass. The requirements for floral initiation and expression appear to be different in soybean; inflorescences were initiated (after induction) under long days, but flowers aborted if the plants were maintained under long days after initiation (Fisher 1962).

With the aid of low magnification of a dissecting microscope, a system of staging floral development has been described for cocklebur (Salisbury 1955), soybean (Borthwick and Parker 1938), and lambsquarters (Kasperbauer et al. 1964). Photoinduced plants of these SDPs readily produce floral initials under

Fig. 12.6. Localized effect of photoperiod on orchardgrass plants (center plant was divided). Shoots left of the divider received cold, 9-hr days (an environment favoring floral induction); shoots on the right received cold, 18-hr days during the induction period. Flowers were expressed after the transfer to long, warm days. Each pot represents a single plant with several tillers (Gardner and Loomis 1953).

long days, the degree of induction reflected in degree and/or rate of floral development. As previously indicated, soybean aborted flowers and fruits under long days (20 hr) despite photoinduction under an earlier experience of short photoperiods (Fisher 1962). Consequently microdissection and staging of the development of floral initials are necessary for study of floral induction per se in a SDP like soybean.

Maintenance under short days would cause initiation but would also increase the floral induction stimulus.

Minimum Age. Most species are nonresponsive to photoperiods during juvenility. A minimum age, size, or stage of development is required to become responsive. The juvenile period has been termed the basic vegetative phase (BVP) (Vergara and Chang 1976). After the BVP is satisfied, the plants then enter the photoperiod-induced phase (PIP) (Fig. 12.2), which is referred to in older literature as the "ripeness to flower" stage. The minimum BVP is usually described in terms of leaf number rather than chronological age; both parameters vary widely in different species and varieties (Table 12.1). The BVP of most tree species is 5 yr or longer. The century plant is not responsive to photoinductive environments until the age of 10 to 12 yr, whereas a lambsquarters ecotype (*Chenopodium rubrum*) can be photoinduced during opening of the cotyledons at germination (Table 12.1). Ripeness to flower seems to be the only

TABLE 12.1. Length of vegetative period (minimum age) for ripeness to flower

Species	Minimum Age	Source
Cocklebur	8 leaves Half to fully expanded; size most effective	Holdsworth 1956 Khudairi and Hamner 1954
Soybean	Up to 6 wk of age	Borthwick and Parker 1938
Tobacco	5–6 leaves	Kasperbauer 1970
Perilla	15 days	Moshkov 1939
Lambsquarters (ecotype)	3.5 days	Cumming 1959; Kasperbauer et al. 1964
Bamboo	5–50 yr	Arber 1934
Wheat, rye	Moist seed in cold	Salisbury 1963
Pine	5 yr	Stanley 1958
Pigweed	30 days, 2 short-day cycles	Zobka 1961
Century plant	5–20 yr; senescence and annual habit after flowering	Hillman 1962
Sweet clover	Only older, larger plants flowered in 100 days under 16-hr days	Kasperbauer et al. 1962
Rice	10–87 days	Vergara and Chang 1976

requirement for flowering of day-neutral plants. They are indifferent to day length and flower only after the BVP, or minimum age, is satisfied. Minimum leaf number is not always constant for a given species or even cultivar. Mineral starvation of plants resulted in a reduction of minimum leaf number by one to two leaves (Holdsworth 1956).

Leaf number rather than leaf area is the operative principle. Photoperiodic studies on highly sensitive species showed that a single leaf of sufficient age on a defoliated plant was adequate for photoperiod reception. With cocklebur, a leaf area of only 2 to 3 cm^2 was effective on a defoliated plant that was in the PIP, but 9.2 cm^2 of cotyledon surface was completely ineffective if the BVP was not satisfied.

Photoinduction Cycles. Floral induction (production of the flowering stimulus) occurs in response to exposure to a specific number of favorable photoinduction cycles. The minimum number of cycles required varies with plant species, cultivars, age, and size. After the minimum number of photoinduction cycles is satisfied, flowering intensity increases (fewer days until flowering) with additional exposure up to a saturation level; that is, the response is quantitative, not absolute. A single dark period of 8.5 hr or longer induced flowering in cocklebur even though the plants were grown thereafter under long days (Hamner 1938; Salisbury 1963). A single dark period of 8.33 hr was not adequate for floral induction, as these plants remained vegetative. Low temperature (5°C) increased the minimum dark period length by 2 to 3 hr. A longer dark period increased flowering intensity of SDPs up to a saturation level, which occurred at 12 to 15 hr (Mann 1940). Other research showed a plateauing response over a range of optimum photoperiods for LDPs and SDPs (Major 1980). A single photoinduction cycle was not adequate for 'Biloxi' soybean; seven cycles were optimum and more had no advantage (Borthwick and Parker 1938). Less sensitive plants required more photoinduction cycles to flower than did sensitive plants. A period of high light intensity was essential prior to and following the dark period (Mann 1940; Lang 1952). At least 4 hr of light were necessary prior to the long dark period at 10°C, but as little as 30 min was adequate at 30°C (Long 1939).

Night Breaks. Length of night rather than day length is the operative factor in photoperiodism, demonstrated by the fact that a very brief interruption of the dark period by white or red light destroyed the long-night effect (Lang 1952). An interruption near the middle of the dark period of an exposure to low-intensity light for 2 min (Salisbury 1963), or even as little as 12 sec, produced the effect of a long day on the SDPs cocklebur and soybean (Table 12.2); that is, the plant remained vegetative. LDPs, on the other hand, were promoted to flower by night breaks (Borthwick and Parker 1938), as might be expected. Interruption was more effective if it occurred after the first 8 hr of darkness in a 10-, 12-, 16-, or 20-hr night. A break before 3 to 4 hr or after 16 to 20 hr of darkness was less effective.

Since in nature the dark period is not interrupted by light, the ecological significance of these findings is open to question. Moonlight, about 0.02 FC

TABLE 12.2. Effect of daily interruptions of the dark period with several consecutive irradiations with red (R) or far-red (FR) light in sequence on flower initiation of cocklebur (*Xanthium*) and soybean

Treatment[a]	Mean Stage of Floral Development in Cocklebur[a]	Mean No. Flowering Nodes in 'Biloxi' Soybean
Dark control	6.0	4.0
R	0.0	0.0
R, FR	5.6	1.6
R, FR, R	0.0	0.0
R, FR, R, FR	4.2	1.0
R, FR, R, FR, R	0.0	. . .
R, FR, R, FR, R, FR	2.4	0.6

Source: Wareing and Phillips 1978, by permission, from Salisbury 1955.
[a]Stages from 1 to 10, maximum development.

($14\ W \cdot m^2$), is too low in energy and in red wavelengths (Salisbury 1969). But the fact that a brief night break produces the equivalent of a long day, similar to supplementing the natural day with more hours of light, is practical for energy conservation in lighting commercial greenhouses.

Light Quality. Photoperiodism is driven by light energy of the red (R) and far-red (FR) portion of the spectrum. The most effective light quality for a night break and the resultant photoperiodic response is R (600–680 nm), but the R effect is reversed if the break by R light is followed immediately by another light exposure consisting of FR light in the 720- to 750-nm range (Parker et al. 1946; Downs 1956) (Table 12.2). If the FR treatment did not immediately follow the R break, the long-night effect, or reversion, did not occur (Downs 1956). Researchers at Beltsville, Md., first showed that sensitivity to light interruption of the dark period was highly dependent on light quality and to a lesser extent on energy level (Fig. 12.7). The effect of the action spectrum for flowering can be summarized as follows (P_r = long-day effect, P_{fr} = short-day effect):

Red: 600–680 nm, high phytochrome absorbancy; effective at low energy levels.

Blue: 380–500 nm, peaking at 445 nm, low phytochrome absorbancy; effective at low or high ($1.7\ mW \cdot cm^{-2}$) energy levels; phytochrome reversion, $P_r \rightarrow P_{fr}$; an equilibrium P_{fr} level of 35% is reached in 8 min.

Far-red: 720–750 nm, peaking at 735 nm, high phytochrome absorbancy; effective in production of P_r (the dark effect); the effect of FR exposure is similar to R if time of exposure is long (excessive energy).

The effects of light quality on phytochrome formation (both P_r and P_{fr} forms) can be summarized by the following schematic diagram:

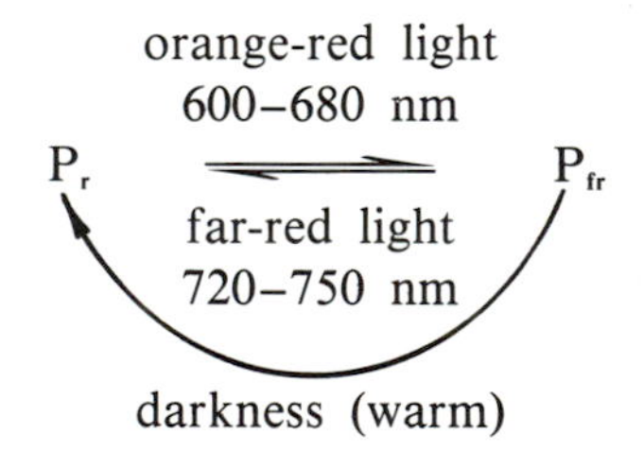

where P_r and P_{fr} are R and FR phytochrome forms, respectively. The P_{fr} form inhibits flowering in SDPs, promotes flowering in LDPs, and promotes germination of light-requiring seeds and certain other developmental processes. The P_r form is the more stable and biologically active of the two forms.

During darkness P_{fr} reverts to P_r at a temperature-dependent rate, which mediates flowering of SDPs and inhibits flowering of LDPs. The response of

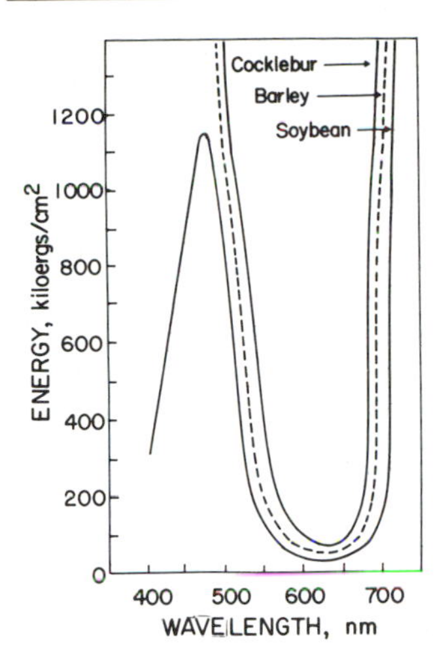

Fig. 12.7. Action spectrum for suppression of floral initiation in cocklebur and soybean (maximum effectiveness 660 nm) and promotion of flowering in 'Wintex' barley (660 nm). The curves give energy requirements at the middle of a long, dark period to prevent or to promote floral initiation in soybean and barley, respectively. At 650 nm very low energy doses as a night break promote flowering of 'Wintex' barley and inhibit flowering of cocklebur and soybean. (From Parker et al. 1946; Borthwick et al. 1956)

'Wintex' barley (LDP) to R at low energy levels was opposite that of cocklebur and soybean (SDPs). A night break with R light promoted flowering of barley but inhibited flowering of cocklebur and soybean (Borthwick et al. 1956) (Fig. 12.7). A brief exposure to FR light immediately following the R exposure reversed the effect of R for both SDPs and LDPs, which suggested that the same pigment controls flowering in both types.

Natural day lengths can be effectively extended with white light (mixture of wavelengths) from incandescent and fluorescent lamps, since both deliver R radiation at a sufficient energy level for the phytochrome response. Recently developed sodium and metal halide lamps are more energy efficient for supplemental lighting. Their effectiveness for greenhouse light is being evaluated.

Factors Modifying Photoinduction. The effect of photoperiod on floral induction is modified by temperature more than by other environmental factors (Thomas and Raper 1982). Low temperature (10°C) increased the critical length of the dark period by 2 to 3 hr (Long 1939), which might be expected because the rate of dark reversion of P_{fr} to P_r is temperature dependent.

The critical length of the dark period was influenced by leaf age in cocklebur (Naylor 1941) and soybean (Fisher and Loomis 1954). Mature leaves promoted and immature leaves inhibited flowering in soybean. Flowering occurred when the ratio of the two became favorable. Inhibition was attributed to the higher auxin level in immature leaves. Leopold (1958) reported that CO_2 enrichment of the atmosphere reduced the critical length of the dark period.

Flowering Stimulus. Since the discovery of photoperiodism, the existence of a chemical messenger or stimulus that signals transformation from vegetative growth to flowering has been postulated. Garner and Allard (1925) demonstrated that leaves are the receptors of the photoperiod stimulus. Probably the strongest evidence supporting a flowering hormone theory has been the classic work of Chailakhyan (1936) with chrysanthemum (SDP). The terminal buds of chrysanthemum plants with the upper leaves removed initiated flowers under long days if the lower leaves were exposed to short days. It was suggested that the stimulus produced in the lower leaves under short days was translocated to the terminal buds held under long days. He named the hormone, or flowering substance, *florigen* and suggested that it moved in the phloem or bark. Neither florigen nor the substance hypothesized to result from vernalization, which has since been given the name *vernalin* (Lang 1952), has been isolated or characterized chemically to date, so their existence can only be conjectured. Perhaps both substances are the same chemical.

Translocation of a flowering substance has been verified in experiments that chilled (at 3°C) the petioles of photoinduced leaves of soybean, which prevented translocation of the flowering stimulus to buds (Borthwick et al. 1941). Uninduced plants flowered when photoinduced leaves were grafted onto them, which has given much credence to the theory of a flowering hormone or other substance that moves from leaves to the flowering loci (Lang and Melchers 1947). Despite the fact that tobacco is a SDP, vegetative 'Maryland Mammoth' tobacco was caused to initiate flowers after receiving a grafted scion from black henbane (LDP) induced to flower by long days prior to the grafting.

Chemical Antagonism and Promotion. Salisbury (1955) showed that indoleacetic acid (IAA) is antagonistic to floral induction in SDPs (Fig. 12.8). IAA interfered with the dark period (Salisbury and Bonner 1956). High concentrations can inhibit flowering in LDPs, although auxins have been reported to promote flowering in certain species (Liverman and Lang 1956; Leopold and Thimann 1949). Spray applications of auxin (e.g., 2,4-D) to promote flowering in pineapple is a common practice (Clark and Kerns 1942). The

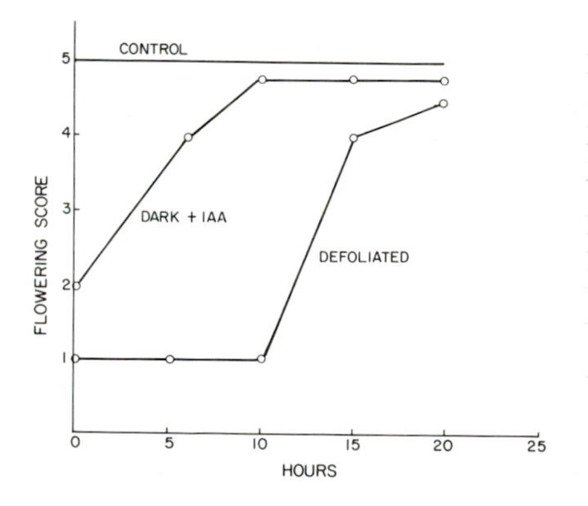

Fig. 12.8. Induction of flowering in cocklebur (flowering score) resulting from IAA and leaf defoliation treatments after the photoinduction cycle. IAA was applied during a second dark period. (From Lockhart and Hamner 1982. Reprinted by permission. © University of Chicago Press. All rights reserved.)

antagonistic effect of auxin apparently depends on the time of application in relation to completion of photoinduction and to effectiveness of the stimulus translocation. Auxin inhibited floral initiation if applied prior to the production and translocation of a sufficient amount of stimulus. Auxin promoted flowering if applied after enough time for translocation of sufficient floral stimulus (Salisbury 1955).

Auxin in young leaves of soybean presumably inhibited flowering until enough mature leaves were produced, that is, until an adequate mature-to-immature leaf ratio was achieved (Fisher and Loomis 1954). Buds and young leaves are rich in auxin.

Unlike auxin, gibberellins (GAs) generally promote flowering. Gibberellic acid can often substitute for all or part of the required number of photoinduction cycles and cold period. GAs induce flowering in certain tropical species that are believed to be insensitive to photoperiod. Treatment with GAs hastened the expression of macroscopic flowers in cocklebur (Greulach and Haesloop 1958). From these data it was concluded that GAs do not replace short days but can substitute for additional cycles, yielding a quantitative effect. GAs completely replaced the cold requirement for floral induction in black henbane but could not substitute for the long-day requirement, however, in initiation after induction (Wittwer and Bukovac 1958). GAs replace the long-day requirement in certain winter annuals and are widely known to change sex expression in certain *monoecious* plants (those with male and female flowering structures in separate flowers on the same plant).

FLORAL INITIATION

After floral induction, the morphological transition of meristems from a vegetative to a floral state is termed floral initiation. Floral initiation has received less emphasis in flowering studies than has floral induction, due in part to the fact that the two stages often have similar environmental requirements, making the distinction between them difficult or impossible. However, in temperate grasses the induction and initiation phases are usually discrete and have distinctly different photoperiodic and thermal requirements. They are naturally separated by the winter season (Fig. 12.5). As with biennials and winter annuals, initiation occurs in the long days of spring following autumn vernalization; the long days are required for initiation. After induction cocklebur (SDP) readily initiates flowers under long days (Salisbury 1963). Soybean (SDP) can initiate inflorescences in long days after two to eight photoinduction cycles, but soybean fails to express flowers if grown thereafter under long days (Hamner 1969). In general the photoperiodic sensitivity of further floral development is considerably less than that of floral induction and initiation, the chemical and morphological transition stages, respectively.

FURTHER FLORAL DEVELOPMENT

Further floral development (expression of visible flowers) may not necessarily occur in conditions favoring induction and initiation (Hamner 1969). For this reason flowering response to photoperiodism has often been reported on initiation of inflorescences observed under low magnification. Orchardgrass plants, with floral initials already present, produced only a few abnormal inflorescences under 9-hr days but flowered profusely under normal spring day lengths or 20-hr days in a greenhouse (Gardner and Loomis 1953). Certain cultivars of soybeans aborted flowers and failed to set fruit under long (20-hr) days (Fisher 1962), due to a failure in anthesis.

In orchardgrass, inflorescences in the 9-hr plants generally aborted before emergence from the leaf sheaths and had abnormally long floral bracts. Initiation was more sensitive to short days than was further floral development, so the long-day effect was qualitative (absolute). In grasses, nitrogen fertilization favored panicle production during spring; competition between inflorescences for nitrogen may result in lack of expression of panicles.

Fruiting

A *fruit* is defined as a mature ovary. *Seeds* are mature fertilized ovules. In fleshy fruits, such as those for table consumption, seeds are a nuisance or of little practical significance. In most field crops seeds are the desired end product, and the fruit is usually of no practical significance. The fruits of grasses (caryopses) are single seeded and dry. In agricultural terms, grass fruits are considered seeds (see Chap. 9).

Generally pollination is the cue to fruit growth, and fertilization triggers ovule growth and seed formation under the influence of growth hormones (Nitsch 1952, 1953). In certain cases fruits can develop to maturity without fertilization or seed formation, a phenomenon known as *parthenocarpy;* in a few instances fruit formation can occur without pollination, a more restricted type of parthenocarpy.

The ecological significance of fruit development is 3-fold: dispersal of seeds, induction of seed dormancy mechanisms, and nourishment and protection of seedlings during establishment.

Adaptive mechanisms for fruit or seed dispersal are manifold, ranging from fleshiness to wings on fruit or seeds. Seed dormancy mechanisms are often in the pericarp (of fruit) (e.g., tomato and the two-seeded cocklebur fruit), but true seeds themselves may also have dormancy mechanisms. The fruit is often rich in nutrients to nourish the seedlings. Some fruits contain growth inhibitors that prevent germination of their own seed for a period (to synchronize with favorable environmental events) and also to inhibit germination and growth of competitor species. For example, the fruit of black walnut contains an inhibitor, juglone, that imparts such inhibition (allelopathy) (see Chap. 7).

POLLINATION

Pollen grains are formed from microspore mother cells in the anthers. Following a meiotic and a mitotic division a tetrad of cells is formed, which matures as pollen grains (Fig. 12.9). Pollen grains may be either bi- or trinucleate (Brewbaker 1959); that is, they may have one or two generative nuclei in addition to the vegetative nucleus. Members of the Gramineae and Compositae families are characteristically trinucleate as a result of a second mitotic division in the microspore. The binucleate pollen also undergoes a second mitotic division upon germination of the pollen grain and becomes trinucleate in effect.

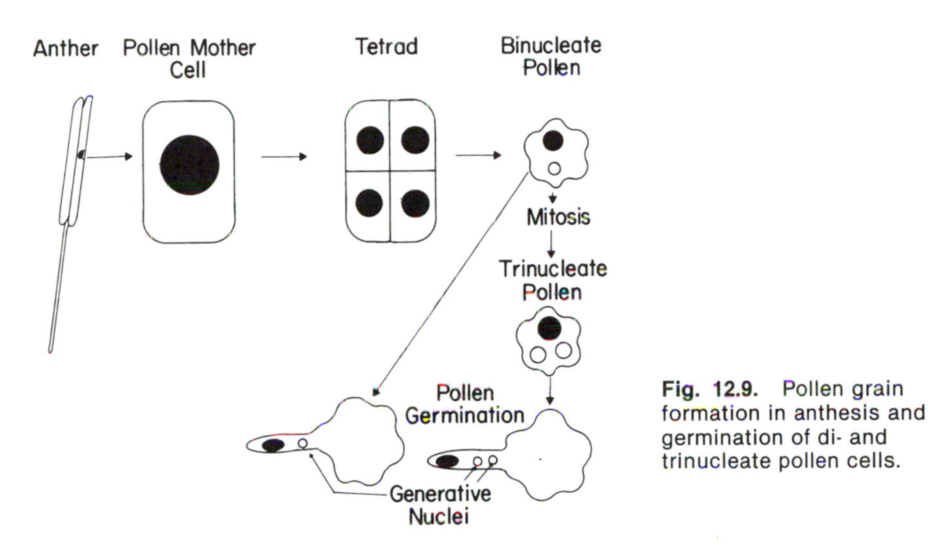

Fig. 12.9. Pollen grain formation in anthesis and germination of di- and trinucleate pollen cells.

Pollen cells germinate almost immediately if brought in contact with a receptive stigma, which provides a suitable substrate stimulation. Germination occurs in vitro using agar or a sugar solution (plus certain minerals, for some species). Germination of pollen is evidenced by a rapid increase in respiration and synthesis of RNA and protein. These processes begin in a span of 2 to 30 min (Leopold and Kriedemann 1975).

A number of factors stimulate pollen germination (Vasil 1960), including favorable concentrations of sucrose, CO_2, boron, and calcium. Boron seems to enhance sucrose utilization. From in vitro studies it appears that growth regulators in the stigma or medium are not requisite to germination, although pollen is rich in auxin and GAs. Increased germination and pollen tube growth from a mass of pollen, compared with that of a single grain, is evidence for growth hormone stimulation. Water extracts of pollen had similar stimulatory effects (Brewbaker and Majumder 1961).

Pollen may fail to germinate on its own or on alien stigma (incompatibility) even though conditions are ostensibly favorable. Numerous species are self-incompatible, such as most members of Leguminoseae, and bee visitation is necessary to rupture the stigmatic membrane for pollen receptivity. Self-incompatibility may also result from differential ripening of male and female gametes. Enzymatic or chemical inhibitors in stigma are believed to cause incompatibility in some species.

FRUIT SET

The onset of fruit growth and development as normally initiated by fertilization is referred to as *fruit set*. Fruit set is associated with a number of physiological events, including rapid fruit growth and flower senescence. In a study of 'White Sim' carnation (Nichols 1971), two growth substances, 2-chlorethylphosphoric acid (ethephon) and 2,4-dichlorophenoxyacetic acid (2,4-D), accelerated senescence of flowers in conjunction with fruit development, probably by stimulating the production of IAA, which frees bound ethylene in the flower. A fertilized strawberry flower produced four to five times more ethylene than an unfertilized flower (Bleasdale 1973). Ethylene apparently is produced in the stigma and style. Unpollinated flowers double ethylene production if sprayed with an auxin. Other substances can affect senescence of unfertilized and fertilized ovaries (Nitsch 1953) (Fig. 12.10).

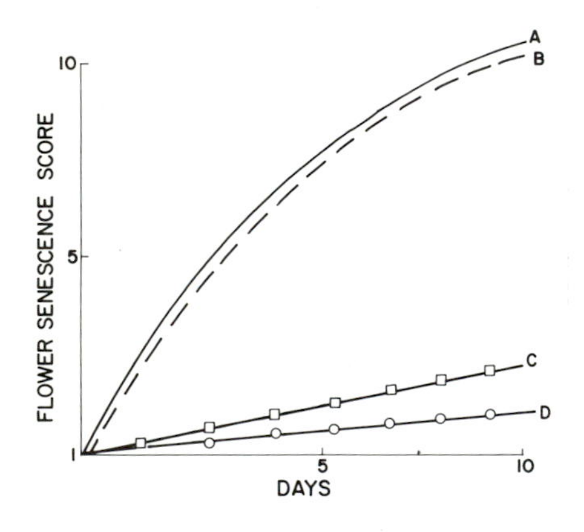

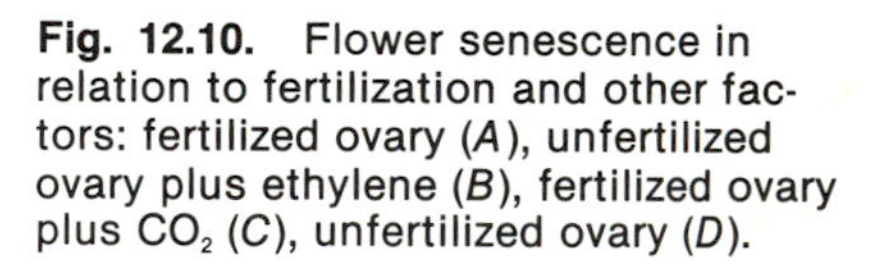
Fig. 12.10. Flower senescence in relation to fertilization and other factors: fertilized ovary (*A*), unfertilized ovary plus ethylene (*B*), fertilized ovary plus CO_2 (*C*), unfertilized ovary (*D*).

Pollen contains auxins that trigger the reactions involved in fruit set. A growing fruit is its own source of auxin (e.g., banana). Synthetic auxins (Gustafson 1936) can stimulate fruit set in a range of species, particularly in Solanaceae and Cucurbitaceae, but not in others, such as apple and cherry. In another group of species the stimulation of fruit growth by synthetic auxin is temporary or lasts only as long as the supply of auxin lasts. Additional auxin must be supplied for fruit growth to continue. GA is used commercially to increase fruit growth of certain seedless types of fruits, such as 'Thompson's Seedless' grape, and can improve the shape of fruits.

Pollination without fertilization triggers parthenocarpic fruit growth in certain species, such as citrus and seedless grape. Types of parthenocarpy can be defined as follows:

1. Fruit growth and development without pollination
2. *Apomixis,* the process of fruit growth and development with pollination but without *syngamy* (fertilization, or union of male and egg nuclei)
3. Fruit growth and development with pollination and syngamy but without seed formation (due to abortion)

Tomato, pepper, pumpkin, banana, and cucumber are representative of the first type. Apomixis in Kentucky bluegrass and citrus is representative of the second type. Seeds of Kentucky bluegrass are viable and usable for propagation. Seedless fruits of the third type, aborted seed parthenocarpy, are common in peach, cherry, and grape. Except for apomixis in a number of grass species and citrus, parthenocarpy is neither found nor desired in crop production, since seed production is the commercial objective of most field crops.

FAILURE IN FRUIT SET

Failure of most flowers to set fruit is the rule rather than the exception. A 50 to 75% failure in soybean and wheat is not uncommon. An ear of maize has about 1,000 potential seeds, but this number is seldom produced. There are three reasons for fruit set failure.

1. Lack of pollination. Anthers and pollen in grasses frequently abort due to heat and desiccation, especially in grasses, a condition referred to as *blasting*. Bees may fail to visit legume flowers requiring cross-pollination.
2. Lack of fertilization because of weak pollen or incompatibility
3. Abortion of flowers and fruits. Abortion is common among *cleistogamous* (self-pollinated) legumes (e.g., soybean). Flowers are prolific, but a majority of them (and even the lower pods) may abort. Pods may abort at a young age, especially on diseased plants in dense, tall canopies. In grasses the whole inflorescence or 50% or more of the florets of an inflorescence may abort. This abortion is believed to be caused by a deficiency of organic nutrients resulting from intraplant competition with flowers and fruits of a head or panicle with a greater competitive advantage. The plant can only set and mature seed to the extent of its assimilate supply. Environmental stresses reduce the assimilate supply and seed number.

FRUIT GROWTH

Growth of an organ, including fruits, is characterized by the standard sigmoid curve (see Chap. 8). But a number of fruits, typically *drupes* (those

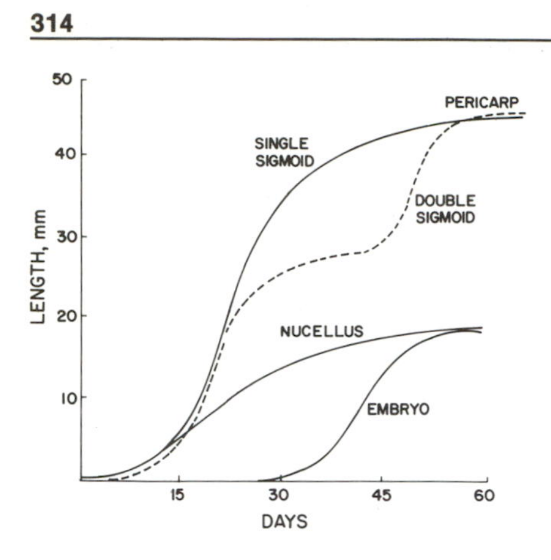

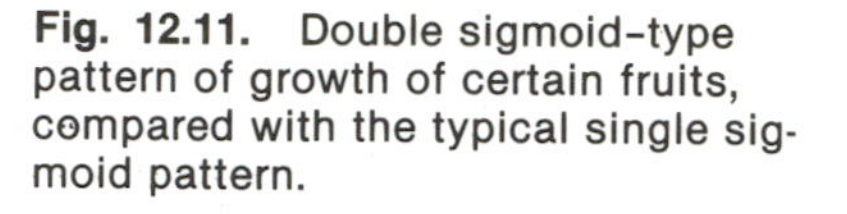
Fig. 12.11. Double sigmoid-type pattern of growth of certain fruits, compared with the typical single sigmoid pattern.

with pits), are characterized by a double sigmoid (Fig. 12.11). Generally the first sigmoid corresponds to seed growth, the second to pericarp growth (fruit growth).

Growth of the ovary is initiated by pollination. Without pollination the flower forms an abscission layer and drops, probably due to lack of the proper growth hormones (Nitsch 1952). Pollination provides a source of growth hormones sufficient for initial fruit growth. The stimulus from pollination is temporary; apparently the endogenous supply of pollen GAs is soon exhausted (Carr and Skene 1961). The second peak of fruit growth occurs with a new supply of hormones from the fruit.

Nitsch (1951) outlined three distinguishable phases of fruit growth:

1. Preanthesis. Growth of the ovary, primarily by cell multiplication
2. Anthesis. Pollination and fertilization of ovules, stimulating ovary growth; unfertilized flowers abscise or are aborted
3. Postfertilization. Size increase of fruits, occurring primarily by cell enlargement

The principal hormones in fruit growth are auxins and GAs. Maize pollen is a rich source of both (Fukui et al. 1958). Extracts from pollen can stimulate fruit growth temporarily, but a new source of hormones, as with seed formation, is usually necessary for continued growth. Fruit growth places heavy demands on mineral nutrients, resulting in mobilization and transport from vegetative to fruit and seed development (Fig. 12.12). The percentage of nitrogen, phosphorus, and potassium in maize stems and leaves peaks shortly after silking and declines with rapid grain formation as nutrients are transferred from vegetative parts to fruits. The *rachis* (cob) of the maize inflorescence is also a source of nutrients during seed formation.

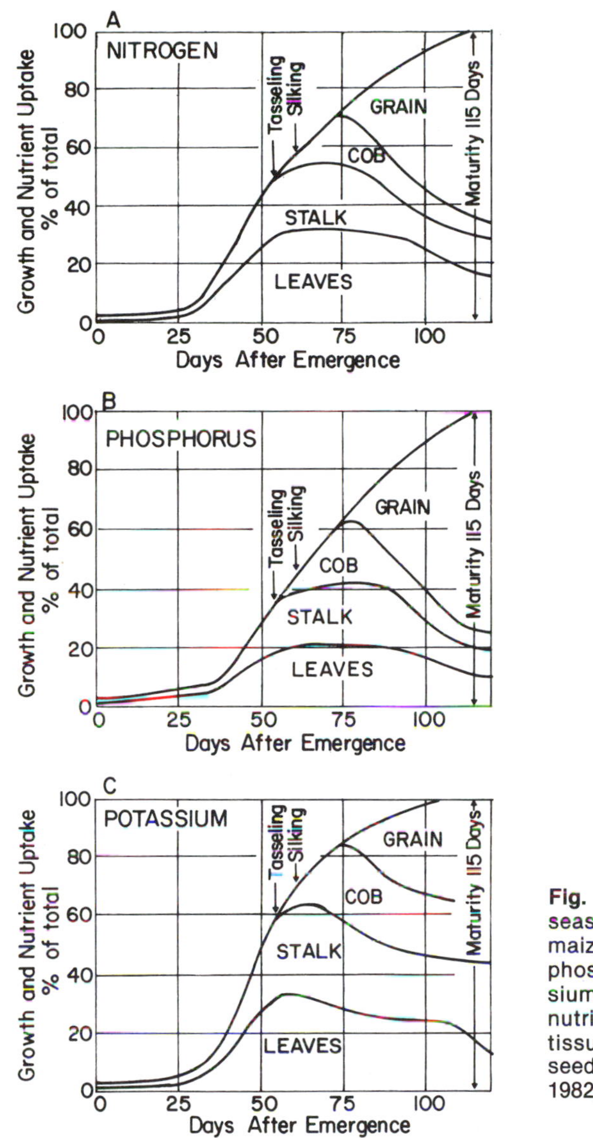

Fig. 12.12. Uptake and seasonal distribution in maize plants of nitrogen, phosphorus, and potassium. Note the loss of nutrients in vegetative tissues to the developing seeds (Ritchie and Hanway 1982, by permission).

SEED GROWTH

Like pollen, seeds have been found to be rich in growth-promoting substances, including auxins, GAs, and cytokinins (Leopold and Kriedemann 1975). Fruits and seeds are the sources of supply after the temporary pollen contribution. For example, the cytokinin zeatin was discovered in the milky endosperm of maize seeds. IAA has also been extracted from maize seeds.

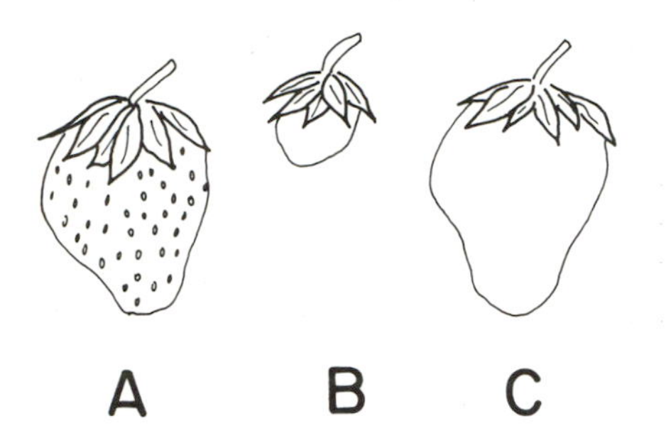

Fig. 12.13. Influence of seeds (achenes) on strawberry fruit (receptacle) development. *A.* Fruit with seeds. *B.* Seeds removed at early development stage. *C.* Seeds removed (as in *B*) and replaced with an auxin paste on the young fruit. (From Nitsch 1950)

Numerous GAs have been isolated from seeds (Letham 1963). Seeds produce ethylene as well during germination.

Generally, except for parthenocarpic fruit growth stimulated by pollination alone, seeds are necessary for fruit development. For example, the receptacle of the strawberry does not develop into a fruit without *achenes* (seeds) (Fig. 12.13). Poor pollination or nutrition resulting in failure of seed set causes dwarfed or deformed strawberry fruits (Nitsch 1950). The role of seed in fruit development of dry fruits is less readily observable. Injury, maturity, and other factors can induce ethylene formation and fruit abortion in most species (Lipe and Morgan 1972), as is common on soybean with stem rot.

MATURATION AND RIPENING

A fruit is mature when full size is reached and the rate of dry weight gain becomes zero. Mature fruits *ripen* by passing through a chain of enzymatic and biochemical events that result in changed chemical composition (Leopold and Kriedemann 1975). In ripening, old enzyme systems senesce and new ones are produced that cause softening and conversion of starches to sugars in fleshy fruits (e.g., apple). The acid level in citrus fruits declines. Chlorophyll is lost, while xanthophyll and carotene increase or are no longer masked out by chlorophyll. Certain of these reactions also occur in grass seeds and legume fruits (pods), although they are less pronounced.

Changes in ripeness are associated with a relatively high respiration rate in *climacteric* (rapid-ripening) fruits. Metabolic activity declines in nonclimacteric fruits, including those of agronomic crops.

Loss of chlorophyll and acceleration of senescence is characteristic of dry, dehiscent fruits (e.g., soybean pods). Carotene increases in yellow maize. Ethylene and abscisic acid play an important role in abscission and dehiscence of pod and capsule dry fruits (e.g., soybean or castor bean).

The number of fruits (e.g., pods of soybean or grains of maize) per plant is established relatively early in the ontogeny of crop plants. In well-watered and well-fertilized plants of adapted cultivars, it is a function of photosynthetic rate or assimilate supply. For a given genotype the number of fruits per plant is a function of spacing and the resulting intercepted light; therefore the number of fruits per unit of land area is related more to intercepted light than to the number of plants. Hence, within limits, light interception and assimilate production per unit of land area determine seed number per unit, regardless of plant number. Once the fruit load is fixed, seed yield becomes a function of seed size. Seed size for a given cultivar is usually relatively constant, but severe stresses during grain filling can cause a reduction in size. This reduction is a consequence of reduced assimilate supply and/or reduced leaf nitrogen. Leaf nitrogen has been considered a principal factor in soybean seed yield (Sinclair 1981). Seed size of peanut has also been shown to be controlled by fruit (pod) size (Nimbkar 1981). Peanut cultivars with small pods produce small seeds because of pod wall constriction, which results in fewer cells and smaller cell size. Peanut cultivars with small pods and seeds usually produce more pods and seeds per plant.

Summary

Flowering, fruiting, and seed set are essential events in crop plant production. These processes are controlled both by environment, particularly photoperiod and temperature, and by genetic or internal factors, particularly growth regulators, photosynthate, and mineral nutrient supply (e.g., nitrogen).

Based on their response to day length (more accurately, length of night), most crop plants can be classified as short-day plants (SDPs), long-day plants (LDPs), or day-neutral plants (DNPs). Soybean, wheat, and tomato, respectively, are examples of these three types. In general late summer to fall flowering indicates SDPs, and spring to early summer flowering LDPs. Tropical plants are usually SDPs but may be photoperiod-insensitive DNPs.

A brief interruption of a long night by red (R) radiation (maximum effectiveness, 660 nm) at low energy levels produces a short-night or long-day effect, preventing flowering in SDPs and promoting flowering in LDPs. An interruption by far-red (FR) radiation (maximum effectiveness, 735 nm) has the opposite effect, promoting flowering in SDPs and preventing it in LDPs. At a warm temperature, FR radiation is equivalent to darkness; the R effect is reversed by the equivalent of darkness. The pigment phytochrome is the photoreceptor of both R and FR radiation. The two forms of phytochrome, P_r and P_{fr}, are photoreversible by R or FR light. The equilibrium concentration of P_r and P_{fr}, and hence the stimulus to flower, is dependent on exposure to length of uninterrupted darkness. P_r is the most biologically active form.

Evidence from grafting experiments shows that the photoperiod stimulus is received by the leaves and translocated to the meristems, causing transformation from a vegetative state to flowering. Evidently the stimulus is not transferred to new or unexposed shoots, or tillers. Grass plants usually have numerous vegetative shoots with flowering shoots on the same plant. Pod initiation of indeterminate varieties of soybean occurs first at the axes of lower nodes and progresses basipetally and acropetally from this point. After the lower axes of the main stem fruit, the branches flower and fruit.

Some plants must have low or near freezing temperatures (i.e., require vernalization) for a period of weeks to complete floral induction. Winter annuals (e.g., wheat) and biennials (e.g., sugar beet) normally require vernalization to flower. Soaked seeds of winter annual cereals can be vernalized, but with most biennials and perennials only the green plants are receptive to the cold period. After vernalization long days are necessary for flowering; in other words, vernalization predisposes a long-day response. Evidently the locus of vernalization is the meristem rather than the leaf.

Flowering consists of three discrete phases—induction, initiation, and further flower development—each of which may have a different set of photoperiod and temperature optima. Floral induction, which occurs naturally in fall for temperate perennial grasses such as orchardgrass, requires a low temperature for a period of weeks along with short days. Initiation, which occurs naturally in spring, requires a warm temperature and long days. Further floral development requires a warm temperature and long days, as for initiation, plus a higher nutritional requirement. The photoperiod requirement for floral initiation in soybean differs from that required for further floral development.

Pollen contains growth hormones and stimulates fruit growth. Fruits and seeds are also rich in growth hormones. Most fruits do not grow without seed growth, apparently because seeds are required as a hormone source, but some are parthenocarpic and grow with only a pollen stimulus to initiate fruit growth. Ripening of fruit involves a different set of hormones from those required for growth. Ethylene is highly active in ripening, especially of climacteric fruits.

References

Arber, A. 1934. The Gramineae: A Study of Cereals, Bamboo, and Grass. New York: Macmillan.
Bleasdale, J. K. A. 1973. Plant Physiology in Relation to Horticulture. Westport, Conn.: AVI.
Borthwick, H. A., and M. W. Parker. 1938. Bot. Gaz. 99:825–39.
Borthwick, H. A., M. W. Parker, and P. H. Heinze. 1941. Bot. Gaz. 102:792–800.
Borthwick, H. A., S. B. Hendricks, and M. W. Parker. 1956. In Radiation Biology, vol. 3, ed. A. Hollaendar. New York: McGraw-Hill.
Brewbaker, J. L. 1959. Indian J. Genet. Plant Breed. 19:121–33.
Brewbaker, J. L., and S. K. Majumder. 1961. Am. J. Bot. 48:457–64.
Carr, D. J., and K. G. M. Skene. 1961. Aust. J. Biol. Sci. 14:1–12.
Chailakhyan, M. K. H. 1936. Proc. Acad. Sci. USSR [Dokl.] 3:433–37.
Clark, H. E., and K. R. Kerns. 1942. Science 95:536–37.
Cooper, J. P. 1950. J. Br. Grassl. Soc. 5:105–12.
Cumming, B. G. 1959. Nature 184:1044–45.
Downs, R. J. 1956. Plant Physiol. 31:279–84.
Evans, L. T. 1969. The Induction of Flowering: Some Case Histories. Ithaca, N.Y.: Cornell University Press.
Fisher, J. E. 1962. Can. J. Bot. 41:871–73.
Fukui, H. N., F. G. Teubner, S. H. Wittwer, and H. M. Sell. 1958. Plant Physiol. 33:144–46.

Gardner, F. P., and W. E. Loomis. 1953. Plant Physiol. 28:201–17.
Garner, W. W., and H. A. Allard. 1920. J. Agric. Res. 18:553–606.
______. 1923. J. Agric. Res. 23:871–920.
______. 1925. J. Agric. Res. 31:555–66.
Greulach, V. A., and J. C. Haesloop. 1958. Science 127:646–47.
Gustafson, F. G. 1936. Proc. Natl. Acad. Sci. 22:628–36.
Hamner, K. C. 1938. Bot. Gaz. 99:615–29.
______. 1969. In The Induction of Flowering, ed. L. T. Evans. Ithaca, N.Y.: Cornell University Press.
Hensel, H. 1953. Ann. Bot. n.s. 17:417–32.
Hillman, W. S. 1962. The Physiology of Flowering. New York: Holt, Rinehart, and Winston.
Holdsworth, M. 1956. J. Exp. Bot. 7:395–409.
Jennings, P. R., and R. K. Zuck. 1954. Bot. Gaz. 116:199–200.
Kasperbauer, M. J. 1970. Agron. J. 62:825–27.
______. 1973. Agron. J. 65:447–50.
Kasperbauer, M. J., F. P. Gardner, and W. E. Loomis. 1962. Plant Physiol. 37:165–70.
Kasperbauer, M. J., H. A. Borthwick, and S. B. Hendricks. 1963. Bot. Gaz. 124:444–51.
______. 1964. Bot. Gaz. 125:75–80.
Khudairi, A. K., and K. C. Hamner. 1954. Plant Physiol. 29:251–57.
Lang, A. 1951. Zutcher 21:241–43.
______. 1952. Annu. Rev. Plant Physiol. 3:265–306.
Lang, A., and G. Melchers. 1947. Z. Naturforsch. 26:444–49.
Leopold, A. C. 1958. Annu. Rev. Plant Physiol. 9:281–310.
Leopold, A. C., and P. E. Kriedemann. 1975. Plant Growth and Development. 2d ed. New York: McGraw-Hill.
Leopold, A. C., and K. V. Thimann. 1949. Am. J. Bot. 36:342–47.
Letham, D. S. 1963. Life Sci. 8:569–73.
Lipe, J. A., and P. W. Morgan. 1972. Plant Physiol. 50:759–64.
Liverman, J. L., and A. Lang. 1956. Plant Physiol. 31:147–50.
Lockhart, J. A., and K. C. Hamner. 1954. Bot. Gaz. 116:133–42.
Long, E. M. 1939. Bot. Gaz. 101:168–88.
Major, D. J. 1980. Can. J. Plant Sci. 60:777–84.
Mann, L. K. 1940. Bot. Gaz. 102:339–56.
Moshkov, B. S. 1939. Proc. Acad. Sci. USSR [Dokl.] 22:456.
______. 1947. Proc. Natl. Acad. Sci. 33:303–12.
Naylor, A. W. 1941. Bot. Gaz. 103:342–53.
Nichols, K. 1971. J. Hortic. Sci. 46:323–32.
Nimbkar, N. 1981. Cell Number in Relation to Seed Size in Peanut (*Arachis hypogaea* L.). Ph.D. diss., University of Florida, Gainesville.
Nitsch, J. P. 1950. Am. J. Bot. 37:211–15.
______. 1951. In Plant Physiology: A Treatise, ed. F. C. Steward. New York: Academic Press.
______. 1952. Q. Rev. Biol. 27:33–57.
______. 1953. Annu. Rev. Plant Physiol. 4:199–236.
Parker, M. W., S. B. Hendricks, H. A. Borthwick, and N. J. Skully. 1946. Bot. Gaz. 108:1–26.
Peterson, M. L., and W. E. Loomis. 1949. Plant Physiol. 24:31–43.
Purvis, O. N., and F. G. Gregory. 1937. Ann. Bot. n.s. 1:569–92.
Ritchie, S. W., and J. J. Hanway. 1982. Iowa State Univ. Spec. Rep. 48.
Salisbury, F. B. 1955. Plant Physiol. 30:327–34.
______. 1963. The Flowering Process. New York: Macmillan.
______. 1969. In The Induction of Flowering, ed. L. T. Evans. Ithaca: Cornell University Press.
Salisbury, F. B., and J. Bonner. 1956. Plant Physiol. 31:141–47.
Schwabe, W. W. 1957. J. Exp. Bot. 8:220–34.
Sinclair, T. R. 1981. Personal communication.

Stanley, R. G. 1958. In The Physiology of Forest Trees, ed. K. V. Thimann. New York: Ronald Press.
Thomas, J. F., E. D. Raper. 1982. Personal communication.
Vasil, I. K. 1960. Nature 187:1134–35.
Vergara, B. S., and T. T. Chang. 1976. The Flowering Response of the Rice Plant to Photoperiod: A Review of Literature. 3d ed. Los Banos, Philippines: International Rice Research Institute.
Wareing, P. F., and I. D. J. Phillips. 1978. The Control of Growth and Differentiation in Plants. 2d ed. Oxford and New York: Pergamon.
Wellensiek, S. J. 1962. Nature 195:307–8.
Wittwer, S. H., and M. J. Bukovac. 1958. Econ. Bot. 12:213–55.
Zobka, G. G. 1961. Am. J. Bot. 48:21–28.

INDEX

Abscisic acid, 86, 89, 174–77, 184, 238
Abscisin II, 175. *See also* Abscisic acid
Abscission, 163
Acetyl-coenzyme A, 26, 223
Achene fruit, 213, 242, 316
Acid rain, 114. *See also* Sulfur
Acropetal transport, 58, 160
Aerobiosis, 107, 108
Actinomycetes, 134, 137
Active absorption, 107
Adenine purine, 114
Adenine,6-aminopurine, 170
ADP and ATP, 9, 26, 61, 107, 116, 134, 222
Agent orange, 160. *See also* 2,4,5-T
Albumin, 219
Aldehydrogenase, 223
Aleurone, seed, 210
Alfisol soils, 102
Allantoin, 140
Allelopathy, 176, 262
Allometry, 195, 249
  coefficient, 195, 207
  constant, 195
Aluminum, 264
Amino acids, 59
Ammonification, 110
Amo-1618, 174–75
Amylase, 170, 222
Amylopectin, 214, 243
Amylose, 214, 216, 243
Anabaena, 134, 136–37
Anion uptake, 107
Apomixis, 213, 242
Apoplast, 61, 62, 74, 108
  transport, 160
ATPase, 109
Auxins, 157–64, 262, 265
  assay methods, 161
  bound forms, 159
  concentration, effects, 161
  metabolism, 160
  natural and synthetic, 159–60
  responses, 161–63, 308
  transport, 160
  uses, 163–64
Avena coleoptile test, 161
Axillary bud, 279, 293
Azolla, 136
Azospirillum brasilense, 134–35, 145–47
Azotobacter, 134–35, 138

Bacteria, photosynthetic, 135–36
Bacteroid, 139–40
Barley aleurone test, 167
Basic vegetative phase, 298, 303–4
Basipetal transport, 58, 160
Beijerinckia, 134
Benzoic acid, 174
Biennial bearing, 164
Biomass
  duration, 206
  production, 45
Biosphere, 100
Blackman response, 190
Blue-green algae, 136–37, 153
Bolting, 301
Boron, 102–3, 125–26, 176
Boundary layer effect, 22. *See also* Carbon fixation
Brassinalide, 157
Bud
  adventitious, 271
  quiescent, 290
BVP. *See* Basic vegetative phase

Calcium, 102–3, 118–20, 128–29
Callus, 157, 163–64
Calvin cycle, 12–13
Canopy photosynthesis
  leaf area duration, 44–46
  leaf area effect, 31–37
  light attenuation, 37–44
  plant density and pattern effects, 48–56. *See also* Carbon fixation
Carbohydrates, 110, 290
Carbon dioxide. *See also* Canopy photosynthesis
  atmosphere concentration, 20, 100
  leaf resistances, 22
  root growth, 263–64
Carbon fixation, 12–17, 20–23
  C_3 vs. C_4 species, 14–16, 17, 29
  Calvin cycle, 12–13
  CAM, 16–17
  dry matter production, 12
  Hatch-Slack pathway, 12–13
Cardinal temperatures, 226, 265
Carotenoids, 9, 11
Carrier theory, 109
Caryopsis, 210, 213

Casparian bands, 108
Cataphyll, 282
Cation
  exchange capacity, 101, 174–75
  uptake, 107
Cell
  differentiation, 187, 197, 251
  division, 187, 251
  enlargement, 87, 187
Chelates, 103
Chlorine, 127–28
Chlormequat, 174–75
Chloroflurecol, 174
Chlorogenic acid, 174, 220. *See also* Seed, dormancy
Chlorophyl, 7–9
Chloroplast, 7–9
Chromatium, 134
Cloning, 173
Clostridium pasteurianum, 134
Cold test, seeds, 230
Companion cells, 61
Contact exchange, 106
Copper, 102–3, 126–27
Core sampling, roots, 247–48
Coronal roots. *See* Roots, coronal
Cotyledons, 211
Coumarin, 220, 238
Crassulation acid metabolism, 16–17
Critical percentages, theory, 192
Crop growth rate, 30, 40–42, 48, 197, 202–5
Cross-inoculation, 140, 145
Crown
  buds, 288
  plants, 196
  roots, 257–59, 262, 269
Culm, 278
Cyanobacteria, 134
Cytokinins, 63, 86, 148, 157, 170–73, 183, 262
  assay methods, 171–72
  cytokinesis, 170
  metabolism, 171
  natural occurrence, 170–71
  responses, 172
  root growth effects, 86
  transport, 171
  uses, 173

Daminozide, 174
Day length, 6, 297–99
Day-neutral plant, 299, 317
Deamination, 223
Defoliants, 180
Denitrification, 111, 112
Determinate and indeterminate species, 91–92
Dicamba, 159
Differentiation, 187, 197–98, 251
  climate and soil effects, 198
  fall hardening, alfalfa, 198
  nitrogen and water effects, 197–98
  zone, roots, 187
Diffuse light, 37
Diffusion
  gradient, 81
  ions in uptake, 106–7
DNA and RNA, 114
Dolomite limestone, 120
Dominance, apical, 172, 287
Donnan free space, 108
Dormancy. *See* Seed, dormancy
Dormin, 175
Drupe fruit, 313

Edaphic growth factors, 188. See *also* Growth and development
EDTA and EDDHA, 103
Einstein unit, 4
Electromagnetic wave theory, 3
Electron, transport, 9
Embryo axis, 212
Endodermis, 108
Epigeal emergence, 240, 289–90
Epinasty, 163
Erucic acid, 217
Escherichia coli, 148
Essential elements, 98, 99
Ethylene, 86, 157, 180–84, 262. *See also* Seed, dormancy
  climacteric fruits and seeds, 180–81
  ethephon, 183
  geotropism, 182
  metabolism, 181
  natural occurence, 180–81
  responses, 181–82
  uses, 183
Etiolation, 280
Evapotranspiration, 76, 83–84, 96
  factors affecting, 82–84
  potential, 84
Exponential growth phase, 199. *See also* Growth, dynamics
Extinction coefficient, 38

Ferromagnesium, 121
Field capacity, 77, 78
Florigen, 308
Flowering, 296–310
  development, 309–10, 318

expression, 302, 318
Garner and Allard's studies, 296, 297, 298
induction, 301–9, 318
initiation, 309, 318
minimum age, 304
ripeness to flower, 303
stimulus, 296
transition to, 297
Flurecol, 174
Food reserves, 290–93, 294
Frankia, 134
Free space, 108
Fructosan, 292
Fruit
climacteric, 180, 316, 318
growth, 313–14
maturing, ripening, 316–17
pollination, 310–11
seed growth, 315
set, 312
set, failure, 313

GA. *See* Gibberellins
Geotropism, 167, 254, 282
Gibbane skeleton, 165
Gibberellins, 157, 164–70, 183–84. *See also* Flowering; Fruit; Seed, germination
assay methods, 167
chemistry, 164–65
naming, 165
natural synthesis and occurrence, 165–66
responses, 167–69
transport, 167
uses, 169
Globulins, 219
Glucosan, 292
Glucose, 216
Glutamine, 113
Glutelins, 219
Glycerol, 217
Glycolysis, 27, 222
Grain fill, 70–71
Ground cover, 31
Growth
analysis, 200–207
correlation, 195–98
dynamics, 199–200
factors, 188–89
hormones, 156–57, 286
inhibitors, 157, 174–80, 183
limiting factors, theories, 189–92
regrowth. *See* Vegetation, regrowth
respiration and, 26–28
retardants, 167
vegetative. *See* Vegetation, growth
Guard cells, 89
Gynophore, 278

Haber-Bosch process, 110, 133
Harvest index, 66, 67, 195, 207
Hatch-Slack pathway, 12–14
Heliotropic movement, 43
Hemicellulose, 216
Hilum, 211
His, gene, 148
Hydrogenation, 218
Hypocotyl, 212
Hypogeal emergence, 240

IAA. *See* Indoleacetic acid
IAA oxidase, 123
Indoleacetaldehyde, 159
Indoleacetic acid, 63, 137, 148
Indoleacetonitrile, 159
Indolepyruvic acid, 159
Inoculation
acid soil, effects, 142
methods, 144
preinoculation, 144
Rhizobium, effective strains, 142
sticking agent, 144
Intravaginal tillers, 281
Inulin, 216
Ion pump theory, 109
Ion interaction, 109
Iron, 102–3, 121–23
Isoprene, 166

Juglone, 176, 236
Juvenile factor
branching, 285
flowering, 303–4

Kaurene, 166
Kinetin, 170
Klebsiella, 136, 148
Kranz anatomy, 14, 61
Krebs cycle, 26, 222, 223
Kylar, 179

Lactone ring, 165
Lactones, 174, 220
LAD. *See* Leaf area duration
LAI. *See* Leaf area index
Laminar resistance, 22
LAR. *See* Leaf area ratio
Laterite soils, 102
Law of diminishing returns, 105, 191
Law of the minimum, 190
Leaf
factor, 163

Leaf (*continued*)
  flag, 70, 274
  growth, 272–76
  inclination, 42–44
  lamina, 272, 293
  nodulating, 135
  primordium, 272, 273
  pseudostem, 272, 278, 294
  resistances, 22–23
  rolling, 83
  sheath, 272
Leaf area duration, 44–45
Leaf area index, 31–37, 45–46, 49–50, 56, 90, 204
  critical vs. optimum, 34–37
Leaf area ratio, 203
Lecithin, 218
Leghemoglobin, 138
Lesquerolic acid, 217
Lichens, 135
Liebig's theory, 190
Life period, seed, 47
Light quality, effects
  on flowering, 306–7, 317
  on germination, 226–27
  on photosynthesis, 3–6, 11
  on stem growth, 278
Linear growth phase, 199–200
Linoleic acid, 217
Linolenic acid, 217
Lodging, 50
Long-day plant, 299, 317
Long-short-day plant, 299
Luxury consumption, 105, 192
Lysine, 219

Macy's theory, 192
Magnesium, 102–3, 120–21
Maleic hydrazide, 180
Maltose, 216
Manganese, 102–3, 123–24
Mannans, 216
Mass flow hypothesis, 63, 106
Megasporogenesis, 209
Meristem, 193–94, 207, 297, 317
  activity, 140, 142, 153
  apical, 194, 207
  diffuse, 194, 207
  intercalary, 194, 207, 272, 277, 293
  lateral, 194
  massed, 194–95, 207
  quiescent buds, 194
  subapical, 251
  vascular cambium, 194, 271
Mesocotyl, 257, 277, 293
Mesophyll resistance, 22–23
Methionine, 219
Mevalonic acid, 166
Micropyle, 211
Microspore mother cell, 311
Microsporogenesis, 209
Middle lamella, 216
Migration coefficient, 66
Mineral nutrients. *See* individual minerals; Nutrients
Mitochondria, 222
Mitscherlich equation, 191
Mollisol soils, 102
Molybdenum, 102–3, 127
Monogerm seeds, 213
Morphactins, 174
Mucagel, root, 147
Mycorrhiza, 147, 188

NAD and NADPH, 7, 9, 26, 116
Naphthaleneacetic acid, 159, 163, 164
Net assimilation rate, 33–34, 56, 203–4
Nif, gene, 148, 150
Night break, photoperiod, 305
Nitrate reductase, 113, 127, 287
Nitrification, 110
  inhibitors, 112
Nitrogen, fixation
  atmospheric, 134
  biological, 134–35
  environmental factors, 150–52
  by free-living organisms, 135–37
  genetic factors, 148–50
  by grasses, 145–48
  industrial, 133
  by nodulating organisms, 137–45
Nitrogenase, 148–49, 150, 151
Nitropyrin, 112
Nonreducing sugars, 59
Nostoc, 134, 136–37
Nucleotides, 59
Nutrients
  availability, 102–4, 128
  concentration, 104, 192–93
  essential elements, 98–100
  functions, 110–29
  ion interactions, 109
  quantitative requirements, 104–6
  recycling, 100, 101
  sources, 100–102
  uptake, 106–9

Oleic acid, 217
Optima and limiting factors theory, 190
Organic acids, 110
Organic gardening, 100
Organic matter, biodegradation, 101
Osmotic adjustment, 87–88, 96

Osmotic potential, 77, 96
Outer space, root, 108
β-Oxidase, 223
Oxisol soils, 102

PAR, 38
Parthenocarpy, 163, 167, 313
Partitioning
  coefficient, 205
  during grain fill, 70–73
  harvest index, 66–67
  remobilization, 71, 73
  during reproductive phases, 65–66
  source-sink relationship, 62–63
  during vegetative phases, 65–66
  yield components, 68
PBH, 135
Pectins, 216
Peduncle, 277
Pentosans, 216
Pentose phosphate pathway, 222
PEP, 12–14
Perennation, 282
Pericarp, 211
Pericycle, 252, 253, 269, 271
Permanent wilting percentage, 77, 78
Peroxidase, 223
Phloem, 58
  loading and unloading, 59, 61, 74
Phosfon-D, 174
Phosphoenol pyruvate carboxylase, 12–15
Phosphokinase, 123
Phospholipids, 218
Phosphorus, 102–3, 115–17, 128
Phosphotransferase, 123
Photoinduction cycles, 304
Photons, light, 4
Photoperiod, 66, 297–99
Photoperiod-induced phase, 298, 303–4
Photorespiration, 14–17
Photosynthesis
  apparatus, 7–17
  apparent vs. true, 23
  carbon dioxide fixation, 12
  factors affecting, 19–25, 30
  leaf as organ, 17–19
  light properties, 3–7
  light reaction, 7–12
  nonlaminar, 33–34, 56, 72
  pathways among species, 12–17
  rates, 24–25, 28–29
  solar radiation, 5–7
  utilization, 25–28
Photosynthetically active ratiation, 38
Photosynthetic photon flux density, 38
Photosystems I and II, 127–28
Phototropism, 156
Phyllochron, 272
Phyllosphere N fixation. *See* Nitrogen fixation, biological
Physiological maturity, 200, 211
Phytic acid, 116
Phytochrome, 278, 306, 317
Phytohormones, 157
Phytomer, 293
Picloram, 160
Pigment excitation, 4
PIP, 298, 303–4
Planck's constant, 4
Plant
  competition, 53
  density, 48–56, 286
Plant growth regulators
  auxins, 157–64
  classification, 156–57
  cytokinins, 170–73
  ethylene, 180–83
  gibberellins, 164–70
  inhibitors, 174–80
Plant growth substances, 156
Plasmalemma, transport across, 107
Plasmodesmata, 58, 107
Plastochron, 272, 273, 289, 293
Podzol soils, 102
Pollen, 311–12
Poly-β-hydroxybutyrate, 135
Porphyrin ring, 11
Potassium, 89, 117–18, 128, 266–67, 289
Potential
  gravitational, 77, 96
  matrix, 77, 96
  pressure, 77
  solute, 77, 96
Poverty-adjustment range, 192
Preinoculation, 144
Prolamins, 219
Proline, 86
Prophyll, 284
Protease, 223
Pseudostem, 272, 278, 294

Quantum efficiency, 11
Quantum theory, 4

Raceme, 283
Ratoon, 282
Remobilization, 68, 74
Respiration
  growth and, 26–28
  seed, 221
Rhizobium
  legume association, 138–45, 153, 162
  meliloti, 120

Rhizobium (*continued*)
strains, 142–44
survival in soils, 144–45
Rhizome, 284, 289, 294
Rhizosphere, 246
Rhodospirillum, 134
Ribulose bis-phosphate, 12–14
Root
adventitious, 257
apex, 251
brace, 257
branching, 252–53
cap, 251
cation exchange capacity, 108
comparison with stem growth, 251
coronal, 257, 262, 269
density, 250, 252, 268
depth, 83
efficiency, 258–59
functions, 246
hairs, 246, 249–50, 253, 258
initiation, 250
length, 250
mucigel, 106
nodal, 257, 269
nutrient transport, 246
prop, 257
rhizosphere, 246
seminal, 257, 262, 269
study methods, 247–50
surface area, 250
systems, 253–58
zones, 250
Rooting
defoliation, 262–63
factors affecting, 200–268
genotype, 260–62
plant competition, 262
soil environment, 263–68
Row spacing, 53, 55

SADH, 174
Saponification, 218
Scarification, 236, 239
Schizocarp, 242
Scutellum, 212, 257
Seed
after-ripening, 238
ball, 209, 213, 242
cotyledon, 211
density, 240
dormancy, 231–43, 310
food reserves, 213–20, 243
germination, 221–43
longevity, 229–30
ontogeny, 211–12
photoblastic, 227
proteins in, 218–20
quiescence, 211, 224, 231
size, 240
structures, 239
treatments to break dormancy, 234–35, 238, 239
types, 232–40
yield, 90
Seedling vigor, 230–31
Shoot-root ratio, 195, 207
Short-day plant, 298, 317
Short-long-day plant, 299
Sigmoid growth curve, 199, 313
Silicon, 99, 128
Sink-source, transport, 62
Soil
bulk density, 262–68
compaction, 267
nutrients, 101–2
parent material, 101
pH, 264
strength, 267
types, 102
Solar energy, utilization, 44–48
Solar radiation
angle, 6
constant, 5
direct and indirect, 37, 43
energy spectrum, 3
factors affecting, 5–6
flecks, 37
output, 6
relation to crop yield, 7
relation to heat, 7, 46–47
seasonal variations, 7
spectrum, 5
Specific mass transfer, 59
Stachyose, 216
Starch, 214–16
Stem growth, 277–89
branching, 281
bunch growth, 284
crown formation, 279
ear shoot, 285
factors affecting, 279–80
internode elongation, 277–78
light, effects of, 278
rhizome, 284
tillering, 282–89
Stolon, 284
Stomata
control, 79, 86
number, 82
resistance, 22
size, 82
Stratification, 166, 177, 236, 300
Sucker, control, 180
Sucrose, 216

Sulfhydryl groups, 114
Sulfur, 102–3, 114–15
Symplast transport, 61, 160
Synergism, 167
Syngamy, 313

Tannin, 220
Testa, 211
TIBA, 179, 184
Tillering, 49, 278
Tordon, 160
Transamination, 223
Translocation, assimilate, 58–61
Transpiration, 16
Transpiration, stream, 108
Triacontanol, 157, 183
Triiodobenzoic acid, 179, 184
Tryptophan, 219
Turgor pressure, 77, 96
2,4-D, 157, 159, 184, 312
2,4,5-T, 160, 184

Ultisol soils, 102
Urides, 59

Vacuole, storage, 107
Vegetation, growth
  branching and tillering, 281–84
  crown development, 279
  factors affecting, 273–76, 279–81, 284–89
  leaves, 272–76
  regrowth, 289, 294
  shoot morphogenesis, 271
  stems and internodes, 277–79
Vegetation, regrowth
  food reserves, 290–93
  food-reserve storage, 290–91
  grasses, 289–90
  LAI, basal, 291
  legumes, 290
Vernalin, 308
Vernalization, 166, 299, 300–301, 318
  locus, 301, 318
  loss, 301
  plant responses, model, 300
Verticillium wilt, 118
Viets effect, 109
Vitamin B_{12}, 100

Water
  availability, 77–78
  evapotranspiration, 80–84
  potential, 76–77
  stress, 84–92
  uptake, 78–79
  use efficiency, 78–79
Waxes, 218
Wein's law, 5

Xylem, 58, 252

Yield
  biological, 50–51, 66
  components, 68
  economic, 7, 66

Zeatin, 170, 184
Zein, 219